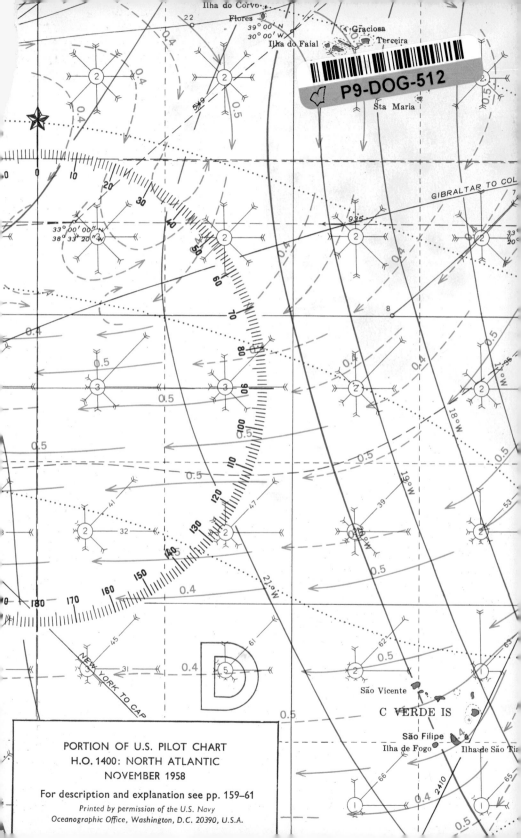

Ilha do Corvo

Flores

39° 00' N
30° 00' W

Ilha do Faial

Graciosa

Terceira

Sta Maria

GIBRALTAR TO COL

33° 00' 00" N
38° 33' 20" W

33°
20°

935

8

NEW YORK TO CAP

18° W

19° W

20° W

21° W

São Vicente

C VERDE IS

São Filipe

Ilha de Fogo

Ilha de São Tia

PORTION OF U.S. PILOT CHART
H.O. 1400: NORTH ATLANTIC
NOVEMBER 1958

For description and explanation see pp. 159–61

*Printed by permission of the U.S. Navy
Oceanographic Office, Washington, D.C. 20390, U.S.A.*

H. Peirce Brawner, Jr.

VOYAGING UNDER SAIL

BY

ERIC C. HISCOCK

Author of *Cruising Under Sail*
Around the World in Wanderer III
Beyond the West Horizon
Atlantic Cruise in Wanderer III, etc.

WITH 150 PHOTOGRAPHS BY
THE AUTHOR
AND 65 DIAGRAMS

SECOND EDITION

LONDON
OXFORD UNIVERSITY PRESS
NEW YORK   MELBOURNE

Oxford University Press, Ely House, London W. 1

GLASGOW NEW YORK TORONTO MELBOURNE WELLINGTON
CAPE TOWN IBADAN NAIROBI DAR ES SALAAM LUSAKA ADDIS ABABA
DELHI BOMBAY CALCUTTA MADRAS KARACHI LAHORE DACCA
KUALA LUMPUR SINGAPORE HONG KONG TOKYO

ISBN 0 19 217527 0

First edition 1959
Second edition 1970
Second impression 1971

PRINTED IN GREAT BRITAIN

To

TOM & ANN WORTH

able and courageous voyagers

PREFACE TO THE SECOND EDITION

SINCE the first edition of this book was published in 1959, I have been able to make with my wife a second circumnavigation of the world and a trip to America in the 30-foot sloop *Wanderer III*; I have also had the pleasure of meeting many more voyaging people and visiting their yachts. With the wider and more mature knowledge so gained to help me, I have revised the whole book, rewriting much of it, and including new material. I have also taken the opportunity to increase the number of plates and diagrams.

I wish to thank my friend Colonel H. G. Hasler for checking my remarks on wind-vane steering gears, and the many voyaging friends who have given me information and allowed me to photograph their yachts. In addition my thanks are due to the following who have kindly given their permission for the reproduction of the yacht plans which appear in Chapter 6: Messrs. John G. Alden of Boston, Mass., U.S.A., designers of *Rena*; Messrs. Laurent Giles & Partners of Lymington, Hants, designers of *Beyond*, *Trekka*, and *Wanderer III*; Mrs. J. G. Hanna of Dunedin, Florida, U.S.A., for the plans of *Little Bear*, designed by her late husband; Mr. J. Muir of Hobart, Tasmania, designer and builder of *Waltzing Matilda*; Mr. Arthur C. Robb of New Cavendish Street, London, W.1, designer of *Kochab*; Mr. Eric Ruck, owner of *Saoirse*; and Mr. Van de Wiele, joint designer with the late F. Mulder of *Omoo*.

E. C. H.

Yacht Wanderer IV
San Diego, California, U.S.A.
January 1970

FROM THE PREFACE TO THE
FIRST EDITION

THE interest taken in ocean voyaging, and the number of long deep-sea passages made in small sailing yachts, have increased greatly in recent years. In the 1920s and 30s a crossing of the Atlantic in a yacht, even one of 20 tons or more, was regarded by most people as a great adventure, or even as a hazardous undertaking, and such voyages were news. Today a number of yachts make the Atlantic crossing each year; few of them are 20-tonners, the majority are considerably smaller than that, and some are quite tiny, 20 feet or even less in length. Not only the Atlantic but the other great oceans, particularly the Pacific, where numbers of Americans, Australians, and New Zealanders cruise, have their quota of little ships making, almost as a matter of course, passages which would have been considered highly dangerous or even impossible only a few years ago. Some people, and my wife and I are fortunate enough to be able to count ourselves among them, have sailed round the globe in their own small vessels; but there are never likely to be many circumnavigations, partly because of the time involved, and partly for the reason that most people choose to make a long voyage because they wish to reach some specific place. However, all small-boat voyagers have two things in common: a love of freedom—for they can go where they will almost unhampered by rules or restrictions, except those which are a part of the seaman's lore—and a desire to pit their skill, wits, and courage against the oceans in every mood. The mainspring of this activity is, I believe, not the desire to be well thought of by others, but the desire to think well of oneself.

It might seem to the uninitiated that a long passage out of sight of land in a small vessel must be a dull business; but that is not so. Everyone who has made such a passage will surely agree that it was one of the most memorable and satisfying things he has ever done. Day and night, perhaps for several weeks, he and his companions drove their little vessel as fast as she could be made to go with safety, often with the warm, powerful trade wind humming in the rigging, and a broad wake, brilliant at night with phosphorescence, hissing out astern. They watched the sea-birds, the large and small inhabitants of the ocean, and saw them at closer quarters than is possible from the deck of a liner. Their vision was bounded by the empty horizon, but by taking observations of the sun or stars they found their way surely across the featureless sea until one day—and what a day that was!— the mountain tops or the palm trees of the land for which they had been steering for so long lifted out of the sea ahead. Presently their vessel, which for many days had been their world, lay peacefully at anchor with the unfamiliar sounds and scents of the land drifting out to her across the still water of the harbour. No doubt they then had a feeling of satisfaction and perhaps of relief, but certainly they enjoyed a sense of worthwhile achievement.

So far as I am aware, there is no book which sets out to be a comprehensive guide to the would-be ocean voyager, or a reference book for the voyager who has got beyond the planning stage; *Voyaging Under Sail*, which is based largely on my own experiences, is an attempt to meet those needs. When I wrote *Cruising Under Sail* my aim was to make it as complete as my knowledge of coastwise cruising allowed. As any good cruising yacht is capable of making an ocean passage, and as much of the knowledge of seamanship and navigation needed for coastwise cruising is equally applicable when making a deep-sea voyage, there would be no point in repeating here the basic and general information which is to be found in that book. *Voyaging Under Sail* may therefore be regarded as a sequel to *Cruising Under Sail*, and aims at supplying the specialized knowledge sought by ocean voyagers. Perhaps the reader who may be doubtful regarding any details of standard cruising practice, gear, or pilotage will permit me to refer him to the earlier book.

When Mr. T. C. Worth returned home at the conclusion of the fine voyage which he and his wife made round the world in *Beyond*, he intended to write a book about it, but pressure of work forbade. He has generously given me the technical notes which were to have been included in that book, together with permission to make what use I like of them here. For this most handsome action I wish to thank him sincerely. I also wish to express my gratitude to my wife for so patiently drawing and lettering the diagrams.

E. C. H.

Yarmouth, I.W.
January 1959

CONTENTS

Part I

THE OCEAN-GOING YACHT AND HER GEAR

Part II

VOYAGING

APPENDIXES

LIST OF PLATES

Endpapers

Part of American pilot chart of the North Atlantic
(Reproduced by courtesy of the U.S. Navy Oceanographic Office,
Washington, D.C. 20390, U.S.A.)

LIST OF DIAGRAMS

ocean or two, or what they bought, believing them to be as near their ideal as finances would permit; and here it is worth noting that until recently very few long-distance voyagers were well-to-do people, indeed, the majority had some difficulty in making both ends meet, so that they were not able to afford to make all the alterations or refinements they believed to be desirable. Yet a large proportion of them succeeded in doing what they set out to do, which emphasizes once again the truth of the saying: 'It's not the ships but the men in them.'

Nevertheless a suitable vessel suitably rigged and fitted does increase the chance of success, and will undoubtedly make the voyage less anxious and more enjoyable, and as today more small sailing yachts than ever before are being prepared, or are even being specially designed and built, for ocean voyaging, it may be well to examine here some of the basic requirements. But in doing this one must remember that sailing men are individuals; each has his own ideas, his likes and his dislikes, and these are as varied as the craft one sees; some of what follows will therefore be unacceptable by some readers.

Size

I say without hesitation that for ocean voyaging the larger the vessel, provided her management is within the capabilities of her crew, and within their financial scope, the better. Apart from illness or incompatibility of the crew, the only thing likely to spoil the pleasure of a voyage is the exhaustion caused by violent motion and its accompanying lack of sleep, and, other things being equal, the larger the vessel the more comfortable will she be. But with the present high cost of building, very few people can afford to build as large a vessel as they would like, and rather than build a very small one, the would-be voyager might do well first to investigate the second-hand market. It will cost him nothing to find out what is available if he places his requirements in the hands of one or more of the reputable firms of yacht-brokers which advertise in the yachting press; but if he does not want to be inundated with particulars of unsuitable craft, he must be precise in listing his requirements, and honest in stating the price he is prepared to pay.

It is incorrect to assume that a very small sailing yacht is easier to handle than one of moderate size, for she will be quicker on the helm and her motion will be more violent. I doubt whether it is sufficiently appreciated that two competent people can manage a yacht of between 40 and 50 feet in length provided the gear is efficiently arranged, no sail exceeds 400 square feet, and an adequate windlass, winches, and

PART I

THE OCEAN-GOING YACHT
AND HER GEAR

〜〜〜〜〜〜〜〜〜〜〜〜〜〜〜〜〜〜〜〜〜〜〜〜〜〜〜〜〜

1

THE HULL

Size—Seaworthiness—Comfort—Self-steering—Speed—Multi-hulls

LONG sea passages have been made and oceans have been crossed successfully by small sailing vessels of all shapes and types. Three of the most famous and contrasting were Slocum's *Spray*, an immensely beamy saucer of a vessel, Voss's *Tilikum*, a narrow dugout canoe rigged as a three-masted schooner, and *Kon-Tiki*,[1] a raft of balsa logs rigged with a squaresail only. Few would imagine that the owners chose such vessels, the first two for circumnavigating the globe, and the last for crossing the eastern half of the Pacific, as being ideal for such purposes. The wreck of the *Spray* was given to Slocum, and was completely rebuilt by him. Voss wished to sail a vessel smaller than *Spray* round the world to win a sum of money, and at the same time to prove his contention that any type of craft could survive the worst of weather provided use was made of a sea-anchor. *Kon-Tiki* was an exact copy of the ancient Inca rafts, and was built and sailed by Thor Heyerdahl and his companions for the purpose of showing that the voyage from South America to the Polynesian islands could be made by such a craft, to strengthen the theory that the Polynesians originally came from the eastward. Unlike some more recent raft voyages, this was no stunt, but a carefully planned and executed scientific expedition.

A study of small-boat ocean voyages shows that remarkably few of the vessels used were built for that specific purpose. The majority were old, or at least second-hand, and were either what the would-be voyagers happened to possess at the time they decided to cross an

---

[1] Brief particulars of most of the yachts mentioned in this book will be found in Appendix II. When length is given in the text, this is the length of the hull overall unless stated to be otherwise.

auxiliary engine are provided. The late Tom Worth and his wife brought their 43-foot cutter *Beyond* (see pages 100-4 and Plates 18 and 20) back to England from New Zealand without undue difficulty; Sten and Brita Holmdahl sailed their engineless, heavy-displacement ketch *Viking* (Plate 7, *bottom*) round the world without assistance; Bill and Phyllis Crowe performed a similar feat in their 39-foot schooner *Lang Syne*; and Miles and Bea Smeeton were alone in their 46-foot ketch *Tzu Hang* for the greater part of their remarkable east-about circumnavigation, much of which was against prevailing winds and currents. I have mentioned man and wife couples here because I wish to emphasize that it is not weight and strength that count when handling a sensibly arranged vessel, but intelligence, skill, and seamanship.

Macpherson, however, considered that the small yacht had advantages, and having sailed to the Mediterranean and back in the 40-foot cutter *Driac*, decided she was too large for his purpose, and had the 32-foot cutter *Driac II* designed and built, and in her voyaged to every place in the world that he had any desire to visit. Incidentally, Macpherson did not commence his remarkable sailing career until he was nearly 60 years of age. Goldsmith also changed to a smaller yacht, replacing his 55-foot cutter *Madalèna* with the 30-foot cutter *Diotima*; but I believe *Madalèna* called for a powerful crew because of the weight of her gear. Both yachts made the round trip from England to the West Indies while owned by him. Each year sees an increasing number of tiny craft making great voyages, such as *Trekka* (see pages 126-9) which John Guzzwell sailed single-handed round the world; the 17-foot Shetland fishing boat *Venus*, which Paul Johnson decked and sailed single-handed across the Atlantic, and the 16-foot dinghy *Wanderer* in which Frank Dye made a number of voyages in high northern latitudes.

Seaworthiness

The basic requirements for an ocean-going yacht are seaworthiness, comfort, and speed, in that order of importance; but if she will be short-handed, ease of handling and the ability to steer herself should be high on the list, though the latter will not be so important if she is fitted with some form of automatic steering gear (see pages 75-82; 96-8).

To achieve a correctly balanced combination of these requirements is a matter for the architect, and even then is not always solved successfully. I feel strongly that the design of a sailing yacht calls for such a high degree of skill and art that few amateur designers can make a good job of it, and that a great deal of money and effort has been wasted building yachts to indifferent (and very often hideous) designs, when, for a not exorbitant fee, men who have made a lifetime study and practice of naval architecture can be employed. However, although most hull forms have been tried in the past, as a visit to the model room of the New York Yacht Club will show, any new design is bound to be experimental in some degree, and even the most skilful designers do sometimes make mistakes; so the man who is thinking of having a yacht built, but is not prepared to take any risk, would do well to consider a stock design to which a number have already been built, for only a design which has proved to be successful is built to repeatedly, and any small failings in the original will probably have been corrected in later models. Also there will be some saving in cost and in time.

The chief requirements, as I see them, for a seaworthy hull are these: stiffness, so that an adequate area of sail can be carried in a strong wind, for this may be needed to obtain an offing or weather some danger; sufficient buoyancy, so that the vessel may ride over the seas without taking heavy water aboard, even when laden with stores and water for a long passage; and balance of hull, so that the forebody matches the afterbody when upright and heeled, in order that the vessel may not carry excessive helm, thus being easy to steer on all points of sailing. In addition sound construction with good materials is essential.

Stiffness is obtained by deep draught and a ballast keel, or by considerable beam and firm sections, or more commonly by a combination of the two. The out-and-out racing machine is an example of the first; if her keel were to drop off she would capsize. The Thames spritsail barge, of which there are now only a few in commission, is an extreme example of the second; although she has no ballast when empty of cargo, her great beam and box-like section make her phenomenally stiff. The righting moment of a ballast keel increases as the angle of heel increases and makes a capsize almost impossible, but a heavy ballast keel, with the narrow beam that so often goes with it, tends to make a vessel roll heavily. The vessel with firm sections and

great beam, but with no ballast keel, is initially stiff, but if heeled beyond a certain angle will lose stability and may capsize, or a heavy sea, bursting under the wide, flat bilge, might cause that to happen. Clearly the ideal cruiser will comprise a skilful blend of all these properties, and although the amateur designer may speculate, he can scarcely hope to hit on the ideal combination; but that is the business of the professional, and provided he can be persuaded that no consideration of racing or of rating rules must be allowed to influence him—and that may not be easy, for good publicity for the designer lies in the racing success of his creations, and little in their success as cruisers—one will probably get as good a combination as is possible. Obviously one will choose a designer whose creations one admires; if, for example, the owner has a preference for considerable beam and moderate draught, he will go to one who has produced successful yachts of such a type in the past.

Buoyancy may be obtained by light construction and/or considerable freeboard. Aircraft building has shown that strength of construction is not necessarily obtained by large scantlings, but rather by correct engineering principles and first-class workmanship. The latter increases the cost, and one has to bear in mind that some small yards which can build a heavily constructed vessel well enough may not be capable of making a good job if light construction is required.

High freeboard is not popular with most designers, possibly because it increases windage and may look ugly on the drawing-board; yet one is for ever putting things into a cruising yacht and hardly ever taking anything out, with the result that many otherwise pleasing yachts float below their designed waterline, the freeboard becomes too little, causing them to be wet, and the boot-top line, cut in by the builder exactly where the designer hopefully put it on his drawings, is submerged, and very soon weed and oil accumulate on the topsides greatly to the yacht's disfigurement. Few things look more slovenly. Good freeboard is particularly desirable at bow and stern because some stores or gear will inevitably be stowed there unless the yacht is very large; so ample buoyancy must be provided. In this respect reverse sheer, or hog-back profile, falls short of the requirement.

PLATE 2

Top: The Canadian trimaran *Caravel II*, sailed by Fred and Kitty Carlisle and their two small daughters, stops at St. George's, Bermuda, on her eastward way across the Atlantic. *Bottom*: Also at Bermuda, the American 30-foot cutter *Elsie*, in which Frank Casper had made a single-handed circumnavigation and was now bound for Europe. The difficulty of installing trim tab steering gear with a Norwegian-type stern has here been overcome by stepping the vane shaft on the rudder head; the vane has been temporarily unshipped.

Most of the yachts built today, apart from those with reverse sheer, have bows with pleasing overhang and ample buoyancy, but in some the stern terminates rather abruptly in a transom. This kind of stern has certain advantages, notably that it is easy, and therefore cheap, to build, and that the rudder can be removed without much difficulty when the yacht is ashore. But aesthetically a forward overhang calls for an overhang aft as a logical ending to the hull, and as so much of an ocean cruiser's time at sea is spent running, the added buoyancy given by overhang at the stern is all to the good, provided one does not fill it up solid with potatoes or other heavy stores. The kind of stern most to my liking is the short, sawn-off counter, and I prefer this to the canoe stern, pleasing though the latter is, because it provides more buoyancy for a given length of overhang as well as more deck space. That I did not have such a stern in *Wanderer III* (see pages 134-9 and Plate 27) was purely a matter of cost. I have heard it said that the pointed stern, either canoe or lifeboat type, is better than the transom or counter because it 'parts the overtaking seas and does not cause them to break'. I am not impressed by this argument because I believe that if a sea is going to break it will do so irrespective of the shape of stern. Lifeboats have pointed sterns because it is sometimes essential for them to manœuvre stern first in broken water, and fishing boats because this type of stern is less likely to suffer damage in a crowded port. Nevertheless, pointed sterns, particularly of the Norwegian type, are popular, and among the many famous yachts which have them are Moitessier's *Joshua*, Dumas's *Lehg II*, and *Elsie* (Plate 2, *bottom*) in which Frank Caspar made a single-handed circumnavigation.

I understand hull balance to mean that no matter what the angle of heel may be the hull will not alter its fore-and-aft trim. A vessel with heavy quarters and a lean bow, for example, will depress her bow when heeled; much helm will then be required to keep her on course, and in certain circumstances she could become unmanageable. Most modern yachts are well balanced, a few of them so perfectly that they carry little helm and the tiller has no 'feel', so that they are not easy to steer on a set course. But by adjusting the rig or the fore-and-aft trim it is usually possible to upset such perfect balance sufficiently to create a little weather helm.

Comfort

No small yacht is comfortable at sea, but some are less uncomfortable than others. Opinions as to why that should be are divided, but it may safely be said that moderation in everything will tend to minimize the violence of the motion. A yacht with all her ballast concentrated in a ballast keel low down and amidships will have a more violent motion than one in which the ballast is spread out, and one that is narrow more than one with a fair proportion of beam. As so much of the ocean cruiser's time will be spent running, possibly under twin running sails, when there may be no fore-and-aft sail set to steady her, it is important that her roll should be slow and without a jerk; the angle to which she rolls is of secondary importance.

Thames measurement is raised more by an increase of beam than by an increase of any other dimension, and I think it was largely the widespread custom of British yacht builders of quoting building costs at so much per Thames-measurement ton, coupled with the hard-dying belief that deep, narrow yachts were faster and better at dealing with the short seas encountered in home waters, which was responsible for the large number of narrow British yachts with concentrated ballast. A ballast keel does not exert a powerful righting moment until the hull has heeled considerably, and then it tends to stop the roll rapidly and fling the hull back the other way; but a yacht which relies for some of her stiffness on beam has some initial stability the moment she starts to heel, and this tends to ease her rolling motion. This fact was impressed upon me when *Wanderer III* was sailing in company with *Moonraker* from Porto Santo to Funchal. The wind was light and on the quarter, and *Wanderer*'s leeward roll was followed each time by a lurch in the opposite direction, during which her mast passed the vertical position. But *Moonraker* was much steadier; she never rolled to windward, and her mast rarely reached even the vertical position. She has 9·7 feet beam on a waterline of 29 feet and all her ballast is inside. On several occasions since then *Wanderer* has made long passages not in company with, but at the same time as, more beamy yachts of her own length, and I noticed how much fresher their people were on reaching port than were my wife and I.

I do not know whether a yacht can have too much beam, but it is certain that if she is properly designed she can have a great deal without spoiling her performance. Carleton Mitchell's American-designed and -built centreboard yawl *Finisterre* has a beam of $11\frac{1}{4}$ feet on a waterline of $27\frac{1}{2}$ feet, an almost unheard-of proportion in Britain, yet she has been highly successful not only as a cruiser with Atlantic crossings to her credit, but also in racing events, with 15 firsts, 6 seconds, and 1 third place in 28 starts during her first season. Much of the credit for this success, of course, must go to her highly skilled skipper and efficient crew, but such a record could never have been achieved if beam were really as detrimental to speed as it is commonly said to be.

Following the example of passenger ships, some power yachts have stabilizing fins to help steady them in a seaway, but considerable mechanical equipment is needed for their operation, and they are not suitable for use in sailing vessels. However, it is possible to rig up a simple form of anti-roll, or roll-damping, gear in any type of yacht, and I have seen examples in *Rena* and *Widgee*, but I believe the latter yacht used hers only while at anchor. A boom is held out over the side by a topping lift (Plate 4A) and has guys to prevent it swinging forward or aft; its outboard end is provided with a sheave or block to take a line, one end of which is secured inboard. A triangular-shaped piece of wood or metal, its size depending on the size of the yacht—that aboard the 30-foot *Widgee* measured approximately 10 inches along each side (Plate 4B)—has a small hole bored at each corner; one corner is weighted with a piece of lead, and if the device is to be used under way, a fin is arranged to keep the weighted corner pointing in the direction of travel. Equal lengths of small line or flexible wire are secured one to each corner of the triangle, and are brought together at a ring or shackle to form a three-legged span, and the line from the boom is secured to that ring or shackle. Boom and line must be of such a length that there can be no risk of the triangle fouling the hull. When the yacht rolls in the direction of the boom, the lead cants the triangle out of the horizontal plane and causes it to sink. As the yacht starts to recover from the roll, the three parts of the span come taut, and the now horizontal flat surface of the triangle offers resistance to being lifted through the water, and acting through the long lever of the boom therefore reduces the speed and angle of roll in the opposite direction. Some yachts make use of two such devices, one rigged at each side.

Self-steering

Unless she has some form of automatic helmsman, it is important that a yacht which is to make a long passage with a small crew should be capable of steering herself for long periods, and thus relieve her crew from the tyranny of the helm. Steering day and night occupies too much of the time that could be spent to better advantage on other occupations, and it soon becomes a bore; while if the yacht has a complement of only two people, the need to steer will cause a loss of sleep which in turn will deprive them of much of the pleasure to be had from the voyage.

Some yachts are said to be good at self-steering, while others are not so good. But before one can compare such yachts and attempt to discover what points favour self-steering, it is necessary to study the outlook of their skippers. One may honestly say that his yacht steered herself across an ocean, but on investigation it may be discovered that he did not bother himself over-much if she was many points off the desired course. If she would not self-steer on the chosen course, then some course on which she would self-steer was found, even though that might add many miles to the trip; another may have found it necessary to take in all sail except the jib or staysail. But swift and satisfying passages are not made like that, and I regard a yacht as a good self-steerer only if she can be made to steer the desired course without a major sail reduction, and not deviate far from it.

It should be possible to make any yacht steer herself close-hauled, and when running dead before the wind if twin sails are used (see Chapter 4); but here we are concerned only with the hull. For many years it was thought that a long straight keel and a vertical sternpost were essentials, but that is not necessarily so: *Kurun*, for example, with very great rake to her sternpost, steered herself quite well, as do some of the yachts with the fin keel and separate or skeg rudder profile, which became widespread in the late 1960s and continues to increase in popularity. One might suppose that the perfectly balanced hull, which carries little helm on any point of sailing, would make the best self-steerer, but this again is not always borne out by the facts. It is an interesting problem which yacht designers have not yet solved, and one of the best known of them, Robert Clark, has stated that, having produced a number of yachts which have almost reached the ideal of self-steering, he is in a good position to confess that he knows very little as to how it may be achieved, but he favours the triangular ✓ profile; if such a profile is decided on, clearly it must be modified to the extent of having sufficient straight of keel to allow the yacht to take the ground safely (this also applies to the above-mentioned fin-keel type) and without pitching forward to an intolerable degree, and to enable her to be slipped safely.

Speed

There is much pleasure and satisfaction to be had from making fast passages. For this reason, and because a slow vessel is rarely a pleasant one to handle, speed does merit some thought as a desirable quality for an ocean-going yacht. But as it is speed down wind that will mostly be required, waterline length, and sufficient sail area for the days when the wind is moderate, should be the chief considerations. Speed when close-hauled is another matter and is not likely to ✓ be of major interest to the voyager; but if it is sought, some sacrifices in other directions will have to be made, and his ocean cruiser will then be more like an ocean racer in hull form, with minimum windage and wetted surface, a high ballast/displacement ratio and perhaps an uncomfortable motion. A slow yacht is usually one with a clumsy hull form, but on an ocean crossing she may make poor runs for another reason: because of her unsweet lines she may become hard to steer under a press of sail, and her steep displacement waves and disturbed wake may interfere with the overtaking seas, so that for safety her speed will have to be reduced, or she may even have to be hove-to, long before such a course would be considered necessary for a more shapely yacht of the same length. It is almost unheard of for an ocean-racing yacht to be hove-to when her course is to leeward, no matter how bad the weather may be, and this, aside from the pluck and judgement of her skipper and the stamina of her crew, which are most important factors, is because her beautifully shaped and easily driven hull causes so little disturbance and is easy to steer.

Multi-hulls

The advantages and disadvantages of multi-hull craft, in which some successful ocean passages have been made, were discussed in *Cruising Under Sail*, but it may be worth repeating briefly here the salient points. With moderate or fresh winds abeam or forward of the beam such craft are usually faster, and because they do not roll in the accepted sense of the term, may be more comfortable than the conventional yacht; but as their speed depends on their light displacement, performance can be spoilt and seaworthiness impaired by taking aboard the gear and stores generally considered necessary for voyaging. The initial stiffness of multi-hull craft imposes considerable strains on sails and rigging, while the strains on the hulls in a seaway are severe, and structural failures are not uncommon. In a squall the windward hull of a catamaran may lift out of the water, and if it lifts beyond a certain point the vessel will lose all stability and capsize. Nevertheless great voyages have been made in catamarans, starting with the early exploration of the Pacific by the Polynesians. In the mid 1930s Eric de Bisschop sailed his *Kaimiloa* from Hawaii to the south of France by way of the Cape, and in 1967 Dr. David Lewis and his family completed a circumnavigation in *Rehu Moana*.

The risk of a capsize in a trimaran is smaller, and this is possibly why that type of multi-hull is more widely used for voyaging. A typical example is the Canadian trimaran *Caravel II* (Plate 2, *top*), which made a west-to-east crossing of the North Atlantic, manned by Fred and Kitty Carlisle and their two small daughters.

2

CONSTRUCTION AND
GENERAL ARRANGEMENT

*Construction—Ventilation—Drainage—Rust and galvanic action—
Protection against worm—Arrangement on deck—Arrangement below—
Freshwater storage—Maintenance*

A TROPICAL climate, where sea and air temperatures are higher, and intense sunshine is encountered, is harder on a wood or steel yacht than is a temperate climate; yet in general the orthodox methods of construction with traditional materials, such as were described in *Cruising Under Sail*, have been found to stand up very well. However, there are certain considerations to be studied, and some modifications that would be desirable, if a yacht is to be altered or built specially with a view to making a voyage in a hot climate.

Construction

Burma teak is the most suitable material for the keel, planking, and deck erections of a wooden yacht; but iroko, a less costly West African timber, is almost as good, for it has most of the properties of teak and stands up admirably to hot weather without warping or splitting; it is the same weight as teak. Genuine African mahogany, by which is meant the pure species *Khoya Irorensis*, is also an excellent timber, and is lighter than teak. But unfortunately the name African mahogany was given to a number of inferior timbers which soon after the war were widely used in Britain for the planking of yachts; these were brittle and liable to a form of decay in way of certain fastenings, notably where copper rivets were used to secure iron floors. Because of the lack of oak with suitable grain, and the space occupied by them, grown frames are no longer used, their place having been taken by steam-bent or laminated timbers, which are highly satisfactory. Composite

PLATE 3

Left: A folding dinghy stows neatly on *Wanderer II*'s narrow coachroof. A bridge deck, seen here with the compass mounted beneath it, should separate the cockpit from the coachroof, its full-length beams, *top right*, strengthening the yacht at a weak point. If the mast passes through or is stepped on the coachroof, this should be strengthened with steel frames and beams combined, *bottom right*.

construction, in which the frames and sometimes other members are of steel, and the planking of timber, is rarely used today, and creates certain galvanic problems.

Sunlight is particularly hard on the deck, and except in large yachts, where the deck can be of a considerable thickness of teak without impairing stability, laid decks are not satisfactory as they generally leak. A light-weight deck of Canadian red cedar—a timber which is not liable to rot—covered with canvas and painted, serves quite well in small craft so long as the canvas does not get torn; in the ordinary course of cruising there should be no reason for this to happen, but in some ports the yacht may be boarded by officials wearing nailed boots, and even the ordinary visitor is not too careful about his foot-wear; I have noticed that women visitors are more ready than men to remove their shore-going shoes when coming aboard.

There are various other deck coverings on the market, and of these Cascover (see page 26) is much more expensive than canvas. A coating of glass reinforced plastics (g.r.p.), which can be given a non-slip finish, does not seem to be affected by the sun, and requires little maintenance, appears to be satisfactory, though I have seen yachts in which it has lifted from the wood after several years of service. Hulls built entirely of this material are, in these days of mass-production, becoming increasingly numerous in both British and American waters, indeed, in the U.S.A. new construction in timber is now rare and very costly; when a sufficient number of g.r.p. hulls can be formed in one mould the cost should be lower than that for wood construction. The outstanding of its several advantages over timber construction for ocean-going yachts are these: No risk of leaks, galvanic or electro-chemical action, more internal space (water and fuel tanks are built in as part of the hull), it requires little maintenance, and there is no possibility of attack by marine borers. Nevertheless the material is comparatively new, and one can only guess at its life expectancy.

The building of small yachts in metal has never become common in Britain, where timber is still the most widely trusted, though not now the most widely used, boatbuilding material. But a few have been built of aluminium alloy, and one of the most notable of these was

---

PLATE 4

A: A simple form of roll-damping gear aboard *Widgee*. A boom is held out one side by topping lift and guys, and takes a line which is secured to a 3-legged span made fast, *B*, to a triangular piece of wood weighted at one corner. *C*: A windsail hoisted on the staysail halyard with its tube passing down through the forehatchway. *D*: A larger windsail occupying the whole hatch-way, as aboard *Altair*, is more effective in reversing the normal aft-to-forward circulation of air below deck.

Beyond; her owner told me at the end of his two-year circumnavigation that he was completely satisfied with it, though since then I understand many rivets have had to be replaced. The ability of alloy to withstand corrosion during long periods of neglect was shown by the yacht *Coimbra*. This 40-footer was wrecked on the island of Tristan da Cunha while bound from England to South Africa. Five years later *Speedwell of Hong Kong* called at the island, and her owner reported that the alloy remains of *Coimbra* showed no sign of corrosion.

I cruised in company with the steel-built American cutter *Altair* (Plate 30) as she was concluding a two-year Atlantic cruise, and she seemed to be in excellent condition; so, too, did the German sloop *Kairos*, also built of steel, when I saw her at the end of her three-year circumnavigation. But the comments of Louis Van de Wiele, after sailing the steel ketch *Omoo* (pages 114-18) round the world, are of interest. He remarked that although much is to be said in favour of a steel hull, such as strength, rigidity, watertightness, and gain of space, he was convinced that a wooden hull is superior for ocean voyaging and long stays in tropical waters. The foremost maintenance problem he had to face was to get paint to adhere, particularly below the waterline, although he tried a variety of paints and methods. He also had a good deal of galvanic trouble. If he were to do it again, he said, his choice would be a wooden ship, preferably planked with teak, and copper fastened. It was therefore with particular interest that I noticed, when I later had the pleasure of meeting Louis and Annie Van de Wiele cruising the West Indies, that *Hierro* (Plate 7, *top*), their new ship, is, after all, built of steel. But, of course, since *Omoo* was built some advances have been made; shot-blasting is now considered to be the best method of preparation for the coats of special paints which are available, and it is now understood by most yards that paint may be applied only when the steel is absolutely dry, which means that it must be at the same temperature as the surrounding air or it will be wetted by condensation. More, too, is now known about galvanic action and its prevention; see page 20. Nevertheless, the maintenance of a steel yacht is greater than that of a wooden one; see page 37.

One might scarcely regard ferro-cement as a pleasant or particularly suitable boat-building material; nevertheless it has been used from time to time and appears to be gaining in popularity among amateur builders; one famous ocean-voyaging yacht to be built of it was the 53-foot Uffa Fox-designed cutter *Awanhee*, in which Dr. Griffith with

his wife and young son sailed west-about from New Zealand south of
the three stormy capes, a tremendous test of ship and crew against the
prevailing westerlies.

Ventilation

In hot climates troubles in wooden hulls are most likely to be caused
by lack of ventilation, imperfect drainage, electrolytic action, and
worm. When a yacht is being built it is possible to attend to most of
these things, and at least some of them can be dealt with in a yacht
which was built with little regard for them.

The aim when planning the ventilation system should be to ensure
that there is no part of the vessel where air cannot circulate. In the
majority of yachts the normal flow of air is in through the after companionway and out through the forehatch; to encourage this flow the
forehatch may be fitted with side flaps (Plate 6B and C) so that it will
form an efficient extractor; the flaps should be hinged so that they
can be folded in when the hatch has to be completely closed. When
Wanderer III was running in the trade winds her hatches were open,
or partly open, most of the time; but when she was close-hauled or
reaching in a fresh wind and her forehatch had to be closed, the ventilation depended on two water-trap vents at the forward end of the coachroof (Plate 6D), and two straight-through vents at the after end within
reach of the helmsman; all four cowls were rather larger than are
fitted to most 30-footers, with 9-inch-diameter mouths and 4-inch-
diameter necks; in addition there was a mushroom vent at bow and
stern, and gratings in the cabin sole forward and aft. But adequate
ventilation of the living space does not necessarily mean that the rest
of the vessel is well ventilated, and trunked ventilation to the engine
space and the bilge may be required. As warm air rises, it is important
that the ceiling should be stopped short of the covering board, or that
sufficient holes be bored there, to enable the air to circulate. In addition
to the normal ventilation it is a good plan to provide an extractor over
the galley. The efficiency of the ventilation may be tested with smoke.

Everything possible should be done to keep the yacht cool and airy,
for she will often be floating in water with a temperature between 70°
and 80 °F., and throughout the day will probably have the sun beating
down on her. Opening ports in the coachroof coamings make a great
difference to the comfort below, and at sea even in moderately rough
weather it is often possible to keep those on the lee side open. Provided they are fitted with soft rubber gaskets, which will need renewing from time to time as the sunlight perishes them, and are never

screwed up so hard as to distort their frames, they should not leak, while any condensation on their glasses, or any drips there may be when they are opened, can be taken care of by a handrail in the form of a drip-catcher, as may be seen in Plate 27, *bottom*. There will be no need to fit a drain pipe to the lowest point of the handrail, for the small quantity of water collecting there can be removed with a sponge-cloth. Skylights also are a great help in promoting freshness below, but as they will have to be closed during rain or when there is much spray on deck, they should not be regarded as a part of the normal ventilation system. So that it may scoop in a good supply of air, a skylight may have hinges with removable pins at its forward and after ends, or perhaps be made to lift off its coamings and turn round, as in *Wind's Song* (Plate 5, *bottom*).

A windsail will do much to make life pleasant when at anchor. Many years ago I made one from cotton spinnaker cloth, but terylene is a better material to use because of its mildew-resisting properties. The upper end of its 14-inch diameter tube was held open by a wooden ring to which it was sewn. The air-scoop was triangular, and measured 5 feet along the foot and 3 feet along each of the other sides: the centre part of the foot was seized to one-half of the circumference of the wooden ring. The scoop was lightly roped and had a lanyard at each corner. With the intention of increasing the natural forward-moving draught below, the windsail was at first rigged with its tube entering the main companionway; but there it was so inconvenient that it was moved to blow down the forehatch, being hoisted on the staysail halyard with its corners hitched to the guardrails (Plate 4c). When the wind was fresh it successfully reversed the natural draught and poured a fine current of fresh air through the cabin; but in light winds it could not overcome the natural flow of air. Possibly its scoop was too small, but I believe it would have worked better if it had occupied the whole area of the forehatch so as to prevent an escape of air there. *Altair* had one, though of much greater size, so arranged (Plate 4D), and it worked splendidly.

A double deck with an insulating material, such as glass wool, in between, will help to keep the accommodation cool, and is essential

---

PLATE 5

Top: At sea the fore part of this terylene awning can be reefed to keep it clear of the runners; its ridge rope is held up by a short spar from the stern of the capsized dinghy, and its after end is spread by a boathook lashed to the boom gallows. *Middle*: A weathercloth rigged to shelter the cockpit when the wind is forward of the beam. *Bottom*: *Wind's Song*'s skylight can be lifted from its coamings and placed with its hinges fore-and-aft or athwartships, so as to extract or blow with the wind in any direction.

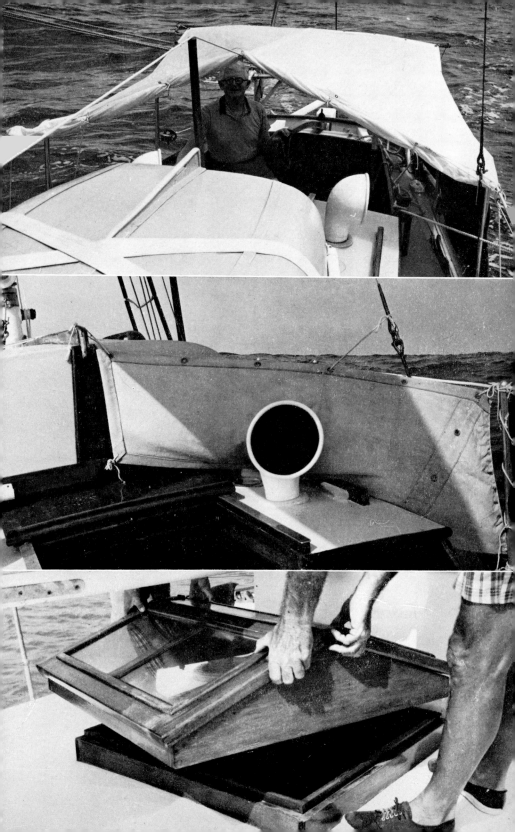

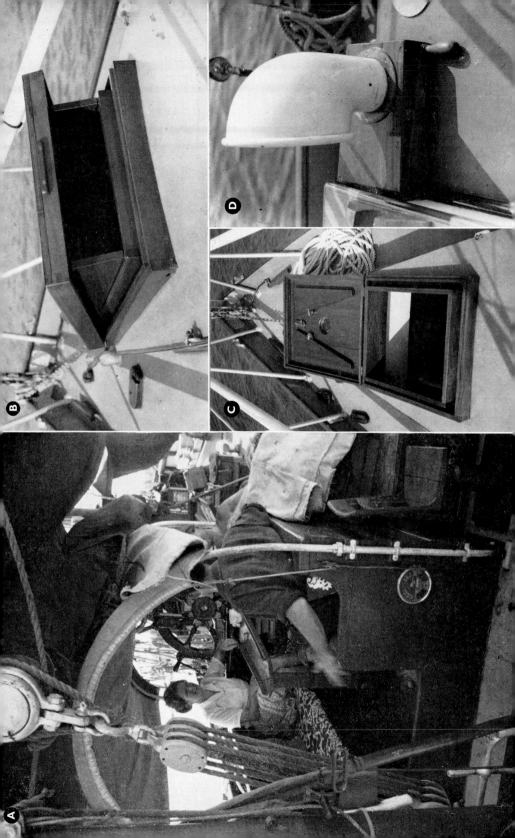

with a steel deck to prevent condensation. The colour of the deck also has a direct bearing on the temperature; the lighter it is the cooler will it and the accommodation be; but a very light colour is hard on the eyes in continuous sunlight. Topsides should certainly be painted a pale colour, preferably white, for that absorbs the least heat, keeps the hull cooler, and tends to reduce the risk of seams opening or paint blistering.

In a hot country an awning is more important than a windsail, and is essential if one wishes to sleep on deck. Ideally it should cover the whole deck, but because of masts and shrouds this is not easy to arrange. An awning (Americans often call it a canopy) of lightweight material can be stowed away in a small space, and is easy to spread or furl, but if it is of white synthetic material it may let too much light through, and in anything of a breeze will flap and rattle. The first awning my wife and I made extended from mast to backstay, and the topping lift was shackled to an eye seized to the middle of the ridge rope to prevent sagging. The after end was spread by the boathook, and a number of lanyards held it to guardrails, runners, and shrouds. As it was spread above the boom one could move about under it with fair ease, but because of its height above the deck, so much sunlight got in one side or the other during the early morning or late afternoon that we fitted it with side curtains to reach down to the guardrails when wanted. With this protection from sun and rain we took all our meals in the cockpit, the evening meal being lit by a paraffin pressure lamp on the main hatch. But that awning was not suitable for use at sea, and as it was so complicated and took so long to spread, our next one was smaller, extending only a short distance forward of the runners, and as it was spread beneath the boom, side curtains were not required. Its after end was spread by the boathook, shipped in chocks on the after side of the boom gallows, and when it was in use at sea its fore part could be reefed (so as to leave the runners clear) and a short pole secured to the stern of the capsized dinghy held the ridge rope from sagging (Plate 5, *top*). However, after we had shipped a vane steering gear there was not often much need for a cockpit awning at sea, but for the odd occasion when a helmsman was needed and required shelter, we carried a 5-foot square of terylene which could usually be

PLATE 6

A: *Arthur Rogers*'s cool little 'deckhouse', which could be fitted with mosquito netting to make a sleeping cabin when in port. *B*: A double coaming forehatch fitted with side-flaps to form an extractor. *C*: The side-flaps are hinged and can be folded flat and secured with turn-buttons when the hatch has to be completely closed. *D*: A water-trap ventilator fitted at the break of a coachroof; a winch handle stows on the side of the box.

C

rigged up in some fashion to provide shade, and in port we rigged it above the forehatch so that the hatch did not have to be closed in rain. A larger vessel might have a permanent wooden shelter over the steering position, such as was used aboard *Havfruen III* during her circumnavigation, or a semi-permanent canvas awning, as aboard *Arthur Rogers*, where a table and seats were arranged on deck close abaft the wheel, and the whole was covered by a small awning on steel hoops. This is shown in Plate 6A, but here the big harbour awning, which covered most of the deck, is also spread. In hot weather all meals were taken in this cool little 'deckhouse', and in port at night mosquito netting was arranged across the open ends, and the owner and his wife slept there.

On the western side of the Atlantic a popular material for awnings is a proofed cotton known as Vivitex; this is resistant to mildew and does not flap unduly. Some awnings have no ridge rope, but are provided with athwartships pockets into which wood or metal battens are shipped; they can therefore be wider than the yacht's beam, thus giving better shade, and can be tilted a little one way or the other to guard against a low sun; but the long battens present something of a stowage problem when not in use.

Drainage

Pockets of salt water trapped in a wooden vessel will do no harm, but accumulations of fresh water may cause rot. With a self-draining cockpit there will be little risk of rainwater getting below, but at times there may be some condensation; this will be most marked when the sea temperature is considerably lower than the air temperature, as, for example, in the Humboldt and Benguella Currents, or after crossing the Gulf Stream. The most likely places for this water to lodge are on top of the butt straps at the turn of the bilge (Fig. 1A). In building, this can be prevented by making the straps a little shorter than the distance between a pair of frames, so as to leave waterways (Fig. 1B), or by shaping their tops to slope; but in an old yacht where the straps are rectangular and fit tightly between the frames, a triangular-section graving piece can be glued on top of each strap (Fig. 1C).

The bilge should have a clear run so that all water will find its way to the pump-well. To ensure this with the usual method of construction, in which each frame is bound to its opposite number and to the keel by wrought iron or bronze floors, limber holes are cut in the keel beneath each floor; to keep these clear of fluff and other rubbish, which will inevitably find its way into the bilge even aboard the cleanest of yachts, a small chain should be rove through the limber holes so that a pull one way or the other will clear them; this chain should be of stainless steel. The alternative is to put cement in the bilge to make a clear, unobstructed run for the bilge water over the throats of the floors. If this is to be done, it will be wise to have a surveyor's report

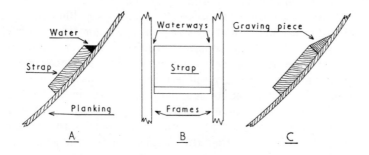

FIG. 1. BUTT STRAPS

A: The usual arrangement where condensed water can lodge on top of the strap.
B: A shorter strap leaving waterways between its ends and the frames. *C*: Strap fitted with a graving piece.

on the state of the hull, and a certificate from the firm entrusted with the work stating why the cement was put in and exactly how it was done, otherwise there might be difficulty in the event of the yacht changing hands at a later date. Pitch is unstable and will move if the yacht is on one tack for any length of time in warm water, while anything placed on it tends to stick or sink in. No matter what steps are taken to promote thorough drainage, it will be as well to sluice plenty of salt water through the bilge from time to time, particularly when condensation is likely; this will also provide an opportunity for testing that important but rarely used item, the bilge pump.

Sometimes the trim of a yacht is slightly altered when she takes on fuel, water, and stores, and then when in port it is possible that rainwater may lodge in the angle between covering board and bulwarks; small holes as additional scuppers should be drilled at those points.

In a steel yacht care should be taken to ensure that neither fresh nor salt water can lodge anywhere below, as, for example, in the angles made between keel, frames, and plating, or on the tank tops. Cement, worked well up the side of the structure, is commonly used to prevent such puddles forming, and offers good protection to steel; but as it is liable to shrink, its edges should be examined, say, once a year, and if a gap or crack is discovered the cement should be chipped away to form a small gutter, and this should then be filled with some soft stopping, such as Selastic, carried well up the steel.

Rust and galvanic action

Rust is a nuisance in spoiling the appearance of a yacht, and is difficult to stop once it has got properly started; it may even be dangerous if allowed to continue unchecked for a long time, although a great deal of rust is produced by a very small wastage of metal, and of itself, so long as it is not being constantly rubbed off, it provides some measure of protection to the steel beneath. The best protection to iron or steel fittings exposed to salt water is galvanizing, whereby after a bath in acid to remove all foreign matter, a coat of zinc is bonded to the metal. If this is well done it will form a protective coat which is serviceable for many years; but in recent years much of this work has often been badly done, so that after a season or two the fittings rust and need to be removed for regalvanizing. This is one reason why stainless steel fittings are so widely used now, and although a brown discoloration often occurs on them, this seems to have no ill effect, and can usually be rubbed off with a plastic saucepan scourer. There are on the market so-called 'cold galvanizing' paints, rich in zinc; these are useful for touching up small bare patches, but the coating they provide is soft and is no substitute for real galvanizing. To protect steel by zinc-spraying is very expensive, and is only effective if the surface is perfectly prepared, i.e. without a trace of scale or rust, and it must be absolutely dry. Clearly it is impracticable to galvanize or to zinc-spray an entire steel hull, and here protection must depend on paint, and any damage to that paint must be made good immediately before rust can form. Steel is more prone to rust than is iron, but the latter is not now much used for yacht fittings because, for a given scantling, it is not so strong, and in yacht construction light weight is generally sought after today.

The greatest structural worry likely to confront the owner of a wood or steel yacht cruising in warm water is the galvanic action which is caused when dissimilar metals are immersed in salt water. An electric current will flow between them causing decay of the less noble metal, and sometimes a decay of the wood surrounding either the noble or the ignoble metal; the warmer and more saline the water the more rapid is the action likely to be.

The obvious solution is to avoid the mixing of metals, but this is not so simple as it might at first appear, and in practical yacht-building is virtually impossible. One of the dangers of galvanic action is that it is not always apparent; the head of a bolt or rivet, for example, may appear to be perfect, but on being twisted or tapped it may break away to reveal that the stem is badly corroded. There is also a type of corrosion of yellow brass in which the interior becomes crystalline and weak, while the outer surface looks perfect. Rapid rusting of steel or iron, a whitish deposit on, or a brightening of yellow metal, or the flaking off of paint, should be investigated as these are often the first signs of galvanic action. In a steel yacht there may be an action between the skin fittings and/or the propeller and the steel plating, to the detriment of the latter. In Britain, so far as I know, skin fittings and sea-cocks are not available in steel in the small sizes needed for use in yachts, and there appears to be no alternative to gunmetal. But at least one firm of electronic engineers, Brookes & Gatehouse of Lymington, make nylon-coated housings for their echo-sounder transducers when these are to be used in steel yachts. In the U.S.A., however, the Jamesbury Corp. of Worcester, Mass., manufactures a range of ball valves which are ideal as sea-cocks, for they require no lubrication and are available in carbon steel, stainless steel, alloy, and monel.

I mention galvanic action not to dismay the would-be voyager, but to remind him of the kind of trouble for which he should be on the alert unless he owns a g.r.p. yacht. Metal fittings should be examined occasionally and a few bolts and rivets driven out for examination; care should also be taken to see that there is no leak from the electric system due to faulty insulation.

The simplest way of guarding against galvanic action is a good coating of non-metallic paint, or perhaps an application of Limpetite (see page 26), which will act as an insulator; but it is almost impossible to keep every part of a steel yacht properly insulated, for the coating on the under side of the keel is certain to get damaged while being launched from a slip, or while taking the ground, and a galvanic couple might then be formed between the keel and the propeller shaft. For many years it has been the custom to fit sacrificial plates (zinc anodes), which of course must never be painted, at the points most likely to be attacked, the intention being that the zinc, which is low in the order of electric potential, should waste away instead of the metal it is supposed to protect. Provided the anodes are of pure (not commercial) zinc, and that they are renewed (making sure the electric contact is good) before they have completely wasted away, they appear to be effective; but if there is any doubt about their quality or their correct positioning, it will be wise to consult one of the firms, such as M. G. Duff & Partners, which have made a close study of galvanic action. There is another but more positive system of protection by means of anodes through which a small current from the vessel's electric supply is caused to flow; this is widely used by naval ships.

Protection against worm

The risk of attack by teredo worm is particularly great in warm water. This creature is so small as to be almost invisible when it attaches itself to the planking and looks for a suitable place at which to enter. Having found an area of bare wood, and this need be no larger than a pinhead, it bores straight in for a short distance, then grows and becomes worm-shaped and proceeds to bore with the grain, rarely or never breaking surface, and it can do most severe and probably unsuspected damage, reducing the wood to a honeycomb. In many warm-water ports I have heard during the silence of a calm night a faint, but almost continuous, clicking and crunching noise coming from below the waterline. For a time I believed this to be made by teredo worms at work on the hull, but I know now that it cannot have been that. I have been told that the noise is made by ✓ small crabs.

It has often been said that teak is not attacked by worm, but I do not believe this; there are many teak-planked vessels afloat in the tropics, but that may be only because they have been kept well protected with paint; it would, however, seem that teak is less likely to be attacked than other timbers. Nevertheless, there are many vessels which are not planked with teak that have survived long periods in worm-infested waters; a notable example was *Arthur Rogers*, in which Tom Hepworth and his wife spent at least ten years cruising and trading in the tropics (see page 271).

Paint is an effective protection against worm, but there is a risk of it being chipped to the bare wood, perhaps by a carelessly handled anchor or by a stranding, and unless the damaged paint is immediately made good the worm could enter. In this connection one should bear in mind that it may not be possible to repaint the bottom of a normal yacht fitted with a ballast keel when cruising in distant waters. There may be no available slip, and the range of tide may be so small that drying out on legs or alongside a quay for repainting is not possible. Among the islands of the central South Pacific, for example, the tidal range is only between one and two feet (it may be of interest to record that when I was at Tahiti in late March and April, high water was always at noon and midnight irrespective of the phases of the moon), ✓ and the only slips, for which there is usually a long waiting list and for which the charge is high, are as far apart as Tahiti, American Samoa, and Fiji. However, even with only a small range of tide a surprisingly large area of the bottom can be painted if the yacht, after taking the ground, is allowed to lie over on her bilge; but of course two tides are required to do both sides.

An old-fashioned yacht, or fishing-boat conversion, having all her ballast inside, can be careened, i.e. hove over on to her beam-ends so as to expose one side of her underwater body from waterline to keel. Before this can be attempted, all the ballast and every movable object on deck and below must be taken ashore, filling caps and air-vents of all tanks must be made leak-proof, and the yacht be securely moored fore-and-aft. A powerful tackle is then secured to the masthead and to some immovable object ashore, or to another vessel, and the yacht is then brought to her beam-ends by heaving on the tackle until the mast is horizontal. I watched this interesting and seamanlike operation being performed by Sten Holmdahl and his wife with their 33-foot ketch *Viking* (Plate 7, *bottom*) in English Harbour, Antigua, a place to which it was nothing new, for in the days when the harbour was used as a British naval base the ships of the fleet were careened there; indeed, three of the old capstans to which the falls of the tackles were led are still standing. Working in their usual quiet and orderly manner, the Holmdahls performed this operation in a day, securing their tackle to a great anchor which lies on the quay.

But as it is not possible to careen a yacht which has a ballast keel, obviously it is desirable that her underwater body should have some form of protection of a more permanent nature than paint. Copper sheathing, which is easily cut and dressed to shape, is the traditional material, and this has the advantage that when it is immersed in water a constant process of exfoliation—formation of soluble salts of copper—takes place, and this discourages marine growths such as weed and barnacles. When copper sheathing is new it has greater antifouling properties than any paint with which I am familiar; indeed, for the first two years of its life on the bottom of *Wanderer III* it fouled so little that my wife and I, wearing diving goggles, had no problem in keeping it clean while the yacht was afloat by giving it an occasional scrub with a hand-brush while bathing; for this purpose we passed a rope from rail to rail under the keel to provide a hand-hold, and moved it along a few feet at a time as necessary. We did this only while at anchor in clear, shallow water where a lookout could be kept for sharks and barracuda. But as oxidization takes place the sheathing becomes less effective; burnishing may give it a new lease of life for a limited time, but eventually it will have to be coated with antifouling paint if the need to scrub at frequent intervals is to be avoided.

Naturally the process of exfoliation slowly destroys the copper, though in British waters sheathing has often had a life of 15 years. But, unfortunately, in some other waters it is subject to a form of corrosion, notably along the waterline, but also on the trailing edge of the rudder and in the vicinity of the propeller. The first time I noticed this was on arrival at Barbados after the 2,700-mile passage from the Canary Islands, when a band of copper about 1 foot deep along the waterline on both sides was seen to be of a pinkish colour, and on investigation was found to be rough and pitted. After some days at anchor the normal dull green colour replaced the pink. On arrival at Panama the corrosion was again noticed, and so on after each long passage, until, on arrival in New Zealand, by which time the sheathing was 20 months old and had spent 12 of those months in warm water, the copper along the waterline was paper thin and holed in many places. The waterline was resheathed at Auckland, and the rudder and propeller aperture sheathing patched as necessary, but on reaching England 20 months later the waterline again had to be resheathed. The copper was surveyed by Lloyd's and pronounced to be 99 per cent pure. Other yachts have experienced similar trouble, including *Diotima* and *Moonraker*. The waterline of the former had to be resheathed at Gibraltar after a year spent on the coasts of Spain and Portugal, and again after crossing the Atlantic to the West Indies. Her original sheathing was the usual 16-ounce, but 18-ounce was used for the waterline in the West Indies and no further trouble was experienced on the homeward passage by way of Bermuda. *Moonraker*'s sheathing had to be patched along the waterline on several occasions, but after six years, two of which were spent in warm water, all of it was in such a bad way that it had to be renewed. That there is nothing new about this kind of corrosion is shown by the fact that the copper sheathing on H.M.S. *Alarm* (one of the first Admiralty vessels to be copper-sheathed) was eaten through on both bows during a voyage to Australia in 1765, and had to be renewed after 15 months.

Although I had made a number of inquiries, no reasonable explanation for this kind of corrosion—which, obviously, is not due to dissimilar metals, for copper is the noblest of any used in normal yacht construction—had been offered until I heard from Gordon Kells, a New Zealander who had made a study of it. His explanation is interesting, and I quote from his letter.

A simple experiment will demonstrate the major points. Boil a quart of seawater to remove entrained air, and put it in a glass jar. Cut two pieces of copper sheet about two inches square, clean them with glasspaper, solder to each a yard of insulated wire and connect these wires to a meter with a reading of from 10 to 20 milli-amps. Immerse the copper plates in the jar of water, then lift one out for a few moments and its wet surface will be seen to become aerated. Return it to the water (electrolyte) and note the current flow on the meter; then repeat with the other plate, and the current will be seen to flow in the opposite direction. The plate with the aerated surface will always be positive. Now shake the container for a few minutes, allowing plenty of air to enter the water, and then repeat the aeration and immersion procedure as before, and it will be seen that there is no appreciable current flow. Faraday's law of electrolysis states that 'the weight of metal removed from the negative of a solution cell is directly proportional to the current flow'. With fully aerated electrolyte no current flows when one part of the conductor is exposed to further aeration, but with unaerated electrolyte continual wetting and aeration of one portion only of the exposed copper creates a big current flow. *Wanderer* and other yachts on the English coasts, in the Gulf Stream and in some parts of other oceans, encountered well aerated water, so the sheathing would not be subjected to differential aeration and corrosion. But when passing through the upwelling unaerated deep ocean waters, the exposed copper was continually wetted and aerated and given positive polarity. Corrosion would take place on the negative area at the boundary, and could be a hundred times more rapid than in home waters.

At least one thing emerges clearly from this; when sheathing a vessel with copper, the sheathing in the neighbourhood of the waterline should be of greater thickness than is considered necessary elsewhere in order to give it a longer life. Alternatively one of the antifouling paints containing a high proportion of pure copper could be used there in the expectation that it might be wasted away rather than the sheathing. (Note that paint containing magnesium must not be used on a copper-sheathed vessel, and that paint containing copper must not be used on steel or alloy vessels, or serious damage may result.)

As well as the disadvantages mentioned above, copper sheathing can cause severe galvanic action with other metal parts, such as keel bolts, rudder pintles, skin fittings, and propeller, and for this reason I would avoid its use in any future yacht I might own. At present there are at least three alternatives to copper sheathing: g.r.p., Cascover, and Limpetite. All provide excellent protection against attack by worm, but none of them have any anti-fouling properties. G.r.p. has already been mentioned as a coating for decks, but some yachts have been sheathed entirely with it, notable examples being *Treasure* and *Rena* (pages 118–22 and Plates 22, 23, and 24, *top*), both of which were built and sheathed by their owners. It can be applied success-fully only to a new yacht with close or splined seams, and before any part of her has absorbed salt water; even then its adhesion may not remain perfect, as it has no elasticity and cannot therefore expand or contract with the wood. Cascover, manufactured by Leicester, Lovell & Co. of North Baddesley, Southampton, is a single layer of tough nylon cloth, bonded to the hull with resin glue, and covered with three or four coats of vinyl paint; it is able to move with the timber to which it is stuck. Like g.r.p., Cascover is best applied to a new hull, and the work is normally done by the firm's operatives, a minimum temperature of 60 °F. having to be maintained during application. Limpetite, a self-curing synthetic rubber, is made by Protective Rubber Coatings Ltd., of Payne's Shipyard, Bristol 3, and is applied in liquid form by brush; skin and rudder fittings do not have to be removed, the owner can do the work himself, and the cost is said to be less than that of copper sheathing.

Arrangement on deck

It is important that the cockpit should be self-draining, for circum-stances may arise in which a considerable quantity of water will be shipped at frequent intervals, notably when sailing with a heavy beam or quartering sea. This water will be unlikely to endanger the yacht, but if it is able to find its way into the bilge, much pumping will be needed. So that the weight of water shipped in the cockpit shall not tend to hold the stern down for long and, by reducing buoyancy, encourage more water to come aboard, the drain pipes must be large, at least $1\frac{1}{4}$ inches in diameter for a small cockpit. Sea-cocks should be fitted at these and all other inlets and outlets in the hull, otherwise a fractured pipe or leaking union could be a serious matter.

If the cockpit of a small yacht is not sufficiently high above the waterline to enable it to be self-draining, its well might be fitted with a waterproof canvas lining, such as I made for *Wanderer II*. The upper fore-and-aft edges of this lining were secured with tacks to two stout wooden bars. With the lining in place, the bars, supported at each end by wooden chocks fixed permanently to the ends of the cockpit, lay snugly under the overhanging edges of the seats, and were wedged securely by crossbars driven home at each end of the cockpit. In the event of the cockpit being filled, such water as was not quickly thrown out by the yacht's motion could be baled out with a bucket. R. D. Graham had a similar lining for *Emanuel*'s cockpit but never had occasion to use it, and I never needed to use mine.

If the yacht has a coachroof this should be separated from the cockpit by a bridge deck (Plate 3, *left*), for the full-length beams of the bridge deck will strengthen the yacht at what would otherwise be a weak point (Plate 3, *top right*). But whether this is done or not, the coachroof coamings are likely to be the weakest points in the whole construction. The rest of her will have closely spaced timbers and deck beams, reinforced by stringers, shelves, and hanging and lodging knees; but the coamings usually consist of wide planks standing on edge and through-bolted to the carlines, relying entirely on their own initial strength and that of their bolts. This weakness is aggravated when the mast is stepped on the coachroof and/or the aft end of the coachroof is raised to form a doghouse. During her westward crossing of the North Atlantic by the northern route, Humphrey Barton's 25-foot sloop *Vertue XXXV* (Plate 1, *right*) was boarded over the quarter by a phenomenal sea during a great gale, and one of the coachroof coamings was split for its entire length. R. McIlvride's 35-foot cutter *Gesture* from New Zealand, also with coachroof and doghouse, was knocked down on to her beam-ends during a hurricane near Lord Howe Island; all the bolts securing the lee coaming were sheared off and the coaming was driven $1\frac{1}{2}$ inches inboard. Although freak seas capable of causing such damage are rare, it is obvious that these weak points do need strengthening, and although it is common practice today to fit steel angle beams and frames combined in way of the mast (Plate 3, *bottom right*), wider use might with advantage be made of such reinforcements at other parts of the coachroof, particularly at the after end and in way of the doghouse. A full-width raised deck amidships with raised topsides is much stronger than a coachroof, and has the additional advantages that it gives more room below, increases buoyancy in the event of the yacht being hove far over, and is not subject to the troublesome leaks which are so common along the carlines of a coachroof. This arrangement appears to be more popular abroad than it is at home. Two notable yachts that have it are the Australian *Waltzing Matilda* (pages 129-34) and the South African *Wylo* (Plate 43, *bottom*). The latter is of particular interest because she is a replica of the 34-foot, hard-chine yawl *Islander*, in which Harry Pidgeon twice sailed round the world single-handed.

Wylo was built at Cape Town by her owner, Frank Wightman, and was sailed by him to Trinidad. The only major difference between the two little vessels is that *Islander* had a conventional coachroof. A much more recently designed and built yacht with a raised deck is *Outward Bound*, which John and Mary Caldwell built in Australia and with their two sons sailed by way of Suez to the Caribbean, where they did chartering work with great success. The impression of space and airiness which this form of construction provides can be seen in Plate 9D.

The central cockpit, as in *Beyond* (Plate 20, *top*), has the merit of enabling the engine, batteries, etc., to be installed in a dry and accessible position amidships, where their weight does not upset the fore-and-aft trim. But with the wind forward of the beam it is liable to be wetter than a cockpit at the stern; also the entry to the after accommodation is vulnerable, so in some yachts which have been developed from the *Beyond* design, the cabin aft is reached by way of a passage running from the forward accommodation beside the engine, one of the cockpit seats providing the required headroom. However, the majority of sailing vessels are steered from aft, which is the traditional position, and from there a better view of the sails can be had. To help keep the helmsman or watchkeeper dry it is usual to rig weather-cloths on the guardrails each side of the cockpit, and sometimes across the stern. These do much to make the cockpit snug and dry, but as the ocean-voyaging yacht spends so much of her time at sea with the wind abaft the beam, it would appear sensible to provide her with a raised stern, or poop, for every inch that the helmsman can be raised above the sea will add to his comfort. The late Conor O'Brien was a staunch advocate of the poop, and anyone who has been aboard the 42-foot ketch *Saoirse* (pages 122-6 and Plates 25 and 26), which he designed and sailed round the world by way of Cape Horn, will at once have realized how wise he was. Disadvantages of a poop may be slightly increased windage aft, though this need be no greater than that caused by many a yacht's doghouse, and the problem of disposing of water trapped on the weather side-deck if there is no break between the after end of the coachroof and the poop; but if an athwartships pipe is fitted between the coamings at deck level, the trapped water can run through to the lee deck and escape through the scupper there; however, such a pipe may hamper movement below. In a large yacht a poop should enable a great cabin with stern windows to be arranged, a delightful thing if skilfully planned by an able designer, but some recent attempts appear to be little more than caricatures.

If much windward work is envisaged, some protection at the

forward end of a cockpit is desirable. A common arrangement is a hood on hoops so that it can be folded flat on deck or coachroof when not required. For occasional windward work I have found that a weather-cloth temporarily rigged across the weather side-deck and part of the coachroof (Plate 5, *middle*) is quite effective.

In planning the layout of a small yacht bound for the tropics, it is important to see that there is a space on deck where her company may sleep. It is not likely that anyone will choose to sleep on deck during a passage in a small yacht, but in port in the tropics the wind often dies at night, and as the yacht will be floating in warm water she cannot cool off, and her cabin and saloon will remain unpleasantly hot and stuffy. All that is needed is a space about 6½ feet long clear of hatches, dinghy chocks, ventilators, etc., for the mattress of each member of the crew, but this must not be close to the side of the deck or the sleepers will not be properly protected from rain by the awning.

Arrangement below

Arrangement below deck is such a personal matter that there would be little point in discussing it in detail here, but the following suggestions might be borne in mind when planning a new yacht, or altering the layout of an old one, for a voyage into warm waters.

An impression of space and airiness is worth striving for, and that can be best achieved by avoiding the division of the accommodation into a number of small compartments; by leaving clear of lockers some of the space above the bookshelves or settee backs; by having plenty of decklights, placed where they cannot be covered by the dinghy, and portlights in coachroof coamings; and by using more light-colour paint than varnish or french polish (Plate 24). But as cooking turns white paint yellow, a lining of Formica for the under side of the deck may be worth consideration.

The galley should be given ample space and locker accommodation, and be situated in an airy and comfortable part of the vessel where the motion is least and where there is good headroom. I would put it as close as possible to the companionway, as on *Widgee* and *Outward Bound* (Plate 9A and C), where fresh air can enter readily, even though the fumes from cooking will then probably pass through the saloon on their way to the forehatch, for the well-being of the cook is of the first importance and his task should be made as pleasant as possible.

A separate chart table is desirable, and although ideally it should measure 42 inches by 28 inches, so as to take an Admiralty chart opened out flat, very few yachts do have one of such a large size;

however, even in the smallest yacht the table should be not less than 28 inches by 21 inches. To economize in space the chart table may have lockers arranged above its outboard part, provided there is a space of 6 inches measured vertically between their under sides and the table. It is convenient to have the fiddle along the inboard edge of the table hinged so that it can be unbolted and turned down out of the way on occasions when the parallel rules would otherwise foul it.

As rolled charts are almost impossible to use, flat stowage must somehow be found for them, preferably in self-locking drawers under the chart table. Or the table itself may be arranged to lift as a school desk lid to provide stowage for the charts immediately required. *Beyond* had this arrangement and on top of her table was hinged a piece of $\frac{1}{8}$-inch perspex, which was used to cover a general chart, usually the appropriate pilot chart; a piece of felt or cloth between this chart and the wooden table kept the chart firmly in position. The work chart was kept on top of the perspex. In a small yacht where no better stowage place can be found, charts may be kept in a rack slung beneath the deck beams, or possibly beneath a bunk mattress. One hundred Admiralty charts folded in the standard manner require a space 28 by 21 by 2 inches deep.

In a small yacht with a quick motion I prefer a sleeping berth only 20 inches wide, otherwise I roll about, and this is a comfortable width when it is used also as a settee. A bunkboard of canvas (Plate 10A) is better than the traditional wooden type; it is comfortable to lean against, and when not in use lies flat and takes no room under the mattress. Bunkboards of netting would probably be more suitable in a hot climate.

Some saving of space can be made by having the feet of the bunks recessed beneath sideboards or lockers, the bedding when not in use being stowed in the recess (Plate 10B) and concealed by an upholstered panel. The disadvantages of this arrangement are that seating space in the saloon is reduced—this is particularly noticeable when there are visitors aboard—and that one's sleeping position cannot be changed end for end; also, if one wishes to lie down, even for a short time, the bedding must be removed from the recess.

Complicated and clever little gadgets are best avoided, for generally they do not work well at sea. Simplicity and strength with reasonable lightness should be the aim everywhere; there should be proper lockers for all the stores and equipment that will be needed, their shelves fitted with high fiddles, and their doors with fastenings that do not permit them to burst open if anything should chance to touch

the release buttons; in this respect there is much to be said for com- ✓
mon bolts or old-fashioned turn-buttons. A large number of moderate-
size lockers is more convenient than a few big ones because it is then
not necessary to remove many articles to reach the one sought. There
is often some wasted space between the settee backs and the yacht's
side, because anything stowed there is apt to fall out when the windward settee back is opened; but if vertical slats are fitted as shown in
Plate 10C, a good stowage place for canned food can be made. All sideboards, galley benches, etc., should have fiddles not less than $1\frac{1}{2}$ inches
high to hold their contents in place (Plate 10E), and this applies also
to the saloon table unless it is of the swinging kind, but here the fiddles
must be removable because they would be much in the way when the
table is used for a variety of purposes in port. Plate 10D shows a non-
swinging table with two hinged leaves; each leaf has its own set of
fiddles, fitted with short metal studs on the under edges to ship into
metal-bushed holes in the leaves; the athwartships fiddles at each end
hold the fore-and-aft fiddles in place, and are themselves held to the
table by threaded studs and wingnuts. The advantage is that when
removed the fiddles take little stowage space, and when in position
one or both leaves can be hinged down with the fiddles still held firmly
in position. They fit the plates, and the space in the middle is wide
enough for mugs, jam-pots, etc.

A swinging table has a lot to recommend it, but only if the saloon
is wide enough to allow a free passage past one side of it, otherwise it
will be lent on heavily sooner or later by someone trying to pass, and
its contents be thrown off. It must have a very heavy balance weight
only a few inches below the fulcrum pins, and will then adapt itself
to any period of roll. If the weight is well below the fulcrums it will
assert its own natural period, and if this nearly coincides with the
yacht's roll, it can get out of hand and throw everything off. The pins,
hooks, or bolts used for securing the table when in port must be as
far as possible from the fulcrums to avoid an unfair strain and to
ensure that the table is rigid. A damp cloth or a sponge-rubber mat
will do much to prevent articles sliding about on the table.

Freshwater storage

Without rationing $2\frac{1}{2}$ gallons of water is sufficient for one person
for one week, provided it is not used for washing, and that sea-water
is used whenever possible in the galley. A saltwater tap will be a help,
and this may be taken off the engine's cooling-water intake. On the
first long trip my wife and I made in *Wanderer II* we checked the water

consumption as follows. We drew all water from the tank into a pint measure. Hanging near the tap was a length of hambro line with eight knots in it, and each time we drew a pint of water we moved a clothes peg up the line one knot; when it reached the top of the line we knew we had used 1 gallon. The yacht had a 15-gallon tank beneath the cockpit; two 2½-gallon cans in the cockpit brought the total capacity up to 20 gallons. As we wished to double this for a trip to the Azores, we bought eight more 2½-gallon cans, and having removed the yacht's little fireplace, stowed them on top of the coal locker in the fireplace recess just forward of the mast. We built a shelf above them to hold the sextant and some other things, and hung a curtain from the shelf to conceal the cans. Today one would probably use the lighter and cheaper plastic containers holding 2 or 3 gallons each (the 5-gallon ones are too heavy when full for bringing off water from the shore), and they should have several 24-hour soakings with water before being put into service. Air-vents are not needed, and may be plugged with slivers of wood. Very small yachts sometimes carry water in plastic bottles, which can be stowed away in awkward parts of the bilge. Water does not keep well in oak breakers.

A larger yacht will have all or most of her water in tanks placed low down, but a stowage space for four of the above-mentioned containers is desirable as they provide a convenient means of ferrying water from the shore; but for cruising in Scandinavian or U.S.A. waters, where the yacht will often lie alongside a pier or quay, a 50-foot length of plastic hose, with provision for fitting the local taps, will be useful; but some of these hoses impart an objectionable taste to the water, so before purchase it is as well to buy a foot or so, fill it with water and cork the ends and leave for an hour or so before tasting.

It is wise to carry water in more than one tank, for then in the event of a tank leaking, or the water in it, perhaps taken from a doubtful source, acquiring a bad taste or smell (all the water aboard the schooner *Director* went bad and poisoned the crew when 1,500 miles from land in the Pacific), the whole supply will not be lost. In *Wanderer III* we had three tanks; one beneath the galley/chartroom floor held 30 gallons, and a pair holding 15 gallons each placed one under part of each

---

settee in the saloon; a shelf in the forepeak held four $2\frac{1}{2}$-gallon cans, bringing the total capacity to 70 gallons—enough for two people for 98 days. Each tank was connected to a common filling pipe (as in Fig. 2) with a screw cap in the side deck, but by means of cocks each could be filled independently of the others, and when the cocks were

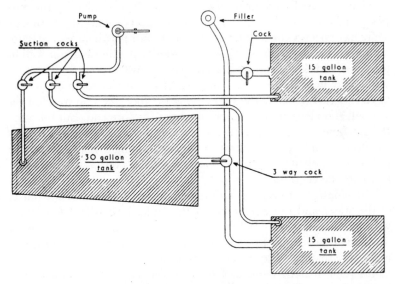

FIG. 2. THREE WATER-TANKS CONNECTED SO THAT EACH MAY BE FILLED OR EMPTIED INDEPENDENTLY OF THE OTHERS

correctly turned there was no intercommunication. The suction pipes from the three tanks led to a small manifold of three cocks from which a pipe went to the pump in the galley. By opening any one of these three cocks water could be sucked from that tank only. The plug type of cock with a tapered barrel turned by a short lever, such as is used at skin fittings, is best for this purpose; the screw-down gate-type is liable to leak and allow water to syphon from one tank to another. The air-vent from each tank (not shown in the Figure) terminated in the oilskin locker under the side deck where the spillage, which usually

---

PLATE 8

Top: The dainty little gaff cutter *Wanderer II* was sailed by Howell and McNulty from England to Tahiti, and then by Howell single-handed to Vancouver, B.C.; although her waterline length is only 21 feet, she achieved the remarkable average of 106 miles a day for 3,600 miles in the Pacific. *Bottom*: *Ice Bird*, one of the Vertue class, on arrival at English Harbour after Dr. Cunningham had sailed her single-handed from Ireland to the West Indies. She subsequently made the return trip, and has since made other notable voyages.

occurs when tanks are being filled, did not matter. Inspection covers should be fastened with studs and nuts, not with round-headed screws, which are sometimes used to save a fraction of space and are difficult to loosen.

Before being put into use, new steel tanks should be cement-washed or coated with one of the special paints provided for the purpose. However, tanks of g.r.p. are being increasingly used, though some have been known to taint the water, and plastic pipes are tending to replace copper or steel. A coachroof or doghouse might well be provided with low coamings, or an awning with a spout, so that rainwater may be collected conveniently. At sea, if the boom of any sail has a recessed track for the slides along the foot, the rain from off the sail will run along it and can be caught at the fore end. But not until the catchment area has had every trace of dirt or salt washed from it should the rainwater be put into a tank containing drinking water, for if it is brackish its keeping properties will be poor and it will contaminate the good water.

The length of time water will remain sweet depends on the purity of the water and the cleanliness of the tank. To purify a tank fill it with water, and while it is filling slowly add Durazone or Parazone at the rate of half a pint to 50 gallons. Allow to stand for an hour, and make sure that the water reaches all pumps, pipes, etc. Flush out with mains (chlorinated) water and then fill for use. This treatment will not harm galvanized tanks. To purify doubtful water see page 250.

The yacht of 40 feet or more is today often provided with a fresh-water shower, and for those who live permanently afloat this may be regarded as more than a luxury. The usual arrangement is to provide the water supply with an electrically operated pressure system, thus eliminating a header tank, and to fit a sump tank with an electric or hand-operated pump. The water for the shower may be heated by bottled gas, or by means of a heat-exchanger on the auxiliary engine. If the shower rose is of small bore and water is used only for wetting and rinsing, two gallons of water should be enough for one shower bath; the hand-held rose on a flexible pipe is the most economical. However, some may well consider it is a lot less bother to go on deck with a cake of soap during a shower of rain, or to use a sponge and bucket.

Maintenance

A belief fostered by the press is that the ocean-voyaging yacht, on reaching port, will look weather-beaten and neglected, her varnish peeling, her paint blistered, chafed, and rust-streaked, and her sails and rigging frayed and disordered, as though to tell of the dreadful weather she has encountered and the hardships endured by her crew. While it is true that a few yachts do have that appearance, they are mostly those whose owners are talking about crossing oceans but have not yet done so, and it is noticeable that those in which neglect is almost studied, usually get into trouble (and into the papers) before they have voyaged very far. The yachts which leave their home ports quietly without publicity, and make uneventful passages, usually arrive at their destinations in trim and shipshape condition, for their owners and crews are efficient and like to keep their little vessels in good order. The attempt to arrive in port at the end of a long passage with the yacht looking as though she has been out for an afternoon sail, can of course be carried to extremes—I tend to err in that direction myself—but normal maintenance, the protection against wear and weather, is plain common sense and pays handsomely in the end.

The ordinary procedure for fitting out for a cruise in home waters, as described in *Cruising Under Sail*, will apply equally well when preparing for an ocean passage, though it may be done with greater care and attention to detail, and it will be wise to use only the best marine paint and varnish, for the little extra that these cost will be well repaid by their longer life and better protective qualities. Sunlight is the most destructive of the elements, and may cause thickly applied paint to blister; several thin coats are therefore better than one thick, and a light colour better than a dark, particularly where the surface is horizontal and is thus exposed to the full power of the sun. If room aboard can be found for it, enough paint and varnish to last for the whole voyage should be carried, as some which is sold abroad, even under well-known trade names, is of indifferent quality and chalks readily. Although some week-end cruising men claim to prefer a paint which soon chalks as it saves some work when rubbing down for the next season, the man who sails long distances, and keeps his yacht in commission for long periods, requires the toughest and most long-lasting protection for his yacht that can be obtained.

So long as the topside enamel is in good condition with a high gloss it will not readily become stained, and an occasional wash down, perhaps with a little detergent, should be all that is needed. The exception to this is when the enamel is fouled by the heavy oil which is often found floating in commercial harbours and sometimes out in the open sea; this should be removed as soon as possible, preferably with paraffin, if it is not to leave a yellow stain. On one occasion while hove-to in the western approach to the English Channel, my wife and

I had the misfortune to encounter a slick of oil, made, presumably, by some tanker pumping the sludge from her tanks. Not only the topsides, but the deck and sails were bespattered with emulsified oil which we removed with toilet tissue and paraffin. Oil is increasingly becoming a menace, not only to the pleasure and comfort of yachtsmen, but to bird and fish life, and even to the livelihood of whole communities ashore, as was emphasized by the loss of the *Torrey Canyon* on the Seven Stones in March 1967.

On two voyages during which the yacht was in commission continuously for a three-year period, and for much of the time was in tropical waters, it was necessary once a year to rub down the topsides and apply a coat of undercoating and enamel. For this the yacht was not hauled out if some well-sheltered berth close to a weather shore, where there was no risk of wash from passing craft, could be found. Commissioners Bay in English Harbour, and Admiralty Bay, Bequia, are two such places in the West Indies. Because the yacht was copper-sheathed the cutting in of the boot-top line was no problem, but for an unsheathed yacht the use of masking tape would be advisable; however, if this is to be used, the boot-top should have been coated with enamel, for the tape is liable to remove ordinary antifouling paint when it is itself removed. The higher the boot-top is the better, for if it is close to the water the yacht will have to be given a list, after mooring her fore-and-aft, by taking a masthead line to the shore or to an anchor laid out on the beam; if she is small a sufficient list may be obtained by letting the boom right out and hanging a weight, such as a couple of containers of water, from its end. The canvas-covered deck and coachroof had to be repainted twice a year. The brightwork was attended to more often whenever it showed signs of losing its gloss, for much less work is involved rubbing down and revarnishing at frequent intervals than in letting the weather get into and discolour the wood, and then having to scrape. Brightwork certainly is a bother because of the short life of varnish compared with paint, and for this reason some people will have none of it, and either leave it bare or

---

PLATE 9

A: *Widgee*'s galley, with its sink and gas cooker, is situated in the best possible place close to the companionway where there is plenty of air and good headroom. *B*: In *Svea*'s galley, with decorative tiles and knife stowage, the pressure paraffin cooker shows an excellent arrangement of fiddles; with the break at each corner pans with low handles can comfortably be accommodated on the hotplate. *C*: *Outward Bound* was built by her owners, and her galley also is aft on the port side; the paraffin-operated refrigerator can be seen behind Mary Caldwell. *D*: *Outward Bound*'s saloon, embellished with Australian wood carvings, shows the space and light which may be obtained by having built-up topsides.

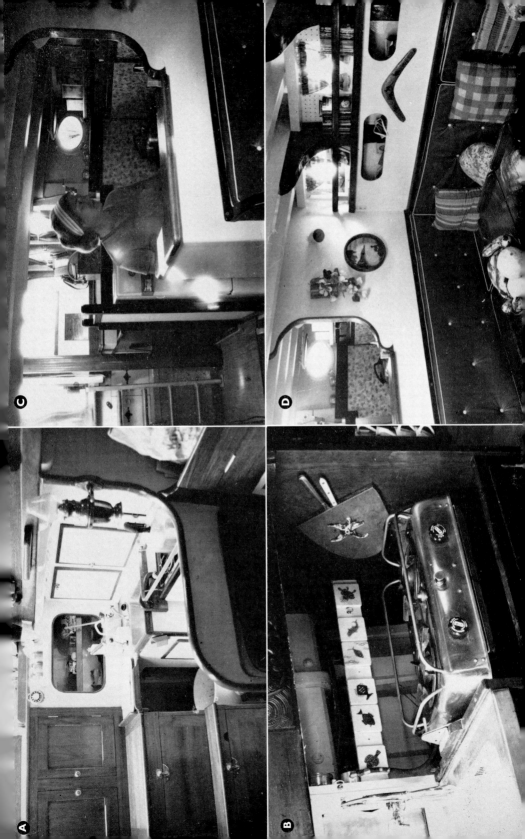

paint everything. In my view this gives a yacht a dead appearance. Cooking and smoking cause white paint below deck to turn yellow, a discoloration which cannot be washed off, and, perhaps because I enjoy my pipe, we usually had to repaint the saloon after two years of living in it. The below-deck varnish, which was occasionally washed down with water and detergent, and then polished with a duster, needed no other attention in the three-year periods, but after that some of it grew sticky and had to be removed. I am told that polyurethane varnish is less liable to this trouble, but in my experience it should not be applied over ordinary varnish or it will peel.

I have found that the maintenance of a steel yacht is very much greater and is less satisfactory than that of a wooden yacht. It is customary for builders to fair up the plating with a thick coat of glazing putty or other form of plaster before painting; such materials appear to have a short life, particularly under water, and when an area lifts or drops off much work is needed to build up a smooth surface. Whereas any minor damage to the paintwork of a wooden yacht can be repaired at one's own convenience with a little undercoat and enamel, with steel such damage must be attended to immediately or rust will form; the bare steel must first be burnished and then be given at least three coats of primer followed by two of undercoat and one of enamel, otherwise the repair will not last more than a few weeks.

It is not possible to give any figures for the effective life of antifouling paints, as this will depend, among other things, on the type of paint, the temperature and salinity of the water, and on the length of time spent in port, for fouling does not occur so quickly on a moving as on a stationary vessel. Moitessier noted that the bottom of his ketch *Joshua* remained remarkably clean throughout his voyage from Tahiti to Spain by way of the Horn, and attributed this to changes of water temperature. A stay of a week or so in fresh water will cause saltwater fouling organisms to die; this does not necessarily mean they will drop off, but will probably make them easier to remove.

PLATE 10

A: A canvas bunk-board is comfortable, and when not in use stows flat beneath the mattress. To economize in space, the foot of a bunk may be recessed beneath a sideboard: by day the bedding is stowed there, *B*, and the recess covered with an upholstered panel, as in *C*; this picture also shows a locker for canned food made with vertical slats behind the hinged settee back. *D*: Each leaf of this table is fitted with its own removable fiddles, and a hand-bearing compass is shipped on a bracket on the table leg to act as a tell-tale. *E*: High fiddles on the galley bench, and beneath it a self-locking drawer, lockers, and sliding bread and meat boards (*Wanderer III*).

3

RIG AND RIGGING

Choice of rig—Jib-headed or gaff rig—Masts and standing rigging—
Running rigging

THE following brief list of yachts which have made great voyages
shows that no rig has a monopoly, and that the choice of single-
or two-masted rig is a personal matter and is not dictated by the
peculiar conditions of long-distance voyaging; but it is probable that
some of the owners mentioned below made do with the rigs their
vessels already possessed when they acquired them.

Macpherson's *Driac II*, Pye's *Moonraker*, Casper's *Elsie*, Tou-
melin's *Kurun*, and Guzzwell's *Treasure* were cutters; Slocum's *Spray*
started as a cutter but was converted to yawl rig. Clark's *Solace*,
Robinson's *Svaap* and Chichester's *Gipsy Moth IV* were ketches.
Pidgeon's *Islander* and Muhlhauser's *Amaryllis* were yawls, while
Fahnestock's *Director* and Crowe's *Lang Syne* were schooners. Many
of the smaller yachts, including Nance's *Cardinal Vertue* and Ann
Davison's *Felicity Ann*, were sloops.

Choice of rig

My preference is for a single-masted rig, and I would not have
two masts unless my vessel were of such a size that with one mast
any sail would be too large to be handled by one person, i.e. in excess
of 400 square feet. The single mast can be well inboard, where its
weight and the drive of the sails set on it will not tend to depress the
bows, and there it can be more efficiently stayed both fore-and-aft
and athwartships. When running dead with the ketch or yawl rig the
mizzen will partly blanket the mainsail unless the sails are spread
wing and wing, but they will not remain comfortably like that for long
in the open sea. Therefore on that point of sailing, unless tacking to
leeward is resorted to, the mizzen will usually be taken in and that
much sail area is lost. Robinson believed that the ketch was the ideal
rig for the small ocean-going yacht, and that was how his *Svaap* was
rigged throughout her splendid circumnavigation. But at Durban
I met the Driscoll brothers from Australia; they had built *Walkabout*
(Plate 43, *top*), a replica of *Svaap*, and had just crossed the Indian

Ocean in her. They had found the mizzen such a nuisance, blanketing the mainsail when running, and causing so much weather helm when the wind was on the quarter—this appears to be a common failing of the rig—that they soon gave up using it, and were then planning to convert the yacht to cutter rig. An argument in favour of the ketch or yawl is that sail can be reduced quickly and easily by handing the mizzen; but it should not be a long or difficult job to reef the mainsail of a cutter or sloop provided the gear for that purpose is efficient, and such a reduction of sail will be less likely to interfere with the balance of the yacht and her steering than will a big reduction at one end of her. But on a reach the sail area of a ketch or yawl can be considerably increased by setting a mizzen staysail, which is a powerful sail and should be of light cloth so that it will hold its wind in light weather. Franklen-Evans in *Kochab* often used mizzen staysail and genoa only when sailing in light weather during his voyages between England and New Zealand, thus saving the heavier working sails from chafe. A fisherman's staysail (sometimes known as a mule) may be set between the masts of a ketch, and will help to steady her; but this is not an efficient sail because its sheet must lead to the head of the mizzen mast, which is too far to windward no matter what the course may be, and the sail therefore creates more heeling influence than forward drive. Another way of increasing the sail area of a ketch is to give her a wishbone rig; but the wishbone is not an easy spar to control at all times, and probably for this reason has never become popular.

The ketch is not so efficient to windward as the sloop or cutter, and although I would not hold that against the rig because not a lot of windward sailing is normally involved when voyaging, there may be occasions when ability to windward is important. The Holmdahls in *Viking* took 51 days for the 4,000-mile passage from Balboa (Panama) to the Marquesas Islands, and on arrival there they tried to beat into Taiohae Bay, a place which is bedevilled with calms and cats-paws. For many hours they struggled to reach water shallow enough to anchor in, but *Viking* was a heavy vessel with a very snug jib-headed ketch rig and they failed. Trader Bob McKitterick watched from his veranda with surprise and dismay, so I have since learnt, as towards evening the Holmdahls reluctantly gave up the attempt and headed away for another island of the group, where they anchored precariously for a short time in a heavy swell, and then, still weary with their long voyage, sailed on for another 17 days to reach Tahiti. I think the moral of this story, and others of the same sort, is that if you intend to rig a heavy-displacement vessel as a ketch you must see to it that she can carry plenty of sail, or else install an auxiliary motor of reasonable power, as indeed almost everyone does today.

The sight of a schooner delights me. There is something glamorous and rakish about that rig, something redolent of palm-fringed anchorages sweet with the scent of vanilla and the heady aroma of copra. I have often thought that I would like to own a gaff-rigged schooner with a poop, a great cabin, and a fiddle head, a couple of graceful squaresail yards crossed on the fore, and a merry Polynesian crew to work her; but she would need a powerful motor to make up for her inefficiency to windward in narrow waters. It is true that not so very long ago sailing vessels which were not very polished performers to windward used to make their way into difficult places; but labour was cheap then, and they carried large crews which were capable of towing or kedging.

I do not consider that the small mizzen of a yawl is worth the cost and complication of the second mast, also it makes it difficult to fit a vane steering gear; but it is worth noting that both the great voyagers Guzzwell and Franklen-Evans chose the yawl rig for their yachts *Trekka* and *Kochab*, but Guzzwell's new yacht *Treasure* is a sloop.

No matter whether they have one or two masts, the tendency among modern yachts is to set a single headsail to the top of the mainmast. This is a pleasant simplification in that only one pair of headsheets has to be attended to, and that so long as the mast is sufficiently stiff it need have only a single pair of crosstrees, and no runners are required. But I question if this is the best rig for a short-handed voyaging yacht of any size, as the No. 1 headsail may be too large to be handed conveniently by one man when a freshening wind calls for it to come in, and when it is in a smaller sail will have to be set in its place; there will then be a gap between the leech of this sail and the mainsail to reduce the slot effect and therefore efficiency is lost.

With the two-headsail rig a freshening wind will call for only one of the headsails, preferably the outer one, to be handed, and it may not then be necessary to set another sail in its place—a considerable saving of physical effort. But as it is not seamanlike to set a headsail from an unstayed part of the mast, as the mast would bow under its pull and the luff of the sail sag to leeward, the two-headsail rig calls for a second pair of crosstrees at the inner sail's point of attachment, and a pair of runners will be needed at that point also. With this rig I would be tempted to have a bowsprit because that would enable a greater area of sail to be carried without increasing the height of the mast, and this would be of value in light winds with the usual ocean swell. Also, when the yacht is close-hauled or is reaching, and she is required to steer herself or to be reasonably well balanced to assist a vane gear to do its work properly, there will be a better chance of achieving this if there are two pairs of headsail sheets to adjust instead of one; and with the outer headsail, the jib, set on a bowsprit, its extra leverage, coupled with the fact that it is less likely to be interfered with by the mainsail, may make it possible for the yacht to steer herself with the wind on the quarter, a difficult feat, but one which the yawl *Spray*, the ketch *Saoirse*, and the cutter *Kurun*—all with bowsprits—managed to perform.

It may be considered retrograde to think of having a bowsprit in these days of the stemhead rig, but for some little time American designers have been alert to its advantages for genuine cruising yachts. The modern bowsprit is not the single round and naked spar of the past, difficult and dangerous to get out along, but is usually a sturdy platform, or sometimes a bipod with a grating across it, and the whole provided with proper 30-inch high guardrails. Such an erection can make a most attractive finish to a beautiful clipper bow (Plate 22), which, like the bowsprit, is showing signs of returning popularity; it has the merit of providing extra deck space forward, and its flare tends to keep the deck drier than does the normal rounded bow; but it is not the best type of bow for a small yacht. When a vessel with a bowsprit is lying at anchor, the chain may grind against the bobstay each time she sheers towards her anchor. Aboard *Rena* Vancill has overcome this in the following manner: On the stem close above the waterline and just below the eye for the bobstay, is a second strong eye backed up by an internal plate, and to this a piece of $1\frac{1}{2}$-inch nylon rope is kept shackled; the rope is in length almost equal to the distance between its eye and the bowsprit end, and its outboard end is spliced to a chain hook (devil's claw). When not in use the hook is lashed to the under side of the bowsprit so that the rope looks like a second bobstay. After the anchor has been let go and sufficient chain has been veered, the hook is disengaged from its idle position and is hooked to the chain outboard of the chain lead; then a fathom or two of chain is paid out until the strain is taken entirely by the nylon rope. Incidentally, this rope serves also as a spring, and enables the yacht to lie more quietly and with little tendency to sheer about.

Jib-headed or gaff rig

Everyone now agrees that the jib-headed (bermudian or marconi) rig is more efficient than the gaff rig when sailing close-hauled, and not one yacht in a hundred built today is given gaff rig. This is under-standable, for efficiency to windward is important for a coastwise cruiser, as much time is likely to be spent on that point of sailing, especially in British waters; also, most modern cruisers are expected at some time or another to take part in races, and their designers, if not their owners, see to it that they are provided with the most efficient rig for the purpose. The jib-headed rig also has these advantages: a permanent backstay can be arranged so that the safety of the mast will not depend on runners or shifting backstays that have to be let go and set up each time a tack or gybe is made; the running rigging is simple, reducing weight and windage aloft, and there is not much risk of chafe.

Nevertheless there are many elderly yachts which do have gaff rig, and if the owner of one of these is planning to make an ocean voyage, there need be no reason for him to feel that he must first modernize his rig, for, provided arrangements are made for setting a large, correctly shaped topsail of light cloth, the rig will serve quite well, and indeed does have certain points to recommend it for off-the-wind sailing. For a given height of mast and length of boom a slightly larger area of sail can be carried abaft the mast, and as this will be in two parts, mainsail and topsail, the mainsail will be of a more manage-able size. Any mainsail must of course be made of sufficiently heavy cloth so that it will not stretch out of shape or otherwise get damaged when used in strong winds. In light winds, particularly in a small vessel, which has a more violent motion, the jib-headed mainsail, because of its weight and height, may not hold its wind properly, and although the boom can, and should, be held firmly by sheet and guy, there will be no means of controlling the upper part of the sail which, by the motion of the yacht, will have the wind thrown out of it again and again; the taller the mast the worse will this be. This trouble is particularly common when running in light winds, but that is just the time when the mainsail will be wanted together with a spinnaker to provide the greatest possible area of canvas. This slamming of the mainsail reduces the yacht's speed, and is hard on the sail, the rigging, and the nerves of the crew. A gaff sail of equal area will, of course, need to be of the same weight, but because its head will not be so high as that of a jib-headed sail it will be less affected by the motion. The important point, however, and this is the chief virtue of the gaff rig, is that above it in light winds can be set a large topsail of light cloth; because of its light weight that sail will retain its wind better, pull more steadily and tend to stop the gaff from swinging; but if the gaff does show a tendency to swing, it can be kept under control by a forward vang, a rope one end of which is secured to the end of the gaff and the other to a cleat or bollard in the fore part of the yacht. The gaff should be a hollow spar and all its gear kept as light as possible, and particular care should be given to the design and lubrication of the jaws, as failure there is one of the commonest troubles.

A disadvantage of gaff rig is that a permanent backstay cannot be used with it, and another is its susceptibility to chafe. This is not insuperable, but great care needs to be taken over the lead of the peak and throat halyards where there should be plenty of drift between the blocks; it may also be necessary to rig an after vang to prevent the gaff from pressing on the lee rigging and crosstree and nipping the topsail there. The jib-headed rig is wonderfully free from chafe, provided the topping lift is not permitted to touch the mainsail (page 51),✓ and that the mainsail is not allowed to move up and down against the lee rigging, on which it is bound to press when the sheet is well eased; this can be ensured by the use of the boom guy (page 50).

Masts and standing rigging

The loss of her mast is the most serious disablement that can happen to a sailing vessel, even more serious than the loss of her rudder, for the stepping of a jury mast might well prove to be impossible, though it is truly amazing what pluck and determination can achieve. Voss rigged a jury mast in *Sea Queen* after she had been rolled over and dismasted in a typhoon near Japan, but a more recent and even more remarkable feat was performed aboard *Tzu Hang*, the 46-foot jib-headed ketch which the Smeetons and John Guzzwell were attempting to sail from Melbourne to the Falkland Islands by way of Cape Horn in 1957. In latitude 52° S., longitude 98° W., about 1,150 miles west by north of the Horn, after 50 days at sea and while running before a great gale (see page 181), the yacht capsized and lost both her masts and suffered other extensive damage, including the loss of skylights, hatches, and rudder. After the water had been baled out, two hollow masts were built from woodwork taken from the saloon, they were stepped, a jury rudder was made, and *Tzu Hang* was then sailed 1,350 miles to Coronel in Chile, her best day's run being 77 miles. On her second attempt to round the Horn she was again dismasted, and once again she returned to Chile under jury rig.

The masts and standing rigging of an ocean cruiser should have an adequate margin of strength to guard against fatigue, deterioration due to wear or weather, or some sudden and unexpected stress, and they should not too closely follow those of ocean racing yachts, which are sometimes kept down to the theoretical minimum of strength in an endeavour to reduce weight and windage aloft.

A solid mast is a little less dependent on its standing rigging than is a hollow one, for in the event of failure of any part of the rigging it will bend farther before breaking. The hollow spar must be regarded purely as a strut and be kept as straight as possible at all times, for if it is permitted to bend it may buckle and break. However, today hardly any solid masts are made, and the only choice remaining is whether to have a new mast made of wood or alloy, and whether to step it on deck or below. So long as they are properly made to correct dimensions, there should be no difference in strength between wood or metal masts, but for a given strength the latter will be the lighter. The lack of the need to paint or otherwise maintain the metal mast is much to its advantage, but I have seen some metal masts which had become badly corroded; however, the anodizing treatment generally used today should guard against this. The noise of halyards tapping a metal mast, even though this may have been lined with sound-deadening material, can be very irritating, particularly to people in neighbouring yachts; but it can so easily be prevented by attention to the lead of the gear (see page 49) that it ought not to be permitted to occur. A wood mast has the advantage that any alteration or addition to its fittings can easily be made, but as it relies entirely on glue to hold it together, I would paint it white rather than use varnish, for this provides better protection and keeps it cooler. Maybe my suspicion of glue is unjustified, but in several hot countries I have seen masts in which the glue had failed, and a pair of British-made dinghy oars with which I started one voyage fell apart at the glued joints after only one year. I thought when I decided to paint the mast and boom of *Wanderer III* white that it might be difficult to keep them looking smart; but except in a few commercial ports where there was some oily soot in the air and a wash-down was needed, that did not prove to be so, and repainting was needed only every other year. But varnish does permit one to see any discoloration of the wood which might indicate damp or decay. A mast stepped on the keel and rigidly wedged at the partners will be

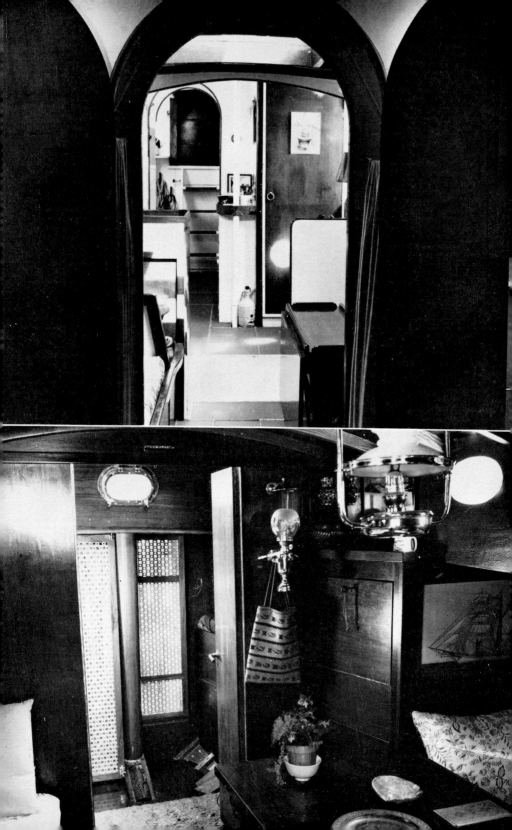

stiffer than one stepped on deck. If it is to be stepped on the coach-roof, this must be specially strengthened not only to take the down-ward thrust but the lateral pressure also; failure to take this precaution has resulted in serious damage to coachroof coamings in heavy weather.

It should be borne in mind when deciding on the sizes of gear and rigging for the ocean-going yacht that she may in a fortnight sail as many miles as would be covered by a coastwise cruiser in a whole season, and that during a single year she may sail as many miles as the other will in ten years. While she is at sea, even when becalmed, she will not be still, and her motion will be constantly working the rigging, on which there will be a strain each time she rolls or pitches. In such circumstances metal fatigue may develop and cause some seemingly sound part of the gear to fracture, or the continuous wear on the pins of rigging screws or other fittings may weaken them by reducing their size. This may sound far-fetched to the coastwise yachtsman, but, as a typical example, I found after 15,000 miles that a steel bolt securing a forestay rigging screw to the stemhead fitting had such a deep groove worn in it that it needed renewing. On another occasion when the yacht was close-hauled for 18 consecutive days on one tack and was pitching into the sea kicked up by the trade wind, the weather runner, although it was set up taut by its lever, was constantly on the move, with the sheave of the fairlead block turning a little one way then the other each time she pitched. The extra weight and windage caused by having all the gear a little larger than is the usual yachting practice is slight, yet it may make all the difference between success and failure.

There is no appreciable action between stainless steel and manganese bronze, and these are the materials most widely used for the rigging and fittings of modern yachts, and they require no maintenance. But all yacht equipment in which these materials are used is expensive, absurdly so in many instances, and where cost is a serious considera-tion there is much to be said for the use of galvanized steel. If this is looked after in the manner suggested below, it may last just as long as stainless steel, and where wire rope is concerned perhaps even longer; also, as there have been failures of stainless steel from time to time,

---

PLATE 12

A: A runner passing through a snatch block and set up with a Highfield lever; by releasing the lever and lifting the wire out of the snatch block, ample slack is provided. *B*: A masthead with three sheaves, one for the main halyard, one for the genoa halyard, and one for the topping lift. *C*: The fall of the topping lift leads through bullseyes on the crosstrees, and the falls of idle halyards are held by thumb-cleats on the lower crosstrees to prevent chafe and noise. *D*: When the sheet is eased the boom will lift and allow the sail to chafe on the lee rigging unless, *E*, the boom is held down with a boom guy.

one may feel that galvanized steel is safer. For her first 70,000 miles of voyaging *Wanderer III* had no stainless gear except the forestays, for which galvanized wire rope is not satisfactory as its zinc coating soon gets destroyed by the hanks of the headsails; all her chain plates, rigging screws, and mast fittings were galvanized and were still in use after 110,000 miles. The splices and hard and soft eyes were parcelled with insulating tape, served with marline, and varnished, and this protection was carried six feet up from the lower ends of all shrouds and the backstay. The naked wire remaining was coated with boiled linseed oil, a new coat being applied at least once a year. The oil tended to fill up the contlines of the rope and make it smooth, and I noticed after changing to stainless standing rigging after ten years that there was a greater tendency for the stainless rope, which of course had no such coating, to chafe the sails. But the rope was the 7×7 type, whereas 1×19, because it is smoother, would be less inclined to chafe; however, cost was a consideration, and the 7×7 rope could be spliced by the yard rigger at no great cost, whereas 1×19 rope would have had to be provided with special terminal fittings, swaged or otherwise attached by a firm specializing in such work. The threads of the galvanized rigging screws, which were of the open barrel type, were smeared with anhydrous lanoline before assembly; then, after the rigging had been set up and the final adjustments made, the exposed screw threads were parcelled with insulating tape, served with marline put on in the form of a west country whipping, and the whole assembly coated with a mixture of boiled linseed oil and varnish in equal parts; after 15 years the threads were still in perfect condition. Anhydrous lanoline is also useful on nuts, bolts, shackle pins, etc., preventing them from seizing up and lasting longer than grease or petroleum jelly, but it is not a lubricant.

The voyager will of course avoid such absurdities as rigging screws with galvanized steel ends screwed into yellow metal barrels, or steel wires running over brass sheaves, for although such self-destructive combinations may last for some time aboard a yacht which is used only for week-end sailing and a fortnight's summer cruise, they will not survive for long at sea where they may be wet for weeks on end. It is wise to use only rigging screws that have been tested and stamped by the makers, for although they will not normally be tested to anything like the working load of the wire rope with which they are to be used, the test should show up any inherent weakness, such as a flaw in the metal. The same applies to shackles, but as these are the commonest weak points of the rigging, they are best avoided by using clevis pins.

Swaged terminal fittings, particularly at the lower ends of the
rigging, are liable to deteriorate under the action of sun and salt water,
and their failure is not uncommon. Hairline cracks in such fittings
should be regarded as a danger sign, for these will probably have
started inside the fittings and by the time they are visible the fault may
be serious. Swaged terminal fittings should therefore be protected
against the elements, and the method mentioned above for the protection of galvanized rigging screws with lanoline, parcelling, and
serving, should do very well. Linked rigging, diamond shrouds, and
sometimes jumper stays have the drawback that a journey aloft is required for their adjustment, and are therefore best avoided. Although
the safety of the mast should not have to depend on runners or shifting
backstays, these, as mentioned on page 40, will be required if a headsail is to set properly from a position on the mast lower than the masthead. The simplest, most efficient, and now almost universal method of
setting up runners is with levers, and even when these are short it is
possible to arrange for the lee runner to have sufficient slack to permit
the boom to be squared right off. Plate 12A shows the arrangement.
The runner (the lower part of which must of course be of flexible wire
rope) leads down through the fairlead block in the foreground, then
aft to a snatch block on deck, and then forward to the lever. When sailing close-hauled sufficient slack is provided by lifting the lever. To
obtain more slack for reaching or running, the bight of the wire is
lifted out of the snatch block, then the runner will be free to go forward
to the lee rigging, where it may be secured by a hook on a length of
shock cord, though if it is to remain there for any length of time, it
will be best lashed securely to prevent chafe. To avoid having to
make a journey forward to deal with this each time a gybe is made,
in some yachts a line is secured to the runner, led down through a
small block at the rail by the rigging, and then aft to a cleat near the
cockpit.

Anyone who is at sea near one of the great thunderstorms which are
common in many parts of the world—when one flash of lightning
rapidly follows another, when there is no noticeable pause between
the flash and the peal of thunder, and the mast is the only upstanding
thing in all that part of the ocean—may well feel apprehensive. However, he should gain some comfort by remembering that since lightning
will take the shortest electrical path to earth, his yacht's mast will not
attract a lightning discharge unless it is almost immediately beneath
the area of the cloud in which the maximum potential is being built
up; in other words the chance of being struck is very small.

To prevent damage being done in the event of the yacht being struck, a straight, low resistance, electrical path must be provided from the masthead(s) to the sea. A steel yacht with a steel or alloy mast does provide such a path, and even if her mast is of wood she should still be safe so long as her topmast shrouds are not insulated, as they sometimes are to break a closed loop for the benefit of radio direction-finding, and that any electrically weak points are bridged; see below. To protect a wooden yacht with an alloy mast, the heel of the mast should be electrically bonded to one of the ballast keel bolts. In a wooden yacht with a wooden mast, a topmast shroud which is secured to a chain plate extending well below the waterline or bonded to a keel bolt will serve, but it will be as well to bridge any shackles or rigging screws which may not provide adequate electrical contact with $\frac{1}{8}$ in. diameter copper wire. But such copper bridges will inevitably form a galvanic couple with the steel, to the detriment of the latter, creating, in my view, a danger far greater than the one we are trying to avoid. I suggest it is better to secure, with many turns, a length of $\frac{1}{8}$ in. diameter copper wire temporarily to the shroud above the rigging screw, and allow several fathoms of it to trail overboard.

The perfectionist may be interested in the specification laid down for yachts by the Bureau of Standards in the U.S.A. This states that at the masthead a pointed metal rod not less than $\frac{1}{4}$ in. diameter should extend 10 inches above the truck, and be connected by a copper wire at least $\frac{1}{8}$ in. diameter to an earthed metal conductor which should have an area of 36 square feet and in wooden vessels be situated under water.

Possibly the risk of damage will be reduced once the rain has started, but this may be late in coming and often marks the end of the storm. St. Elmo's fire, a blurred ball or balls of greenish light which may rest on the truck or other parts of the mast, is harmless, and usually indicates that the worst of the storm is over.

Running rigging

The normal well-tried arrangements for running rigging, such as are commonly used for coastwise cruising, call for little comment here, as most of them will serve equally well on longer passages; but chafe must be guarded against, and the safety factor perhaps increased. I like to have more than one masthead sheave, and in *Wanderer III* had three (Plate 12B), one for the main halyard, one for the genoa, jib-topsail, or other masthead headsail, and the third for the topping lift. A damaged or escaped halyard would then not be a very serious matter as one or other of the remaining ropes could be used. The sheaves might with advantage be of nylon or tufnol on a stainless pin, so that no lubrication will be needed, and the sheaves should be of greater diameter than the diameter of the mast to prevent the falls of the halyards bearing on the mast.

I have not found stainless steel flexible rope to be entirely satisfactory for running rigging which has much use, as the small wires of which the strands are composed appear to be more brittle than ordinary steel, and tend to break where the rope works over a sheave, weakening it and making it unpleasant to handle. Some makes of this rope, particularly American, are less prone to this trouble than others, but I have not often experienced the trouble with galvanized steel rope, though admittedly I always scrapped this as soon as it began to rust. Sometimes the rusting and subsequent destruction of galvanized steel flexible rope is hastened because it is handled by winches with bronze barrels, and when wet a galvanic couple is produced. A similar action sometimes takes place where the wire touches a bronze lever, but this can easily be prevented by parcelling and serving the wire at that point.

However, for the above reasons I gave up using wire rope halyards some years ago, and replaced them with single-part halyards of three-strand, pre-stretched terylene, taken to open-barrel winches on the mast for the final setting up. These are in every respect easier to handle than are halyards of wire rope, and their life is much longer. To prevent halyards from tapping and chafing on the mast when not in use, their working ends can be made fast away from the mast, e.g. on guardrails at bow or stern, and their falls flipped into thumb-cleats on the forward side of the crosstrees, as in Plate 12C.

Headsail sheets will probably chafe on the shrouds on some point of sailing. If new shrouds are being rigged, a few feet of loose-fitting plastic tubing (I have found Alkathene to be good and not affected by sunlight) may be threaded on before splicing or fitting terminals (Plate 15D). On existing shrouds split wooden rollers, the two halves being seized together, will serve to prevent chafe of the sheets just as well. When the mainsheet is arranged in the usual way with its blocks on deck, or one of them on a track, it will, when eased, lie across the lee guardrail and may bear heavily on it. An excellent way of preventing this is to have the sheet blocks on a very high horse, as was, I believe, first fitted aboard Humphrey Barton's *Vertue XXXV*; this may be seen in Plate 1, *right* and Plate 27, *top*. The sheet will then lead clear of the guardrails, and the horse will provide a fine anchorage for

E

the after ends of the guardrails. Places where chafe may take place temporarily, such as where a spinnaker brace or guy touches a runner, can be prevented by parcelling at that point with a strip of sailcloth, and as this may often have to be shifted, it can conveniently be held in position with a couple of household clothes pegs.

I regard the boom guy as an important item of gear, not to prevent the helmsman from gybing all-standing, though it will certainly assist him when running in rough or in very light weather, but to hold the boom down. There is a strong tendency for the boom (when the sheet is eased) to lift with each puff of wind and each roll of the yacht, and chafe of the mainsail on the shrouds may result if it is permitted to do so (Plate 12D). Provided the boom is sufficiently high above the deck, all that is needed is a strong rope secured to its outboard end and set up to a suitable cleat right forward. There is no need to fit it with a tackle or lead it to a winch or windlass, for to get it tight one has only to pay out more sheet than will be needed, make fast the guy forward hand-taut, and then heave in on the mainsheet, which, being itself in the form of a tackle, or having its own winch, will set the guy taut. The trouble with this simple arrangement is that each time an adjust-ment of the sheet has to be made, or the yacht has to be gybed, a journey forward is necessary. Barton's ingenious method of overcoming this was to lead the guy through a block at the stemhead, and bring it aft to the idle staysail sheet winch on the weather side on which it was set up and adjusted. When gybing he had only to ease off the guy, get the boom inboard, cast off the fast end of the guy and make fast the other end (a snap shackle spliced to each end of the guy enabled this to be done quickly), gybe, pay out the sheet, and set up the guy on the other sheet winch. All this could be done from or close to the cockpit. I tried the arrangement and found it convenient, but did not continue with it because the guy imposed too great a strain on the guardrail stanchions on the weather side, outside of which it had to pass.

In a gaff-rigged yacht one cannot prevent the topping lift from rubbing on the sail on one or other tack or gybe except by unshackling it at the boom-end, which is not desirable as it might be needed in a hurry when handing or reefing the sail; so one must either fit two lifts and unshackle the lee one, or else provide the lift with plenty of baggy-wrinkle. But one of the advantages of the jib-headed rig is that the topping lift will go to the masthead, and unless the sail has a lot of roach, the lift will not touch the sail so long as it is kept taut. When the boom guy is in use, i.e. when the mainsheet is eased, the topping lift can be set up to take some of the strain off the leech of the sail and to ✓

keep itself out of mischief. But when close-hauled, or nearly so, and the guy is no longer in use because its angle with the boom would be too inefficient, the matter is not so simple; then the boom will be constantly lifting and falling a little, and each time it rises the topping lift will fall slack and chafe the sail. One way of keeping it taut is to bypass it at the boom-end with 2 or 3 feet of shock-cord in such a way that when there is no strain on the lift the shock-cord will take charge and hold it taut within limits, but should the topping lift be needed, the cord will stretch and allow it to take the strain. But I soon found, as did Robinson in *Svaap*, that a wire lift is not altogether satisfactory on a long passage, as it is weakened by the jerking to which it is subjected and by the sharp nip where it leaves the masthead sheave when the boom is squared off. After the original wire lift had started to strand at that point, I replaced it with a length of $1\frac{1}{8}$-inch nylon rope as a temporary measure, but that served so well that it was retained, and of itself had so much elasticity that the topping lift remained taut without the need for shock-cord. Chafe on the crosstrees and any chance of the fall fouling their ends was prevented by leading the fall through fairleads screwed to the upper and lower crosstrees, as shown on the left of Plate 12C.

Because nylon rope has so much elasticity, it is not suitable for any other items of rigging except perhaps main and mizzen sheets; terylene rope is so excellent for all other purposes that I can see no point in considering the use of any natural-fibre rope, such as hemp, cotton, or sisal. Terylene rope does not absorb water, so when wet it does not swell or become stiff, and it retains its full strength then, which is nearly one-quarter greater than that of the best white hemp. However, loosely plaited terylene rope, such as is widely used for sheets which are handled by winches, does hold a lot of water between its strands, and may take several days to dry out. I therefore prefer to use the 3-strand type for all purposes. Terylene rope is not subject to damage by bacterial growth or insects, but after considerable exposure to sunlight, it (and nylon) hardens, becomes less tractable, and loses some strength. Although the outer surface may quickly become fluffy when chafed, this appears to act as a protection, and the rope will then withstand considerable further chafe without much harm, and if the normal precautions are taken will outlast many times a natural fibre rope and so repay its original greater cost.

4

SAILS AND SELF-STEERING ARRANGEMENTS

Sailcloth—Fore-and-aft sails—Reefing—Heavy-weather sails—
Running sails and self-steering arrangements—Wind-vane
steering gears

ALTHOUGH the fore-and-aft sails needed for voyaging will differ little from those which are used for coastwise cruising, the following remarks may be of interest if any new sails are to be made with a long passage in view.

Sailcloth

For a long time after terylene (dacron in the U.S.A. and dilon in Germany) had become widely used for the sails of racing and inshore cruising yachts, doubts were felt about its suitability for the sails of voyaging yachts, and for these sails of cotton and sometimes of flax continued to be used, and indeed still are used to some extent today. It was not so much the material as the stitching of it that was suspect, the argument being that because the cloth is hard the stitches cannot sink into it for protection as they do in natural fibre cloth, and are therefore subject to chafe where a sail touches any part of the rigging; it was also thought that the hard cloth might even cut its own stitching.

For her first three-year circumnavigation *Wanderer III* was provided with mainsail and working headsails of cotton. The mainsail was hand-sewn, and although I had a replacement sent out to South Africa for the final 7,500 miles of the voyage, this proved to be unnecessary, and the original would almost certainly have brought her safely home, though its foot, which was the part most exposed when

---

PLATE 13

A: The drum of the genoa furling gear aboard *Svea* is enclosed, and in the event of the line breaking or becoming snarled, a lever can be inserted between the two vertical plates above the drum, and furling accomplished. *B*: The furling line is handled by a winch mounted on a pinrail beside the cockpit. *C*: Worm-type roller-reefing gear; on the right is the original handle, which was too short and fouled the rolled sail; on the left is the improved handle. *D*: Enclosed roller gear, with a flange to stop the sail from creeping forward and fouling it. The bolt-rope along the foot of the sail is held in a groove in the alloy boom. *E*: Single-part trysail sheets lead through blocks right aft and then to the staysail sheet winches. Here the sail is set to steady the yacht as she runs under twins with mainsail stowed and coated.

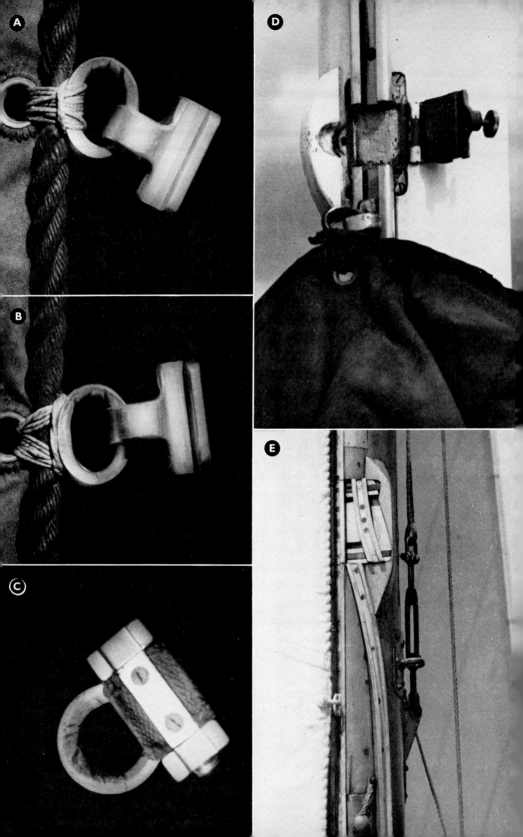

the sail was stowed—even though we kept the sail coated as much as possible—was showing signs of weakness. The seams, particularly near the head where they tended to chafe on the topmast shrouds, had needed a little attention now and then. For her second circumnavigation, which took the same time as the earlier one and covered almost as many miles, she was provided with working sails of tan-dyed terylene, and because my sailmaker and I had the then common suspicion of the stitching, each seam of the mainsail, after being machine-sewn in the normal manner, was reinforced with a single row of hand-made stitches. This sail survived the whole voyage, during which very little more attention to the seams was needed than had been required by the cotton sail, and I concluded that the hand sewing had not been necessary. But the foot, which like that of its cotton forerunner had been more exposed than the rest of the sail, had become hard and degraded by the end of the voyage, so that a sail-needle thrust through it broke the fibres instead of parting them, and left a clearly seen hole. A third sail for subsequent voyaging, mostly in sunny waters, also had a useful life of about three years.

The degrading, and eventual destruction, caused by ultra-violet rays is the only serious drawback of terylene, and all one can do about it is to see that sails when not in use are either coated or stowed below; but as mildew can grow on terylene, white sails should be dried before coating if their appearance is to be preserved; however, this grows only on the surface and does not rot the cloth, and it can usually be scrubbed off. I prefer the appearance of tan-dyed sails and find them easier on the eyes; of course they absorb more heat than white sails, but I understand it is not the heat but the ultra-violet rays which are so destructive; unfortunately the colour fades. Recently manufacturers and one or two sailmaking firms have taken to weaving a softer cloth, but so far as is at present known this is no less vulnerable to ultra-violet rays.

Provided precautions are taken against chafe, there should be little trouble with the stitching, but it is worth having a third row of machine-made stitches put in between the other two rows in each seam. It might be thought that three rows will chafe as readily as two when a seam rubs against a shroud, but this is not so. The zig-zag stitches at

PLATE 14

Slides and gates. *A*: Gibbons Y.22 slide incorrectly seized to the sail; such a seizing will impose a breaking strain on the thimble. *B*: The slide correctly seized on, the marline passing round the circumference of the thimble. *C*: A lubricating slide fitted with wick and oil reservoir. *D*: The gate in a track should be high enough to permit all the mainsail slides to be accommodated below it. *E*: A duplicate track on the lower part of the mast permits the trysail slides to be inserted before the mainsail is lowered, a switch leading them into the main track.

the edge of a seam cross over the edge of the cloth on one side where they pass from the double to the single thickness of cloth, and it is along this sharp edge that they are most vulnerable to chafe. But the stitches of a third row, put in between the other two, do not cross over the edge of the cloth, and therefore are not so subject to chafe. Because of their mistrust of the stitching some owners have all seams resewn every year or two; but I regard that as a mistake because the cloth will have become degraded to some extent so that the needle will punch holes in it rather than part the warp and weft, with the result that the whole seam is weakened; indeed, some sails given this treatment have split right along a resewn seam.

It may be argued that as the working life of a terylene sail is no greater than that of a cotton sail, the extra cost of terylene is not worth while. But terylene has several advantages over cotton and flax and today is not much more expensive. Other things being equal, a somewhat lighter cloth can be used without risk of weakness or stretching out of shape (recommended weights for all sizes and types of sail are given in *Cruising Under Sail*), and the lighter a sail is the more efficient will it be in light or moderate winds; also, it is easier to set, hand, or reef. As terylene of the correct weight does not stretch, a new sail of that material can, and should, be pulled out to its made length on luff and foot the first time it is used, and it will require none of the careful attention in the right weather conditions that a cotton sail calls for in its early days; and as terylene does not absorb water, it does not grow appreciably heavier when wet, nor does it shrink, so once the sail is set no further attention is required to halyard or outhaul; being smoother it offers less friction to the wind blowing over it, and being less permeable it is probably more efficient. Finally, it is not subject to rot if stowed away wet. With terylene offering so many advantages to the voyager, there can be little reason for him to use cotton or flax, unless he wishes to make his own sails, when he will find the natural fibre cloths easier to work with; their cost is a little less, and flax may perhaps be worth considering for heavy-weather sails; see page 60.

Nylon has some of the merits of terylene, but is too elastic to be used for sails which are carried on the wind, and if used in strong winds may stretch permanently out of shape. It is, however, widely used for spinnakers.

Fore-and-aft sails

It has often been said that the cloths of a cruiser's mainsail should be vertical, i.e. parallel to the leech, because if a seam were to run or a cloth split for its full length, one would still be left with two workable areas of sailcloth; whereas if the sail were cross-cut, i.e. with its cloths at right angles to the leech, a cloth or seam split for its full length would render the sail useless. However, today very few sails are made with vertical cloths as sailmakers much prefer to make mainsails and mizzens cross-cut, which results in better setting and more efficient sails; but some mainsails have been made on the mitre, i.e. with the upper cloths at right angles to the leech, and the lower cloths at right angles to the foot, as in a headsail, and their owners spoke highly of them.

For a cruising yacht the leech of a mainsail or mizzen should not have any roach, i.e. its leech should not curve outwards like that of a racing yacht, but should be straight from head to clew. Such a sail does not look so pretty, especially if it has a high aspect ratio; but roach is of value only in giving a slightly increased sail area for a given height of mast and length of boom, and is out of place in a seagoing yacht for the following reasons: Roach calls for battens to hold it out and stop it curling, but these are a nuisance because they are liable to get hitched in the rigging when the sail is being set or handed, unless those operations are performed head to wind, which is inconvenient and sometimes impossible. Battens can also be a danger, for when running with a light wind the rolling may cause the upper part of the sail to move to and fro athwartships in its own plane, and it will then be possible for a batten to insert itself between two parts of the rigging, e.g. shrouds, and with both ends trapped. The sail cannot then be lowered, it is unlikely that a hand sent aloft will be able to clear it, and probably the only way of freeing that batten will be to gybe all-standing, with a broken batten, a torn sail, or a damaged crosstree as a result. So that it will not be necessary to remove a batten when reefing, it is usual for the lower batten pocket to be sewn to the sail parallel to the boom instead of at right angles to the leech; but with roller reefing it is likely that the aft end of the boom will droop a little as the sail is rolled down; the batten will then not be parallel to the boom and will have to be removed after all; anyone who has attempted to do this when the sail is hard-sheeted, as it must be to bring the batten within reach, in a fresh breeze, will know what a difficult and perhaps dangerous job it can be, and that the best chance of success will lie in threading a piece of wire through the holes in the end of the batten and heaving on that.

The luff of a terylene mainsail or mizzen may be fitted with a flexible wire rope, a terylene tape, or a 3-strand pre-stretched terylene rope; the latter is much to be preferred as it is strong and flexible, and the stitches holding it to the sail are buried in the contlines and are therefore not subject to chafe where the slides are secured. Sails, including large headsails, in which tapes made of several layers of terylene were sewn to the luff with several rows of machine-made stitches, had quite a vogue in the U.S.A. at one time, and some remarkable claims were made for them; but their use in seagoing yachts is now not generally approved by leading sailmakers.

Traditionally the luff of a gaff sail is seized to wooden hoops sliding on the mast; these cause chafe, and have to be cut away or unlashed when roller reefing gear is used, and they have nothing to recommend them. A lacing of terylene rope is in every way to be preferred, but it must not be rove round and round in a spiral or it will jam; instead, after passing through an eyelet in the sail it must be brought back the same side of the mast. The luff of a jib-headed mainsail or mizzen will of course be provided with slides running in a track on the mast, but even so a rope lacing may with advantage be provided for the lower part of the luff; it will be found easier when reefing to unreeve some of this lacing rather than remove a number of slides from the track and insert them again later when the reef is shaken out. A gate in the track (Plate 14D) will be required for entering the slides, and it is convenient to have this a sufficient distance from the bottom so that all the slides can be accommodated in the track below it when the sail is stowed, for it is easier to drop slides in than it is to enter them at the bottom and push them up, particularly when the job has to be done in a strong wind, as for example when setting a trysail at sea (page 61). I have used James Gibbons Y22 slides for many years and found them satisfactory. This type of slide has a thimble arranged vertically on a horizontal bar on its after side, and the thimble is seized tightly to the sail so that all movement and wear is between the thimble and the slide. The thimble should not be seized to the sail as in Plate 14A, for that would impose a breaking strain on it, but as shown in Plate 14B. A lubricating slide of the same pattern, but incorporating a small wick and reservoir (Plate 14C), is available, but with an alloy track and slides lubricating oil tends to produce a black, non-lubricating coating, which may disfigure the sail and in wet weather trickle down on to the deck. Therefore slides made of nylon, which are not so strong but require no lubrication, may be preferred. The Gibbons slide is not available in this material, but if slightly reduced in size it can be nylon-coated; however, in my experience the coating tends to chip and peel off after a year or two of use.

With gaff rig I would regard the topsail as a working sail, and the handing of it as equivalent to the taking in of the first reef of a jib-headed sail. To set well it should be of such a shape that its head and foot are of equal length (Plate 8, *top*). It is best made with the mitre running horizontally from clew to luff, and a thimble to take a single rope leader should be seized to the luff at the mitre (Fig. 3, *left*). The sail must be a little smaller than the space available so that it can be

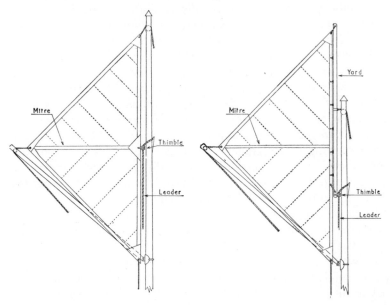

FIG. 3. GAFF TOPSAILS

Left: A jib-headed topsail fitted with a thimble and set on a single leader. *Right*: Topsail bent to a yard, the thimble for the leader being seized to the heel of the yard.

sheeted flat for sailing close-hauled. If the mast is not tall enough to permit the correct proportions, part of the luff can be bent to a light wood or alloy yard; then the thimble should be seized to the heel of the yard (Fig. 3, *right*); this will ensure that the sail sets well, and it will keep the yard under control while the sail is being set or handed. If these suggestions are followed, the gaff topsail should prove to be a valuable and easily handled sail.

Most headsails are fitted with a flexible wire luff rope because this does not stretch and permit the sail to sag to leeward. But galvanized steel wire has only a short life in this capacity, for its lower part, which will often be encased in wet sailcloth for long periods, will rust. Stainless wire is not much better as it is subject to a chemical action resulting from salt water and sunlight being in contact with the terylene into which it is fitted, and this can destroy the wire. It should be possible to overcome this trouble by insulating the wire from the cloth with a plastic coating, but it is far better to use monel rope which is not subject to decay, and this is now the practice of some sailmakers.

Reefing

Whether point and pendant or roller reefing is employed is a personal matter. The former produces a better reef and a better setting sail, but the latter is quicker in use, enables all the work to be done from a comparatively secure position beside the mast, and a reef can be taken in or shaken out on any point of sailing without the need to bring the boom inboard. If roller reefing is decided on, the precaution should be taken of having the sail fitted with reefing cringles and eyelets for a lacing in case anything should go wrong with the roller gear.

An advantage of point and pendant reefing is that if the reefs in the sail are arranged not parallel to the boom, but in such a manner that the cringles on the leech are higher than the cringles on the luff, when the sail is reefed the outboard end of the boom will be higher than normal, and thus less liable to bury itself in the sea as the yacht rolls. I have not seen a sail with which this has been done, but Moitessier (see page 60) had it and thought it essential for a vessel sailing in stormy areas.

I favour roller reefing and prefer the worm to the ratchet type, but a common failing of this is that the handle has too short a crank to give sufficient power, and that it fouls the rolled-up sail after a few rolls have been taken in. Plate 13C shows an open type (Appledore) worm gear; on the right is the original handle, which was useless, and on the left the improved handle I had made in which not only the crank but also the part of the handle which ships on the worm spindle has been made longer; with so long a crank the handle can only be used on the weather side without fouling the mast winches; but that is no disadvantage for it is always best to work to windward. With this gear there is considerable friction between worm and toothed wheel, and in this exposed position neither oil nor grease retains its lubricating properties for long. However, several firms make roller reefing gears in which worm and toothed wheel are totally enclosed and lubrication by oil is efficient. Such a gear is shown in Plate 13D, and it will be noted that a deep flange is incorporated to prevent the luff of the sail creeping forward off the end of the boom. But even with this modern and costly gear the crank of the handle is too short for ease of working.

It is obvious that as the luff rope rolls down on top of itself, and the leech rolls down in a spiral, the after end of the boom will droop increasingly as the sail is rolled down unless it is of greater diameter than the forward end. But this is not the only consideration, for as Claud Worth pointed out in *Yacht Cruising* many years ago, the boom requires an even greater diameter in the middle, otherwise when reefed the sail will have too much belly. As different sails need differently tapered booms, the best that can be done is to experiment with gently tapered lengths of wood rolled in with the sail as it is reefed, and when these have been shaped to give the proper belly without the outboard end of the boom drooping, they can be fitted permanently to the boom, but this will be more easily done with a wood rather than an alloy boom. Note that an alloy boom cannot of itself be given any taper except at considerable cost.

It is not often that a headsail is reefed today, the usual practice being to take it in and set a smaller one, probably of heavier cloth, in its place. In connection with the handling of a headsail, mention should be made of Wykeham Martin furling gear, whereby a sail can be made to roll up on its own luff rope by means of a drum and line. Although this gear appears to have gone out of fashion in British yachts, it is widely used, sometimes for large genoas, in American cruising yachts. Some people even use it for reefing, but I believe a poorly setting sail results, and certainly the strain on the furling line and on the attachments of the sail to its 1×19 wire luff rope is considerable; Ernest Ratsey told me that in his opinion this gear is the sailmaker's best friend, nevertheless it was used by the Kittredges in *Svea* during their circumnavigation, and by Irving Johnson in his ketch *Yankee*, and both speak highly of it. The furling line may with advantage be handled by a small winch (Plate 13B) and it is good insurance to have a place where a lever can be inserted above the drum (Plate 13A), so that in the event of the furling line becoming snarled the sail can be rolled up, though slowly, by hand. But even when used only for furling, the gear would appear to have the following disadvantages: while the sail is furled and is still in position, no other (smaller) headsail can be set hanked to the topmast stay without bearing heavily on the furled sail on one tack or the other, and if the furled sail is lowered to get it out of the way, stowage of the stiff, thin sausage presents a real problem; also, if when in port the furled sail is left aloft, as it usually is because of the stowage difficulty, the edges of leech and foot will be constantly exposed to the sun, which will in time cause a deterioration of the cloth. A coat is not easy to arrange, and I was not much impressed with the only one I have seen.

Heavy-weather sails

It is usual for the ocean-going yacht to carry in her sail-locker a trysail, i.e. a small, strong, boomless, three-cornered sail, to be used in place of the mainsail in event of damage to the latter, or when the area of the mainsail cannot by reefing be reduced sufficiently for very strong winds; she may also have small, strong headsails, a spitfire jib and/or a storm staysail. A common mistake is for such sails to be too heavy, as was pointed out by Moitessier in a letter to a friend published in the *Cruising Club News*, after he and his wife had completed their remarkable voyage in the ketch *Joshua* from Tahiti non-stop to Spain by way of Cape Horn.

Have a set of very small sails at your disposal, but not heavy, stiff sails with enormous boltropes. This would be an error because these traditional foul weather sails are exhausting to use, cumbrous, heavy to handle, difficult to install and difficult to master before managing to furl them. These heavy traditional sails are a calamity in boats our size.

Anyone who has in heavy weather had to stow the mainsail and bend on and set in its place a trysail of heavy flax—a material commonly used because it does not readily tear—will surely agree with Moitessier, and would hope in future to have instead a trysail of terylene, roped all round, perhaps, but lightly and of course with terylene rope, and with all points where chafe might occur reinforced with patches and extra stitching. However, in strong winds a terylene sail can flutter alarmingly, the vibration being transmitted through mast and rigging to the hull, and such fluttering, which usually occurs when the wind is forward of the beam, might over a period of time injure the sail.

In a gaff-rigged yacht the trysail will probably be set on the throat halyard, and its luff will be held to the mast by a rope lacing or separate lanyards, each in the Irving Johnson manner being led from one eyelet round the mast to the next eyelet below, and secured there with a figure-of-eight knot, the resulting diagonal lead preventing the lanyards from binding. With jib-headed rig the main halyard will

---

PLATE 15

Wanderer II's twin running sails, *A* from forward, *B* from aft; the sails were set flying, and their booms had open jaws to fit on the mast, up which they could be pushed to the desired height. *C*: Originally *Wanderer III*'s twin sails were hanked to stays, across which was clamped a wooden bar to take the goosenecks of the booms. The combined area of these sails was only 250 square feet, but in light winds the area could be increased by setting a 330-square-foot ghoster in place of one of the twins, *D*; this sail was of thin nylon, and the sun can be seen shining through it at the top left of the picture.

almost certainly be used for the trysail, though it is held by some people that because of the long drift between the masthead sheave and the head of the trysail, an unfair strain is placed on the track in way of the head of the sail, for of course track and slides must be used for that part of the luff which extends above the crosstrees. I have not heard of a track pulling away from the mast for that reason, but the risk could be avoided by making use of reinforced track, i.e. with flanges at each side to take additional fastenings, in that neighbourhood. When the sail is set its foot should be clear above the main boom when the latter, with sail stowed on it, is in its gallows. A lacing or lanyards may with advantage be used for the luff below the crosstrees; but if slides are to be used for the full length of the luff, to avoid the awkward job of inserting each through the track gate as the sail is hoisted—with the yacht rolling heavily with no sail to steady her—it may be thought worth while to have a duplicate short length of track on the lower part of the mast, into which all the trysail slides can first be fed before the mainsail is lowered, and with a switch to lead them into the main track as the sail goes up (Plate 14E).

Aboard *Wanderer III* the trysail was sometimes carried for long periods, once for 14 consecutive days in the Indian Ocean, during one of which the yacht made a day's run of 150 miles under that 75-square-foot sail only. I reached the conclusion that the simpler the heavy-weather gear is the better, as it needs to be brought into use under difficult conditions and perhaps in the dark when the sense of feel is more valuable than that of sight, and when chafe must be guarded against. Originally the sheet was rigged as a luff tackle; but when setting the sail, or when gybing, the block on the clew was a danger; also, unless the sail was handed for each gybe and its sheet rearranged, the latter bore heavily on the furled mainsail on one gybe. I therefore replaced the tackle with two single-part sheets, rove each of them through single blocks secured to the mainsheet horse, one on each quarter, and thence to the headsail winches. The sheet in use, the lee one, thus passed clear of the furled mainsail (Plate 13E, though here the trysail was set not because of heavy weather but to ease the rolling).

PLATE 16

A: Under her twins *Wanderer III* rolled heavily. Here the original crossbar for the booms is still in use, but the sails are set flying. *B*: After the booms had been lengthened, they were shipped in a fitting on the mast, but the sails were still tacked down to their old positions forward of the mast. *C*: Firm and round, the twins pulled steadily, but they did not work properly when the wind was more than two points out on the quarter. *D*: To achieve self-steering under twin running sails, it is usually necessary to take the braces through blocks at the quarters and hitch them to the tiller, an arrangement first used by Waller in *Imogen*.

The working headsail halyard will be used without modification for the heavy-weather headsail, but to enable the normal headsail sheet leads to be used efficiently if they are of the fixed pattern, the clew of a very small headsail will need to be cut high, otherwise the sail must be provided with a tack rope of such a length as to enable the clew of the sail to lie on the same line as that of the normal working headsail.

Running sails and self-steering arrangements

Purely for the purpose of running, a squaresail with all the gear it needs is not worth having in a small vessel, and it will not of itself make a vessel self-steering. It is a fact that *Buttercup* crossed the Atlantic under a simply rigged squaresail without being steered by hand, but steering was achieved not by that sail but by means of one of the earlier of the wind-vane steering gears. Bill Leng, to whom Macpherson presented *Driac II* when he gave up sailing, told me that during his passage from the Cape to England he became convinced that the squaresail was not worth its locker space, and that the ship ran faster and much more steadily under mainsail and spinnaker. But a squaresail does of course reduce the work when gybing.

To make square rig worth while it must be of ample area, the yard being at least twice the beam of the ship, and it should be possible to set a raffee above the squaresail; also the standing rigging should be so arranged that it is possible to brace the yard sharp up to enable the square canvas to be carried with the wind on the beam or even forward of the beam. This means that the lower shrouds will have to be led well aft, and the crosstrees pivoted so that the lee topmast shroud can be pushed out of the way by the yard as it is braced up. Although this can be done aboard a gaff-rigged vessel without jeopardizing the mast or seriously interfering with the working of the fore-and-aft sails, it cannot be done safely when the mast is tall and is intended for the jib-headed rig. In large vessels making long passages, however, some square canvas may have advantages, as was shown by the 96-foot brigantine *Yankee* (Plate 44) during the four 18-month voyages that she made round the world in 10 years; she even set studding sails. *Yankee* carried a crew of about 20 amateurs, but that squaresails do not necessarily require a large crew was demonstrated by Robinson, who modernized the brigantine rig for his beautiful 70-foot *Varua* (Plate 32, *top*) so that she could be handled by two men, as he describes in his book *To the Great Southern Sea* (Peter Davies). Conor O'Brien was a staunch advocate of squaresails in yachts, and anyone who is interested in the subject, which revives some of the difficulties and delights of the great days of sail, and is seeking sound, practical advice, will find what he needs in O'Brien's *The Small Ocean-going Yacht* (Oxford).

Most people today prefer to use a spinnaker rather than a square-sail for increasing the sail area when running; it requires less gear, and by easing the after guy, it can be used effectively with winds from astern to abeam without any modification having to be made to the standing rigging. As shipping and unshipping a boom can be awkward when the motion is wild, it is a good plan to keep the boom permanently shipped on its gooseneck, and by means of its topping lift, stow it when not in use vertically up and down the mast, with its upper end resting in a chock on the fore side of the mast. So that the boom can be used on either gybe without having to be unshipped, it must of course be short enough to pass beneath the lower forestay (if there is one) when topped up. If two running sails are to be used, both booms can be stowed in this way, or in the manner described on page 74.

A parachute spinnaker, i.e. one with great belly, cannot be regarded as a seaworthy sail because of its tendency to collapse when the yacht rolls. Indeed, except for racing purposes a spinnaker (I prefer the term 'running sail' in this context) which is likely to be carried for long periods in the open sea, is best made without much belly, even though it may then have little lift, and it should be so arranged that it cannot chafe on anything; large area is of secondary importance.

It should be possible to get a reasonably well-balanced yacht to steer herself close-hauled with the helm lashed so long as the wind is steady in force; but if the wind freshens she will luff, and if it eases she will bear away. It is therefore desirable to achieve a balance of sail so that the helm may be left free. As all normal yachts carry some degree of weather helm, a reduction in the after part of the sail-plan may have to be made by taking in a reef or handing the mizzen; but if the yacht normally carries only a little weather helm, self-steering may be achieved by hardening in a headsail, or easing the main or mizzen sheet until balance under sail is obtained. In this respect a bowsprit may offer some advantage (see below). I had plenty of opportunity for experiment with *Wanderer III*, jib-headed sloop, when she was bound north in the Atlantic from Ascension Island to the Azores, for after crossing the equator and passing through the doldrums, she had to make her way close-hauled through the full width of the north-east trade wind belt, a distance of 1,600 miles. With the No. 1 staysail set we had to roll down about 2 feet of the mainsail to get her to balance, and she then steered herself six points off the wind—that was as close as I considered it paid to sail her in rough water—with the tiller free. The only effect a freshening or easing of the wind had was to make her increase or decrease speed; but when the wind changed direction, as it did a point or two most mornings and evenings, she changed course with it and continued to steer herself, making daily runs of between 90 and 100 miles. Mostly the wind was fresh, so the reef was no disadvantage, and when a further reduction of sail was needed as the wind strengthened and a deeper reef had to be rolled in, the balance was maintained by setting a smaller staysail. There were, however, occasions when, because of the steepness or irregularity of the sea, she refused to steer herself unless a piece of elastic was attached to the tiller on one side or the other, and the tension on this had to be adjusted with precision.

In favourable conditions the above method may be made to work with the wind just free, but with some uncertainty when it is abeam. For many of the trade wind passages likely to be made by a yacht, however, the wind will be on the quarter or right aft, and it is for sailing in such conditions that various special running sails and other arrangements have been devised.

It may be possible for a yacht with a long forward overhang or a bowsprit to steer herself with the wind on the quarter if the jib is sheeted flat, for the bowsprit or overhang increases the leverage of that sail and reduces the blanketing effect of the mainsail on it. If, however, the yacht cannot be made to self-steer by flattening the jib, or if she has no jib, the method used successfully in *Kurun*, *Viking*, and *Solace* should be tried. For this the staysail has to be rigged with a boom along its foot, and it is an advantage if it is sheeted to a horse. A steering line, made fast to the boom, is rove through a block to windward, then aft through a second block to windward of the tiller, and thence to the tiller (Fig. 4A). This line exerts a certain pull on the tiller which is counteracted by shock-cord fitted to the lee side of the tiller. Now let us suppose the yacht is steering herself on the desired course with the wind on the port quarter, and she luffs; the angle of incidence of the wind on the staysail increases, the pressure on the sail increases

PLATE 17
Beyond's twin booms were pivoted on the mast 18 feet above the deck, and the sails were hanked to twin stays 5 inches apart. *A*: The starboard sail is already set, and the port sail is being hanked on. *B*: With boom and brace secured to the sail, the sail is being hoisted up its stay. *C*: As the sail goes aloft it lifts the boom, and this is automatically pulled aft by the brace until, with the sail fully set, *D*, it takes up its correct position. (See also Plate 18.)

and pushes it to leeward together with the steering line, which then pulls the tiller to windward against the pull of the shock-cord (Fig. 4B). As soon as the yacht has returned to her course, the pressure on the staysail decreases, and the shock-cord pulls the tiller back to its original position. If the yacht were to bear away off her course, the pressure on the staysail and the pull on the steering line would decrease and the shock-cord would pull the tiller to leeward (Fig. 4C) to return the yacht to her course. But aboard a yacht which rolls violently this

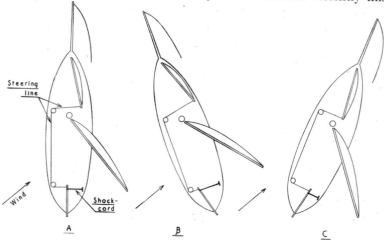

FIG. 4.   THE SELF-STEERING ARRANGEMENTS AS USED IN *Kurun*, *Viking*,
AND *Solace* WHEN THE WIND IS ON THE QUARTER

The staysail controls the tiller by a steering line against the pull of the shock-cord. *A*: On course. *B*: When the yacht luffs, the tiller is automatically pulled to windward by the steering line, and when she bears away, *C*, the tiller is pulled to leeward by the shock-cord.

arrangement will be useless when the wind is much abaft the beam, because the wind will be thrown out of the staysail with every leeward roll, and a course to windward of the desired one will be the inevitable result. In moderate weather it is likely to work best with the wind between two and four points (23 and 45 degrees) abaft the beam, but much will depend on the size and type of yacht and the nature of the sea. In *Viking* shock-cord was rigged on the tiller end of the steering line in the same manner as it may be used on a topping lift (page 51) to ease the jerk on the tiller when the staysail shook.

PLATE 18

The combined area of *Beyond*'s twin sails was 750 square feet, and in a force 5 wind gave her a speed of 7½ knots. I took this and the photographs reproduced in Plate 17 off Viti Levu, Fiji, during *Beyond*'s voyage round the world.

F

John Letcher used some variations of the above arrangement with considerable success in his 20-foot *Island Girl* during his single-handed voyages between San Diego, Hawaii, and Alaska, and these are shown in Fig. 5. At A in a light air with mainsail and genoa set, the steering force is taken by way of a snatch block and line from a bend in the genoa sheet. If the wind freshens, or the yacht tends to luff, the increased strain on the sheet pulls the block to leeward, thus

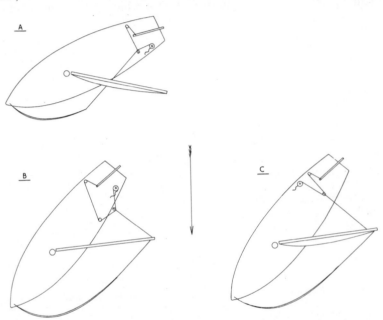

FIG. 5. SELF-STEERING ARRANGEMENTS USED IN *Island Girl*
A. Main and genoa are set, and the steering force is taken from a bend in the genoa sheet. *B*. The genoa only is set, and its sheet, passing through a block at the boom-end, swigs on the steering line. *C*. Mainsheet is entirely slack, the steering line from the boom-end going indirectly to the tiller.

moving the tiller to windward to correct the heading. At B, also in light winds, the genoa only is set and its sheet is rove through a block at the end of the main boom, the boom being held firmly outboard by sheet, fore guy, and topping lift; here the sheet swigs on the steering line, and the increased tension caused by the yacht luffing pulls the tiller to windward. At C, mainsail and genoa are set, but the mainsheet is entirely slack, and the steering force, which is halved by the block arrangement, is in the single steering line which goes indirectly to the tiller. Letcher found that he often needed to control the tiller by a piece of elastic leading to the lee side of the cockpit (this is not shown in the Figure), and that the finer adjustments consisted of changing the position of the steering line block on the weather side and changing the tension in the elastic. Such adjustments, he found, should not be made indiscriminately, but always with careful consideration of how the self-steering is failing to provide the desired forces on the helm. The mainsheet is not shown in any of the drawings.

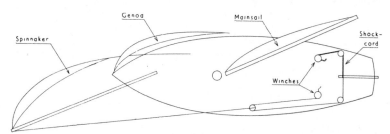

FIG. 6. SELF-STEERING ARRANGEMENT USED BY JOHN GOODWIN IN
Speedwell of Hong Kong

When the wind was light and abeam, a total sail area of 640 sq. ft. could be set.

In *Speedwell of Hong Kong*, one of the 25-foot Vertue class (see also page 74), when the wind fell light and was more or less abeam, John Goodwin, who was a great driver, found it was possible to carry mainsail, genoa, and spinnaker all at the same time, and he discovered that the latter, on its 14-foot boom, did not interfere with the set of the genoa. The total sail area of this rig was 640 square feet, which is a remarkable amount for so small a yacht; a Vertue normally carries a maximum of 314 square feet and is not considered to be under-canvased. Self-steering was achieved in the following manner. The spinnaker guy was led aft to a single whip, one end of which passed through the quarter block and was made fast to the tiller, while the other went to the weather headsail sheet winch so that it could be adjusted easily. Heavy elastic shock-cord secured to the tiller was adjusted by the lee sheet winch to achieve balance (Fig. 6). If the yacht luffed, the spinnaker exerted a greater pull than the shock-cord and the tiller was moved to windward; if she bore away the shock-cord exerted a greater pull than the spinnaker, and the tiller was moved to leeward until she returned to the course.

There have been a few vessels which, their owners claimed, would steer themselves with the mainsail set even when the wind was more than four points abaft the beam, and Slocum's famous *Spray* was one of them. But in the absence of a wind-vane steering gear, see page 75, usually it is necessary to lower the mainsail and run under headsails only, or to set twin running sails, to get a yacht to steer herself with the wind aft. There are several ways of rigging these twin sails, and I believe the first method to work successfully was that devised by

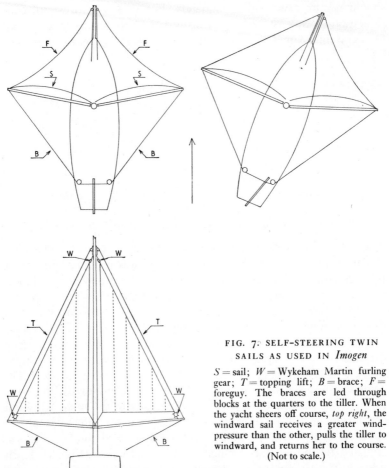

FIG. 7. SELF-STEERING TWIN SAILS AS USED IN *Imogen*

S = sail; W = Wykeham Martin furling gear; T = topping lift; B = brace; F = foreguy. The braces are led through blocks at the quarters to the tiller. When the yacht sheers off course, *top right*, the windward sail receives a greater wind-pressure than the other, pulls the tiller to windward, and returns her to the course. (Not to scale.)

Otway Waller for use in *Imogen* in 1930 (Fig. 7). The sails (*S*) were triangular, their combined area forming an equilateral triangle, and they were spread by two booms shipped into goosenecks on the mast. The sails were sheeted to the foot of the mast and their outboard edges were fitted with Wykeham Martin furling gear (*W*) so that the sails could be easily furled simply by rolling them up. The booms, held by topping lifts (*T*), were normally trimmed about 10 degrees forward of the beam, and their braces (*B*) were rove through blocks each side of the tiller and were hitched to the tiller. Foreguys (*F*) were rove through blocks at the bowsprit end and set up inboard, but their only function was to hold the booms in position while the sails were being set, and in the event of the sails being taken aback; at other times the guys were slack. Let us now suppose that *Imogen* is running under her twins dead before the wind (Fig. 7, *top left*) and she yaws to starboard (Fig. 7, *top right*). The wind-pressure on the starboard sail becomes greater than that on the port sail, its brace exerts a greater pull on the tiller, and the tiller is moved to starboard to put the yacht back on her course.

A year or two later experiments made with model yachts showed that if the twin sails were hanked to stays so arranged that there was a space 3 or 4 feet wide between them at the deck, and that the angle of the booms was increased to about 20 degrees forward of the beam, it was not necessary to take the braces to the tiller to make the model steer itself dead before the wind. But arrangements which work in models are not always a success in full-size yachts, especially when the latter, with no fore-and-aft sails to steady them, are rolling heavily and causing the apparent wind to change direction with every roll. Although some people have made this rig work, others found that it offered no advantage because the braces still had to be taken to the tiller, and the gap between the sails had certain drawbacks when the yachts rolled.

In *Wanderer II* I used the Waller rig, but without the furling gear, which seemed to me unnecessary, and the twin booms had open jaws to fit on the mast so that they could be pushed well up the mast to counteract the tendency of the outer ends of the booms to lift, and there they were lashed. The booms were 11 feet long and the total area of the sails was 220 square feet, which was found to be insufficient during her Atlantic crossing, when the trade wind was light. This rig (Plate 15A and B), with the braces led to the tiller (Plate 16D), worked well enough dead before the wind, but was of little use with the wind on the quarter, because, or so I thought then, it was impossible to brace the lee boom far enough aft, for its swing was restricted by the shrouds. I made some use of the rig when sailing single-handed and during a trip to the Azores with my wife, but it did not really come into its own until the little ship, manned by Bill Howell and Frank McNulty, made the 3,600-mile passage from the Galapagos Islands to Tahiti at an average speed of 106 miles a day, much of that great distance being run under the twin sails; it was a remarkable feat for a yacht of barely 21 feet on the waterline, and it says much for the courage, drive, and seamanship of her young Australian crew.

Marin-Marie had at the time not heard of the Waller rig, but he had made a study of the steering arrangements of model yachts, and for his single-handed crossing of the Atlantic he rigged his gaff cutter *Winnibelle* with twin staysails. These were set hanked to twin forestays a few inches apart (Fig. 8, *top*). It would appear to be important that the arc described by each boom should be centred on the stay on which the sail is set, to ensure an harmonious rotation of the whole. If, as is often done, the boom is pivoted on the mast while the sail is on the forestay, the act of bracing the boom aft will flatten the sail, while if the boom is allowed to go forward the sail will become too baggy. Marin-Marie overcame this trouble by clamping a steel fitting (*A*) across the twin forestays a few feet above the deck, and shipping the goosenecks of the booms in its ends. The thrust of the booms tended to twist the stays together, so he set up chains from the crossbar to the fiferail at the foot of the mast. As in *Imogen*, the foreguys were eased off after the sails had been set. The rig had several advantages over earlier ones. Being farther forward, the centre of effort of the sails was well forward of the centre of lateral resistance, so there was less tendency for the yacht to yaw; the angle of the forestays gave the sails a lifting component instead of the pressing component imparted by sails which are set parallel to the mast; and when sailing with the wind on the quarter, the lee boom could be braced well aft without touching the shrouds, its swing being limited only by the fact that the farther aft it was braced the nearer did its outboard end droop towards the water.

But a trouble with the rig was that because of the low position of the inboard ends of the booms in relation to their outer ends, which had to be kept high if they were not to dip into the sea, the boom-ends lifted in anything of a breeze; therefore the braces, instead of being taken direct to the quarter blocks and the tiller, had first to pass through blocks situated just forward of the shrouds (L in Fig. 8, *top right*) so as to hold the booms down. The strain on these blocks and on the braces was considerable, and not until hemp braces instead of wire, larger blocks, and much tallow had been used could the self-steering gear be made to work. *Kurun*, with an almost identical twin rig, suffered from similar troubles, and *Galway Blazer* even more so because the inboard ends of her booms were even lower than those of *Winnibelle*, being shipped on the forward corners of the coachroof, and while her twin sails were in operation, constant renewals and

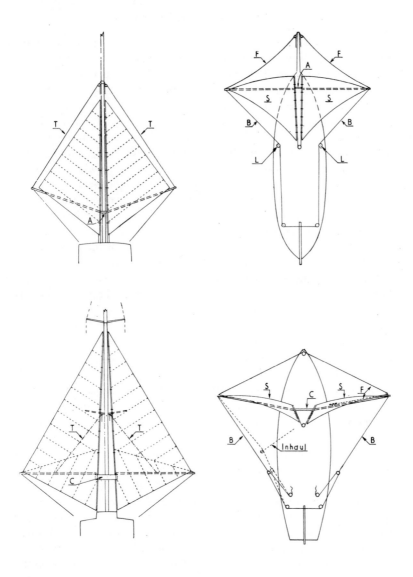

FIG. 8. TWIN STAYSAILS

Top: Winnibelle's twin sails, *S*, were hanked to twin forestays, the inboard ends of the booms being shipped in a steel fitting, *A*, clamped to the stays. To prevent the booms from lifting, the braces, *B*, were led through fairlead blocks, *L*, before being taken to the tiller. *Bottom: Wanderer III*'s twin sails were originally hanked to special stays set up to the coachroof forward of the mast, the booms being shipped in the ends of a crossbar, *C*, clamped across the stays.
   F = foreguy set up to cleat on starboard boom; *T* = topping lift. (Not to scale.)

repairs had to be made to the blocks and braces. Yet the rig served its purpose in all three yachts, each steering herself for 26 consecutive days in the north-east trade wind. A remarkable thing about *Winnibelle* was that when the wind fell light and the big masthead spinnaker was set in addition to the twins, she still continued to steer herself.

When working out the twin rig for *Wanderer III*, I decided to attempt to overcome the lifting of the outer ends of the booms and yet avoid the friction of the forward brace blocks by making constant use of a foreguy; but as the yacht had no bowsprit to which to rig the guy, the sails could not be set on twin forestays set up at the stemhead, as in *Winnibelle* and *Kurun*. I therefore secured two special stays to the mast at the root of the upper crosstrees, and when the twin sails were to be used, set them up with rigging screws to plates on the sides of the coachroof $3\frac{1}{2}$ feet apart and 3 feet forward of the mast. Across the stays 4 feet above the deck I clamped an oak crossbar (c in Fig. 8, *bottom*), with bolts and wing-nuts, and shipped the goosenecks of the booms into sockets at its ends (Plates 15C and 16A). Topping lifts were rove through bullseyes seized to the under sides of the lower crosstrees, and the braces led direct to quarter blocks, thence to the tiller. The foreguy was made fast to the outboard end of the port boom, was rove through a block at the stemhead, thence through a bullseye at the outer end of the starboard boom, and so to a cleat on that boom. It was therefore easy to adjust from the foredeck, and although it was kept taut when the sails were set, so as to prevent the outer ends of the booms from lifting, it did not interfere with the fore-and-aft swing of the booms.

This rig had the advantage of reducing the strain on the braces and their blocks, and it permitted the lee boom to be braced some way aft for a quartering wind before it came against the shrouds. But the self-steering did not work well when the wind was more than about two points out on the quarter, and I believed that was partly due to the wide gap between the luffs of the sails, permitting wind to get through and backwind the lee sail. It does seem, however, that few twin-sail rigs will work properly with the wind more than two points out on the quarter, though it is not easy to obtain accurate information as to the latitude within which the rig will work in other yachts, probably because most of the available accounts concern the crossing of the North Atlantic by the southern route, a large proportion of which is usually a dead run.

An improvement in self-steering with the wind on the quarter can sometimes be made by rigging an inhaul on one of the braces. This consists of a block on the brace between the boom end and the quarter block, with a line secured to it for setting up on board, as shown in dotted lines rigged on the port brace in Fig. 8, *bottom right*. If, for example, one wishes the yacht to steer more to starboard, the inhaul is put on the port brace, and when it has been set up it will be found that, by increasing the pull on that brace, it will keep the tiller a little to port and thus achieve the desired result.

It is not convenient to lead each brace through its quarter block to the tiller, because the only way of adjusting a brace will then be to re-hitch it to the tiller, and except in light airs or perhaps in a very small yacht the strain will be too great to permit this to be done. Instead the inboard end of the brace should be provided with, or rigged in the form of, a whip. If the moving block of the whip is on the brace, as in Fig. 8, *bottom right*, one end of the whip will be hitched to the tiller and the other will go to a cleat or winch; a mechanical advantage of 2 to 1 when trimming is thus obtained; but for a given movement of the boom-end the tiller will be moved twice the distance, and this may be excessive unless a tiller extension is fitted. If the whip is rigged the other way round with the moving block on the tiller, the tiller will move only half the distance that the boom-end moves, and no mechanical advantage is obtained for adjustment; however, this will not matter provided the brace is taken to a winch.

The chief disadvantage of *Wanderer III*'s original twin rig was that a lot of work was involved when changing to it from normal rig; so on arrival in New Zealand the stays and crossbar were scrapped, the booms were lengthened and were thereafter shipped in a double gooseneck fitting on the mast band (Plate 16B) and the sails were set flying. That was an improvement as the booms with their gear ready on them could be kept when not in use topped up and down the mast, from which position they could be quickly lowered into the working position. The self-steering still worked within the same limits, i.e. with the wind from aft to about two points out on either quarter. The combined area of the sails was 250 square feet, which is too small for a yacht displacing 8 tons, and never gave a day's run better than 125 miles, but in light winds the area could be increased by replacing one of the twin sails with the 330 square-foot ghoster, which was hanked to the topmast stay and boomed out in the usual way (Plate 15D). With the new arrangement it soon became clear that there was no need to take the braces to the tiller, and the twins continued to steer within the same limits as before with the tiller left free, though sometimes a piece of elastic had to be used on one side or the other of it, depending on the run of the sea. The braces were led straight from the boom ends to the headsail sheet winches with which all adjustments were made.

One of the most outstanding passages to be made under twin running sails was John Goodwin's single-handed crossing of the Atlantic from Las Palmas to Barbados in the sloop *Speedwell of Hong Kong*. This Laurent Giles-designed Vertue took 25 days to sail the 2,700 miles, an average of 108 miles a day, and she steered herself for the entire passage. This is in marked contrast to A. G. Hamilton's single-handed crossing of the Atlantic by the northern route in *Salmo*, a sister ship to *Speedwell*; to get sleep he hove-to or ran under a headsail only.

In *Speedwell* the inboard end of each boom was notched to fit on the forestay, where a pin held it in position, and a line secured with a rolling hitch kept it at the desired height. Sometimes twin spinnakers totalling 480 square feet were used, but more often twin masthead genoas were set by a single halyard, their hanks being alternated on the topmast stay; the braces were led to the tiller by way of blocks near the shrouds as in *Winnibelle*, and at the quarters. The total area of this rig was 400 square feet, and when the wind freshened one of the genoas was replaced by the working staysail, reducing the area to 310 square feet, and the tiller was balanced by light elastic. Foreguys rigged to a temporary bowsprit were needed only in light weather.

Tom Worth brought a fresh outlook to the twin sail problem in *Beyond* for her world voyage (Plate 18) and was well pleased with the result. He had two pairs of twin sails; the larger, totalling 750 square feet, were of 1½-ounce cotton and were carried in winds of up to force 5, when they gave a speed of 7½ knots; the smaller pair were of 2½-ounce nylon with a total area of 450 square feet. The sails were hanked to twin topmast stays about 5 inches apart, and the 15-foot booms, which were 2½-inch diameter light-alloy tubes, were pivoted not at their lower but at their upper ends, the goosenecks being on the mast 18 feet above the deck. When not in use the booms hung down each side of the mast with their lower ends lashed to the shrouds. There were no topping lifts, foreguys, or outhauls.

The photographs in Plate 17, which, together with Plate 18, I took off Viti Levu, Fiji, show the port sail of the larger pair of twins being set. First it is hanked to its stay and the tack made fast (A); the lower end of its boom is then unlashed, and the clew of the sail and the brace shackled on (B). The sail is then hoisted (C), and as it goes up it lifts the boom into a horizontal position (D). With this rig the high position of the inboard ends of the booms, coupled with the downward pull of the braces, kept the booms from lifting. The sails had wire bolt ropes (these need to be strong and well secured) to take the weight of the booms and the downward pull of the braces, and each had a light line in the tabling at the foot so that the belly could be increased to help them keep full in light airs. To take in the sails, the braces were not touched, and on the halyards being let go the sails ran down their stays neatly on to the deck. Although steering was by wheel, a tiller could be fitted for use with the twins; strong shock-cord was fitted each side of it to steady it and provide fine adjustment, and on occasions this needed a good deal of attention. 'Jack', as the self-steering arrangement was called, did not work well when the wind was more than 20 degrees out on the quarter.

The only trouble experienced with *Beyond*'s twin rig, which was in use a great deal during her circumnavigation, was the breaking of one of the boom goosenecks. This was of the conventional design, i.e. a joint with two hinge pins, that nearest the mast having a vertical axis, and then a second with a horizontal axis, just like a main boom gooseneck. 'If', as Worth pointed out in the *Royal Cruising Club Journal* for 1953, 'the boom is hinged high up the mast and comes down to a vertical position, there is little to prevent it going up the wrong way and subjecting itself to heavy strains. This trouble is easily overcome if the first hinge on the mast has a horizontal axis.' Alternatively a ball-and-socket fitting might be used.

A disadvantage of twin running sails is that under them alone rolling can be violent. *Wanderer III* at times built up a period roll during which she flung herself from 32° one way to 32° the other at 2-second intervals. In an attempt to reduce or slow down the roll, the trysail sheeted flat amidships was tried on several occasions, but it interfered with the self-steering ability of the twins, and so could be used only with a hand at the helm, when of course there was not much point, except freedom from chafe, in continuing to use the twins. The Pyes in *Moonraker* had the same experience, but as they had a third hand aboard they set the trysail only at night so that the two below could enjoy their rest, and let the twins steer only in the daytime. During a more recent voyage to America, for which *Wanderer III* was fitted with a wind-vane steering gear, the twins were still often used to save the mainsail from wear, but it was found possible to use the trysail to steady her without any ill effects on the steering.

Wind-vane steering gears

The idea of using a wind-vane to control the main or an auxiliary rudder and so keep a yacht on course is not new, for such a gear was

devised and used by Marin-Marie in 1936 for his west-to-east crossing of the Atlantic in the motor yacht *Arielle*, and by Ian Major in
1955 for an Atlantic crossing in the opposite direction in *Buttercup*.
Since then the need for vane gears by the participants in the single-
handed Transatlantic races has led to developments and improvements, with the result that such gears are more efficient and are now
widely used by short-handed ocean-going yachts, greatly to their benefit; nevertheless some of the self-steering arrangements mentioned in
the previous section continue to be of value because anything that helps
a yacht to balance under sail, and to be easy on the helm, will assist a
vane gear to do its job better. After we had fitted *Wanderer III* with
vane gear, my wife and I found the business of crossing oceans less
arduous; it was now possible for us to get all the sleep we needed so
that we arrived in port fresher and more relaxed than we ever had in
the past, and at sea there was plenty of time for cooking, navigating,
and reading; indeed, on a passage we made in 1965 from Tenerife to
Barbados we steered by hand only for an hour or two at each end, and
for the rest of the 26-day passage steering was entirely by vane gear.

The general principle on which vane gears work is this: the yacht
is put on course and the sails are trimmed for that point of sailing; the
vane, which is usually made of plywood, though some are of metal, is
then adjusted so that when the yacht is on course it lies head to wind;
if the yacht goes off course the vane receives wind pressure on one or
other of its sides so that it tries to turn head to wind, and in doing so it
moves the main rudder or an auxiliary rudder or a trim tab and brings
her back on course.

The arrangement in which the vane moves the main rudder is the
simplest mechanically, for the vane can be coupled by a clamp or
clutch to a yoke which in turn may be linked to the tiller by lines leading through blocks or over sheaves, and the adjustment for different
points of sailing can be made by altering the length of these steering
lines. This was the type of gear devised by Francis Chichester and used
by him when he won the first single-handed Transatlantic race in
Gipsy Moth III, and it had the merit that no modification was needed
to hull or rudder. But as the vane, which was of sailcloth, had no
mechanical advantage, it had to be so large (45 square feet) for use
in moderate winds that it required reefing in strong winds, and because of its large size it had to be partly furled each time a gybe was
made.

H. G. (Blondie) Hasler, the originator of the single-handed Transatlantic race, has gained practical experience with vane gears from,

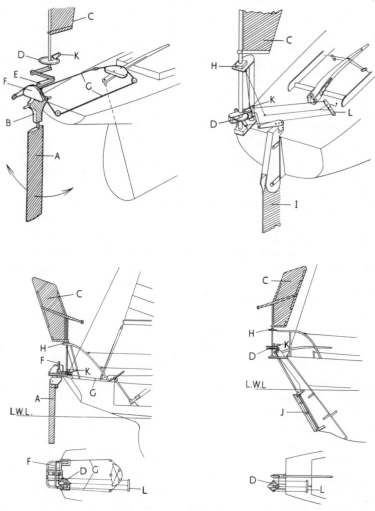

FIG. 9. HASLER WIND-VANE STEERING GEARS

Top left and bottom left: Pendulum-servo. *Top right:* Auxiliary rudder. *Bottom right:* Trim tab. *A* = blade; *B* = servo box; *C* = vane; *D* = toothed wheel; *E* = servo tiller; *F* = quadrant; *G* = steering lines; *H* = vane-shaft upper bearing; *I* = auxiliary rudder; *J* = trim tab; *K* = latch; *L* = brownstick.

among other sources, four single-handed crossings of the North Atlantic in the Chinese-rigged 25-foot Folkboat *Jester*; for a total of 12,000 miles *Jester* was steered by hand for less than 50 miles. Hasler made a close study of vane-gear requirements, and is now widely regarded as the leader in this field. In his later type of gear (Fig. 9, *top left* and *bottom left*), which he calls the pendulum-servo, and which is suitable for a yacht with an inboard rudder stock, i.e. one with a counter or canoe stern, in which a trim tab (see below) on the main rudder is scarcely practicable, he has overcome the disadvantages of Chichester's gear. A long, narrow wooden blade (*A* in Fig. 9) is hung vertically over the stern, and is carried by a servo box (*B*), which allows it to be turned like a rudder by means of a wind-vane (*C*), which is linked to it through a toothed wheel (*D*) and short servo tiller (*E*). The servo box is mounted on fore-and-aft bearings on a tubular metal bumkin, so that, together with the blade, it can swing from side to side like a pendulum. An athwartships quadrant (*F*) is fixed to the servo box, and steering lines (*G*) lead from this through blocks or over sheaves to a similar quadrant mounted on the head of the rudder stock. When the vane receives pressure on one or the other of its sides due to the yacht going off course, it turns the blade out of the fore-and-aft line; the flow of water past the blade's lower part causes it to swing sideways, and this movement is transmitted to the rudder by way of the quadrants and steering lines. Because of its length the blade has considerable leverage, so it needs to be turned only a little out of the fore-and-aft line to be effective; therefore the vane can be very much smaller than that which was used by Chichester, and a little smaller than those needed for some other types of gear. The pendulum blade can be hinged up out of the water for ease in shipping and un-shipping the servo assembly at sea; the whole gear is portable and requires no modification to be made to the yacht or her rudder, and the only necessary additions, which are of a semi-permanent nature, are a pair of base plates to hold the bumkin, a supporting arm for the vane-shaft upper bearing (*H*), and a quadrant on the rudder stock. I have had no experience with it, but this gear is highly spoken of by many owners, and Chichester used it during his voyage round the world by way of Cape Horn. However, his *Gipsy Moth IV* was at the time probably the largest and fastest yacht ever to be fitted with vane gear, and that hers suffered damage on the outward passage to Sydney was not surprising. Alec Rose, who in *Lively Lady* sailed in Chichester's track a year later, successfully used a veteran pendulum-servo gear which had previously made a double crossing of the North Atlantic. Tabarly in *Pen Duick II*, in which he won the 1964 Transatlantic race, used a vane which was pivoted not on a vertical but on a horizontal axis; such an arrangement has the advantage that it does not lose the pressure of the wind as it turns, and it is more sensitive to wind striking it at a small angle; however, linking it to the rudder or a trim tab presents engineering difficulties, and Tabarly's gear failed, so that he had to steer by hand for much of the race. I understand that M. F. Gunning has patented a gear in which the Tabarly type vane is combined with a form of pendulum-servo.

As we have already seen, if a wind-vane is by itself to move the main rudder, which must be of large area as it is needed for manœuvring, the vane will have to be so large as to be impracticable. But if, as in Fig. 9, *top right*, and Plate 19E, a small auxiliary rudder (*I*) is arranged as far aft as possible, a comparatively small vane should be capable of moving it sufficiently to keep a well-balanced yacht on her course, but it may need some help from the main rudder, which can be given a bias one way or the other with tiller lines or elastic.

With the trim tab type of gear (Fig. 9, *bottom right*, and Plate 19A–D), use is made of a narrow flap, the trim tab (*J*), hinged to the trailing edge of the main rudder, which usually needs to be straightened to accommodate it by removing a small part of the blade and providing fairing pieces at top and bottom. Because of its small area, which may be about one-fifth of the total area of the rudder, the tab is easily turned by a small vane. When the tab is turned in one direction it imparts a considerable turning force to the rudder in the opposite direction. So, when the yacht wanders off course, the wind turns the vane, the vane turns the tab, and the tab turns the rudder. Clearly such an arrangement has the disadvantage that it can be used easily only in yachts with outboard rudders, i.e. those with transom or pointed sterns, but in this connection it is worth noting that it cannot be fitted without some modification to a Norwegian-type stern because the upper part of the curved rudder stock moves in an athwartships arc as the rudder moves. Also, because the tab moves one way and the rudder moves the other way, some drag is inevitable. Nevertheless this type of gear is popular, particularly with people who make their own, and is widely used.

With all vane gears provision has to be made so that the vane can be set correctly for the course that is to be steered, and so that it may be instantly disconnected in order that the watchkeeper can take over the steering in an emergency, such as when risk of collision seems likely. Sometimes a clutch is used, but the more usual arrangement is to provide a toothed wheel on the vane shaft and a latch (*K*) to engage with this (Plate 19B and C), the latch being lifted from a distance, if desired, by pulling on a line. Such wheels usually have 36 teeth (if they had more they might not be strong enough), so the finest steering adjustment possible is 10°, which for practical purposes in the open sea may be regarded as good enough; but if a finer adjustment is required, some slight alteration of a sheet or brace will probably provide it, or the tiller's movement may be slightly restricted with a piece of elastic. In the standard Hasler latch gear the arm which carries the latch is provided with an athwartships yoke with a hole at each end; small lines from these holes are led forward and secured to the ends of a brownstick (*M*), named after Neville Brown who invented this arrangement, placed within reach of the watchkeeper. A third line, from the latch, is secured to a central hole in the brownstick. When the vane is in operation all three lines are slack. A forward pull on the stick will lift and disengage the latch, and if a strain is kept on the stick, a movement of the stick about its own centre in a horizontal plane will steer the yacht by means of the two lines, the yoke, and the tab or servo blade; or, if preferred, the brownstick can be secured with a strain on it, and steering be done by tiller. When the emergency is over the yacht is put back on her course, and when she is steady on it the pull on the stick is released so that the lines fall slack and the latch engages and the vane resumes steering. In the above circumstances, or when an adjustment to the vane is needed because of a change in wind direction, it might be helpful if the teeth on the wheel were numbered, but such numbers would be too small to be seen without going aft. Instead, I have found it satisfactory to paint two adjoining teeth in each group of four, as may be seen in Plate 19c; then if originally the latch lay between two painted teeth, or two unpainted teeth, or either of the two remaining combinations, the position is readily recognized; also, when moving the latch one tooth either way this marking is a help. As an optional refinement Hasler later introduced a worm-drive latch gear, in which the toothed wheel and latch are superseded by a bronze worm which drops into mesh with a worm-wheel when the latch-line is released. To this extent it can be used in exactly the same way as the normal latch, but an endless line (which replaces

PLATE 19

Wind-vane steering gears. *A*: With the Hasler trim tab gear the vane is offset, and *B*, in this example its shaft is mounted on a platform secured to the bumkin, its upper bearing being held by a bracket built out from the after guardrail. *C*: the vane is coupled to the trim tab tiller on the rudder head by means of a latch and toothed wheel (for adjustment) and a link. The latch line and brownstick lines can be seen as well as the painting of alternate pairs of teeth. *D*: The trailing edge of the rudder has been straightened to accommodate the tab, which is about one-fifth the area of the rudder. *E*: In this gear, designed for smaller yachts by John Flewitt, the vane is coupled to an auxiliary rudder, which is independent of the main rudder, so no modifications need to be made to the latter. Here the vane shaft has been lengthened a little so that the vane will swing clear of the backstay.

the yoke and yoke-lines) leads from the cockpit round a sheave on the worm spindle, and by pulling either side of this line the worm is turned, thus altering course by infinitely variable stages. As the worm never has to be unlatched for making alterations of course, there is no need for markings on the worm-wheel. A large prototype of this latch gear was used by Chichester during his circumnavigation.

The size of vane will to some extent depend on the type of yacht and her steering characteristics, and its shape may have to be dictated by the position of a backstay and/or the end of the main or mizzen boom, or some other obstruction to its necessary 360° swing. But if, when studying the sail plan, it may at first seem to be impossible to arrange a vane of a worthwhile area, it should be remembered that with most vane gears the vane may be, or may have to be, placed off the fore-and-aft centre line, and the farther off it is the more easily will it clear the obstruction and the wider can it be. The only limit to the length of the link connecting it to the servo tiller, or quadrant, is that a vane positioned outside a continuation of the deck edge will be vulnerable when the yacht lies alongside a quay or another vessel.

A tall and narrow vane is not the best shape, nor is it possible to support such a vane properly, for clearly the upper shaft-bearing must be below the lower edge of the vane. In *Wanderer III* a tall, narrow, vane had to be used because of the limited space between the boom-end and the backstay; as is usual because of weight considerations, it was made of thin marine plywood, and in strong winds its upper part vibrated. One windy night in the mouth of the Tagus, when we were beating up for the weather shore in a gale, the vane, which was not in use at the time, broke right across just above the top of its shaft—a rare mishap. Portuguese friends made a new one of thicker material to my dimensions, but this turned out to be too small, so I made an addition of about $1\frac{1}{2}$ square feet, and in port this could be unshipped to prevent vibration. This vane, Plate 19A, besides being too narrow, was not a good shape because its leading edge was too nearly vertical (a much better shape is shown in Fig. 9, *bottom*); nevertheless it worked

---

PLATE 20

Beyond. Top: The cockpit, unlike that of most sailing vessels, is amidships, where it is well protected by the deckhouse, in which is the chart table and compass. The port cockpit seat lifts for access to the engine-room, and the hatch in the foreground leads to the after cabin. There is a saddle for the helmsman. *Bottom right*: The ship has all manner of brilliantly conceived and carefully worked-out details, such as the combined mainsheet horse, boom gallows, and bollards seen here, and forward, *bottom left*, the projecting lead for the anchor chain, and the lifeboat type fairleads for warps. Some of these ideas have been incorporated in more recent examples of this type of yacht.

G

remarkably well for the rest of the voyage, including two crossings of the Atlantic.

Any vane requires a counterweight to reduce friction, which must be kept to a minimum, and to check the vane's tendency to hang to leeward when the yacht is heeled; it is also needed to steady the vane in a seaway. It seems that with a vertical vane-shaft the weight may be about half that which would be required to balance the vane completely.

Vane gears can be used in yachts fitted with wheel steering. Sometimes steering lines from the pendulum-type gear can be taken to a drum on the steering-wheel axle, but usually it is preferable to disconnect the steering near the rudder head and take the steering lines to a quadrant or auxiliary tiller on the rudder head, which may have to be extended above the deck for the purpose.

There are several firms which make and sell the required hardware for some of the patented vane gears; for example, M. S. Gibb, Ltd. of Warsash, Southampton, manufacture and sell all the Hasler gears, and provide customers with information regarding the size and shape of vane, blade, or tab, which they do not supply; but such ready-made equipment is not cheap. Some amateur-designed and built gears have proved to be far from satisfactory, and anyone who may be thinking of trying his hand at this would do well first to study the Amateur Yacht Research Society's booklet on the subject.

Some brief remarks about the automatic helmsman, whereby a yacht under sail or power can be made to steer any desired course by means of a magnetic device and an electric motor or an hydraulic ram, will be found on pages 96-8.

5

MECHANICAL AND
ELECTRICAL EQUIPMENT

Auxiliary engines—Electricity—Cooking stoves—Refrigerators—
Radio—Automatic helmsmen

UNLESS some member of the crew is a skilled mechanic and has with him the tools of his trade, it will be advisable to keep the mechanical equipment of the ocean-going yacht as simple as possible, and to fit only such items as have proved trustworthy in the damp and corrosive conditions which prevail at sea, especially if her voyage takes her into the tropics; and so to arrange matters that the mechanical equipment is needed only for convenience, comfort, or pleasure, and not for safety. If any vital job, such as emptying the bilge, pumping fresh water from the tanks, or weighing the anchor, is done mechanically, there should be an alternative manually-operated system available for use in an emergency; and if electricity is used for lighting, paraffin-burning lamps should be provided as a stand-by. As an extreme example of over-mechanization, I recall an American yacht which I met at Tahiti. She relied on electricity for lighting and cooking as well as for pumping fresh water and flushing the heads, and had no alternative system for doing any of those things. A breakdown of her dynamo had converted what started as a pleasure cruise into an existence of inconvenience and squalor; if the breakdown had occurred at sea the result might have been very serious.

Auxiliary engines

The keen sailing man will probably wish to make little or no use of his auxiliary engine when passage-making, and he may consider that its only purpose will be to charge a battery and get his ship into or out of port in the event of the wind failing, and that one of very low power will therefore serve. That was my own view when *Wanderer III* was being built, and it was strengthened by considerations of weight and space. I therefore had installed in her a two-stroke engine of only 4 h.p. That was a mistake, for the engine gave her a speed of 3 knots in smooth water and a calm, and could scarcely hold her head into a moderate breeze. Some of the lagoons in the South Pacific were barred to her because the engine had insufficient power to push her in through their narrow passes against the wind and outflowing current; also on a number of occasions when berth had to be shifted a short distance in a fresh wind, it was necessary to make sail. The engine's low power and lack of astern gear were also the cause of some anxiety when passing through the Panama Canal, and when entering or leaving small, crowded ports. For her second circumnavigation I therefore had her fitted with an 8 h.p. engine of the same reliable make (Stuart Turner) driving a 14-inch 2-bladed propeller, and fitted with astern gear. This was a marked improvement, though by today's standards 1 h.p. per ton of displacement would be regarded as meagre. Once again we experienced awkward moments in the locks of the Panama Canal with a strong trade wind blowing from astern, and we could well have done with greater power both there and in the Suez Canal, and later while traversing 1,000 miles of the Intracoastal Waterway on the east coast of the U.S.A.

The range under power was about 100 miles, but by carrying extra fuel in cans in the cockpit lockers, as we did for the passage up the Red Sea (and never used any of it), the range could be doubled. However, for an ocean crossing I did not care to load the yacht down aft to that extent, and therefore reckoned to use the engine only when near the shore or for battery charging.

I am glad to have had the opportunity of making a passage through the doldrums under sail, for I feel that is an experience which every sailorman ought to have once in his lifetime, so that he may learn to be patient and to appreciate some of the difficulties with which his forefathers had to contend. But once is enough, and if I ever have to pass through that area of calms, squalls, rain, and heat again, I hope to have an engine of greater power and a plentiful supply of fuel for it. There is also this point to consider: a powerful engine running slowly is longer-lasting, quieter, and cooler than a less powerful unit giving its maximum output.

For use in the ocean-going yacht the diesel engine has the merit that because of its economy in fuel it will give more miles to the gallon than a petrol engine of similar power. Also fuel is cheaper, the risk of fire is greatly reduced, and because the diesel engine works on the principle of compression-ignition, no electrical equipment is required for its operation. However, today most marine engines do rely on electricity for starting, and many of them have no provision for hand-starting in the event of battery or starter-motor failure. In this connection it should be remembered that a battery in a partly exhausted condition can cause damage to the starter, which is not a continuously rated machine and may burn out if it has to crank slowly and excessively. Although there is little excuse for battery failure if a separate battery is kept for engine starting and there is means of charging it on board, some may prefer to be completely independent of electricity and install hydraulic starting. With this in normal operation the hydraulic accumulator is charged by the engine through a belt-driven pump; but if through repeated attempts to start the engine the pressure drops to a useless level, the pump may be operated by hand-lever and pressure restored within about a minute. However, the makers of some engines in which the starter has to be fitted at the opposite end of the crankshaft to the flywheel do not approve of this equipment because it puts an unfair strain on the crankshaft.

In eliminating plumbing and risk of corrosion the air-cooled diesel has advantages. If its air trunking, which takes quite a bit of room, is correctly arranged, and the engine space is bulkheaded off from the rest of the accommodation, it should not cause the yacht to become overheated, and the bulkhead will serve the additional purpose of preventing the smell of diesel oil from pervading the rest of the accommodation; but in a small vessel such a bulkhead is not easy to arrange if there is to be reasonable access to the engine, and I know of several yachts in which the air-cooled diesel raised the cabin temperature to an intolerable degree in hot weather.

More engine failures are due to dirty fuel than to any other single cause, and as fuel bought in some places abroad may have been transported or stored in dirty or rusty containers, it is even more important than usual to strain it through a fine gauze funnel when filling the tanks. But this alone is not sufficient, for some dirt is sure to get through, and condensation in the tanks when they are not completely full, due to changes in temperature, will almost certainly take place; the motion will prevent all of this dirt and water from settling in the tank sump, so two good filters, of the replaceable paper cartridge type, should be fitted on the fuel line between tank and fuel pump of a diesel, and tank and carburettor of a petrol or paraffin engine, and the cartridges be renewed at frequent intervals. With gravity feed an airlock in the fuel line may cause the engine to stop; a choked air-vent, a badly seated filter bowl, or the fuel pipe passing too close to the exhaust system, or having kinks or uphill bends in it, may be the cause. With petrol engines a frequent cause of difficult starting is the use of old sparking plugs with defective insulation, or plugs in which the gap is set too wide (particularly when the engine is hot) or leakage from the high-tension cables. I have found it no bad thing to spray from time to time the entire ignition system with one of the aerosol waterproof insulators sold for the purpose. The modern diesel is a remarkably trouble-free machine provided the maker's recommendations are followed.

Petrol, diesel oil, and paraffin (often known abroad as kerosene) are usually available in small quantities even in very remote parts of the world, but if much use is to be made of the engine it may be best to make arrangements in advance with the Shell Company, which will supply a document authorizing the supply of fuel at a number of named ports along the route. Probably these bunkering stations will be at large commercial ports which one does not wish to visit, but it will be found that Shell often have sub-sub-agents, of which they may know little, in remote places, and these will gladly supply fuel against the document and charge it to the owner's account in England. By this means fuel may often be obtained duty-free in places where this would otherwise be impossible, and provided the point is mentioned on the document, lubricating oil and paraffin may also be obtained in this way.

Electricity

The most convenient form of lighting is by electricity, and this has the advantages, which are not so obvious until one has cruised and lived aboard in a hot climate, that—unlike paraffin or gas—it gives off practically no heat, and as it is not affected by draughts, the ventilation system can be kept working without interfering with it. Also of course other electrical equipment, such as a radio set and fans, can be run off the same supply.

There are two main types of storage battery; lead/acid and alkaline. Each cell of the former provides about 2 volts, and this does not drop appreciably until the battery is nearly discharged; but such batteries have a short life, possibly only 2 years or so, and are damaged by excessive heat, as for example when being charged at a high rate; also they give off an inflammable gas when on charge. They should therefore be stowed in a well-ventilated place and perhaps be cooled by fan when on charge, but if this is installed in the outlet vent it must be of the gas- or flame-proof type to avoid risk of explosion. The alkaline cell provides only 1·2 volts so that a greater number of cells is required to provide a given voltage; also the first cost is much higher; but this type of battery is very robust, it remains almost unharmed by ill treatment, and should have a very long life. A disadvantage is that there is a greater drop in voltage during discharge, so that, ideally, additional cells are needed to be brought in by some form of voltage control to maintain the ship's supply at the correct level.

If the current consumed by some item of equipment is known (this is usually marked on it in watts) together with the capacity of the battery, it is possible to tell for approximately how long the battery will keep that item of equipment working. The capacity of a battery, in ampere/hours, is generally marked on it. In theory a battery with a capacity of 100 a/h could provide a current of 1 ampere for 100 hours, or 2 amperes for 50 hours, etc.

As one volt $\times$ one amp. $=$ one watt,

$$\frac{\text{watts}}{\text{volts}} = \text{amps.}$$

Suppose, for example, the yacht has three navigation lights, each 12 watts, running off her 90 a/h 12-volt battery. The total consumption will be 36 watts.

$$\frac{36 \text{ watts}}{12 \text{ volts}} = 3 \text{ amps.} \qquad \frac{90 \text{ a/h}}{3 \text{ amps.}} = 30 \text{ hours.}$$

If, therefore, the lights are switched on for a total of 30 hours they will discharge the battery completely.

But it must be realized that no battery has an efficiency of 100 per cent, and that ageing reduces the capacity. A 100 a/h battery will in fact not give 1 amp. for 100 hours even when new, and it will be found that when recharging the 100 a/h battery a charge of 140 amp. hours will be needed, and this is the same for both lead/acid and alkaline batteries. Lead/acid batteries slowly lose their charge when not in use, and the higher the temperature the greater is the loss; at 65 °F. the loss is approximately 1 per cent per day, but at 100 °F. it is as much as 3 per cent. If this type of battery remains uncharged for 2 or 3 months sulphation of its plates will result, thereby reducing efficiency; regular charging is therefore necessary.

There are two types of generator: dynamo and alternator. Both generate alternating current, which must be rectified (converted to direct current) for the purpose of charging batteries; the dynamo does this mechanically by means of its commutator and brushes, while the alternator uses diodes for the purpose. For use afloat the modern alternator appears to have every advantage: it is smaller and weighs less than a dynamo of equal output; it requires hardly any maintenance, and as it is effective over a very wide speed range, a high output can be obtained at low engine speed. With this quick and efficient means of charging at his disposal, there no longer seems to be much need for the voyaging man to bother about other methods of charging, such as windmills and free-spinning propellers, which have so often been tried with limited success, but a few words about them might be of interest here. When *Beyond* sailed into English Harbour during her world voyage, all who watched her enter thought for a moment that she had been converted from cutter to yawl rig, but as she drew closer, the 'mizzen' proved to be a tall pillar fitted with a three-bladed propeller coupled to a generator. The next time I saw her was at Suva, Fiji, and the windmill had vanished. When I asked Tom Worth about this he told me that it had not been a success. Running before the wind, the apparent wind was often insufficient to drive it, and in harbour there was usually too much wind or too little. At the best it gave 2 to 3 amps. from time to time. Aboard *Waltzing Matilda* (see pages 129–34) a dynamo was coupled to the de-clutched propeller shaft, but it only charged when she was sailing at 7 knots or more, and one similarly fitted aboard *Kochab* (see pages 104–10) failed to work at all; but success is more likely with the alternator because of its higher output when turning slowly. However, the matter of wear and of drag should be taken into consideration. With the engine's clutch disengaged the spinning propeller will keep the gearbox components turning, and if reduction gear is fitted this speed may be high, while if the gearbox is hydraulically operated, lubrication of it without the engine running may be insufficient. So a clutch may have to be fitted in the shaft between gearbox and propeller, and if this is of the dog type a brake will be needed on the shaft to enable the clutch to be engaged and disengaged when the yacht is sailing fast. A fixed propeller, unless it be of the two-bladed type installed on the centre line and locked with its blades vertically abaft the sternpost, can increase a vessel's total resistance by as much as 20 per cent. But if, as was shown by experiments made at Burnham-on-Crouch by P. Newall Petticrow, the propeller is free to rotate, but is not able to rotate at a high enough speed—and that may be the result of coupling a generator to it—its drag may be increased by a further 25 per cent. The greatest drag appears to be created with a propeller speed of around 100 r.p.m. It would therefore seem that even if this method of charging a battery proved to be satisfactory in other respects, it might have a serious effect on the performance of a yacht under sail.

In countries such as the U.S.A., where the use of marinas is widespread, it is usual to provide the yacht with a socket into which a shore lead carrying 110 or 240 volts can be plugged. This may be wired by way of a charging unit to the batteries so that they are kept fully charged while in port, and at the same time lights and other equipment are run by the float-over. Alternatively, the shore supply may be taken direct to several specially wired points into which 110 or 240-volt equipment, such as fans, radiators, lights, etc., can be plugged. If high-voltage photographic equipment or power tools are to be used, a dynamotor to step up 12 or 24 volts to 110 or 240 volts will be useful, but only motors capable of being run on direct current can be used with this. Wiring—the lower the voltage the thicker the copper core needs to be—should be twin, not single and earth return, and every circuit ought to be provided with a fuse or circuit-breaker to guard against a short circuit; even so it is prudent to fit a switch on the positive battery lead so that the current can be turned off in an emergency or when the yacht is to be left unattended for any length of time. Damp, which causes short circuits and corrosion, is the chief enemy of electric equipment, so wherever possible the fittings should be of waterproof type, and all terminals, contacts, and moving parts should be kept lightly coated with petroleum jelly, or treated with one of the special aerosol sprays.

Cooking stoves

Some notable ocean-going yachts, among them *Idle Hour*, *Southseaman*, and *Saoirse*, had wood- or coal-burning cooking stoves; but in tropical waters the heat radiated must have been a hardship for the cook, and stowage space for a sufficient quantity of the bulky fuel must have been a problem; indeed, *Saoirse* put to sea on her first ocean crossing with 'a deck cargo of coal'. Nevertheless some very experienced people still prefer to use solid fuel cookers, among them the circumnavigator Peter Tangvald, who was installing one and a large bunker for charcoal in the schooner which in 1969 he was building with his own hands in French Guiana.

Aboard a few American yachts, among them the Vancils' *Rena*, electricity is used for cooking; but because of the very high consumption, either a generator must be run or a shore supply used when cooking is to be done. Because of the low fire risk, methylated spirit (commercial alcohol) is approved by the U.S. Coastguard, and cookers burning this fuel, either gravity-fed or, more suitably, under pressure, are widely used in American yachts. They are clean and silent, and the flame can be as easily regulated as that of any gas jet. But this type of stove will not burn properly if more than a few degrees from the vertical, and in some places the fuel is such a high price that it is cheaper to burn duty-free rum.

British yachts, and those of most nationalities other than American, generally use cooking stoves burning either bottled gas or paraffin (kerosene). The gas cooker is simple and efficient, it requires no pre-heating, the heat is easily regulated, and roasting, and sometimes grilling and toasting, can be done without other equipment; but it has two drawbacks: Abroad one cannot be sure of obtaining replacement cylinders having the same screw thread, or of being able to get empty cylinders refilled, so a selection of thread adaptors should be carried; also, unless the precautions mentioned in *Cruising Under Sail* are taken, there could be a grave risk of explosion. As there is no means of knowing, except by weight, how much gas remains in a cylinder, and as the cylinder is bound to become empty while the cooker is in use, it is a good plan where space permits to mount two cylinders side by side and connect them to the regulator by means of a two-way cock so that one can switch over to the full cylinder immediately and change the empty one at leisure.

Cooking stoves burning paraffin under pressure—the Primus type, and it is important to use only genuine Primus burners made in Sweden, as are standard fittings in Taylor's excellent Para-Fin cookers, for there are some worthless imitations on the market—have several disadvantages: The burners tend to carbon up if run at low pressure for a long time, and the flame does not lend itself to simmering without the use of asbestos mats. Also the burners require preheating with methylated spirit; for this, if one of the cans which will measure the exact quantity into the cup is not available, a clip-on preheating wick, such as is supplied for use with Tilley and other pressure lamps, can be used, and this, incidentally, is the best means of preheating a cooker which is not free to swing. Unless the cooker is a large one, roasting or baking can only be done by placing an oven temporarily over one of the burners, and such ovens are not so efficient as gas ovens which have the burners inside. However, the fire risk is small, there is no risk of explosion, and if a few spare parts are carried any failure can easily be rectified; unlike gas, paraffin is readily available even in the smallest island, for it is widely used for lighting purposes. But even when bought under well-known trade names, it is today common for the fuel to contain a little water; when a drop of this reaches the hot burner it is converted into steam, the burner goes out, and a match applied to it may be blown out several times before fuel comes through again. I have found that even when filling the tank through a funnel with a 'water-proof' gauze some water still gets through, and the cure for the trouble is to fit a bowl-type filter with a paper cartridge in the fuel line between tank and stove. An alternative when in the U.S.A. is to use not paraffin but mineral spirits, paint reducer, or even cleaning fluid, all of which I am told are refined forms of paraffin.

With the modern frying-pan the handle is low and almost horizontal, and cannot easily be used if the cooker has the old style of fiddle, i.e. a continuous rail about 2 inches high right round the hotplate. Instead the fiddle should be dipped or broken in several places, perhaps at each corner as in Plate 9B, to accommodate the frying-pan handle.

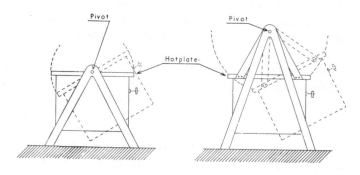

FIG. 10. SWINGING GALLEY STOVE

Left: The pivots should be on the same plane as the hotplate so that the motion is that of a see-saw; if the pivots are above the hotplate, *right*, the motion will be that of a pendulum.

No matter what type of galley stove is fitted in a small yacht, it must be slung on fore-and-aft pivots, so that it will remain upright whatever the angle of heel; athwartships pivots are not required. The pivots should be on the same plane as the hotplate, so that as the yacht rolls the motion of the hotplate will be that of a see-saw (Fig. 10, *left*); if the pivots are above the hotplate, the motion of the hotplate will be that of a pendulum (Fig. 10, *right*), and then although the pots will be kept in position by the fiddles, their contents may be thrown out by centrifugal force; also the stove will require more swinging room. A weight may be needed at the bottom of a shallow stove to counterbalance the weight of the pots and their contents.

The fuel for a swinging gas cooker will have to pass through a flexible pipe, which should be of the metal type. This is the most likely source of danger in the whole installation, and it should be examined for leaks at frequent intervals. A swinging paraffin cooker with a remote tank will also need a flexible pipe, and a leak here could be highly dangerous because the fuel in the pipe will be under pressure even when the cooker is not in use.

Refrigerators

For coastwise cruising in a warm climate a refrigerator may be regarded as something more than a luxury, for it will enable the crew to live more off fresh food than would otherwise be possible, and will help to prevent waste; but unless room can be found for one with a capacity of 2 cubic feet or more its value on a long passage will not be great.

If a liquid under pressure is permitted to expand, it will absorb heat from its surroundings, and this is the principle of refrigeration. In a mechanical refrigerator the refrigerant is compressed by a compressor driven either by a thermostatically controlled electric motor, as is used in most household models, but which inevitably puts a considerable drain on the yacht's batteries, or by the auxiliary engine; if the insulation of the cold chamber is really efficient and enough holding plates are provided, it may be sufficient to run this compressor only for half an hour once or twice a day. But the running time will also depend on the outside temperature, the required inside temperature, and the frequency with which the cold chamber is opened; the top-loading chamber, though not so convenient in use, is far less wasteful than the front-loading type, from which the heavy cold air will escape the moment the door is opened.

The absorption type of refrigerator has no moving parts, and the refrigerant in a coiled pipe is caused to expand by the direct application of heat, the source of which may be bottled gas or paraffin. If gas is used for cooking it may seem reasonable to use it for this purpose also provided proper safety precautions are taken, for with up-to-date models the gas supply is automatically turned off in the event of the flame being blown out. Gas is cleaner and a little more efficient than paraffin, and with the latter fuel regular removal of carbon from the wick must be attended to. The coil is usually in the form of a stack of flattened S's, and the unit needs to be installed in such a way that the S's lie fore-and-aft; if they lie athwartships it is possible when the yacht heels a few degrees for a gas-lock to form in one of the bends, and refrigeration will then cease. But even when correctly installed the absorption type of refrigerator tends to be temperamental, and is highly susceptible to angle, motion, and vibration; also it raises the temperature in its vicinity, and makers of these units are reluctant to sell them for use afloat because of fire risk. Nevertheless this type has been used successfully in a number of yachts, including John Guzz-well's *Treasure* during her voyage from England to New Zealand; but her owner had scrapped the shallow, unseaworthy paraffin tank and provided gravity feed through a float chamber.

Radio

Any cruising yacht will of course need a radio receiver provided (as most are) with the medium-wave broadcast band—approximately 540 to 1,600 kilocycles per second[1] (555 to 188 metres)—which will be used for obtaining weather forecasts, time signals, news, and entertainment when within about 200 miles of stations broadcasting on that band. If the set is of British make it will probably have also the long-wave broadcast band—150 to 225 kc/s—enabling the B.B.C. Radio 2 programme and a few continental stations to be received perhaps up to a distance of about 400 miles. The radio requirements of the foreign-going yacht differ in that she will find the long-wave broadcast band of little value outside European waters, but she will need a set capable of receiving not only the medium-wave stations when cruising coastwise abroad, but also stations broadcasting on the short-wave bands, because only short-wave signals are capable of being received with certainty at great distances.

Aboard a British yacht probably the B.B.C. World Service will be the one most often required, and if the set is to be capable of being tuned in to all of the 20-odd frequencies on which that service is broadcast (see below) it needs to cover the short-wave bands from 5 to 22 mega-cycles (mc/s). A great many stations of all nationalities broadcast very close to one another on most sections of the short-wave bands, and their number continues to increase; so if anything approaching satis-factory reception is to be had (it is far from perfect even in the best conditions) a set with selective tuning by micrometer, and widely spaced, clearly marked scales, is essential. The British Eddystone and the American Zenith are good though costly examples of such sets; they are fully transistorized and protected against climatic conditions, and work off their own internal batteries, an arrangement I consider preferable to using the ship's supply, as no wiring is called for and the set is portable, which may be an important consideration if its own telescopic or internal aerial is to be used. However, sometimes reception is improved by using an external aerial and an earth (as described in *Cruising Under Sail*), but when in an area where there is

[1] Since a cycle per second is now known as a Hertz (Hz), the abbreviations for kilocycle per second and megacycle per second become kHz and mHz respectively.

great congestion on the medium-wave band, such as along the east and west coasts of the U.S.A.—where many towns have their own commercial stations, and some have several—it may be an advantage to disconnect the aerial to cut out some of the interference when listening-in to a near-by station. In fully transistorized sets the batteries have a remarkably long life, but if these are of the common carbon-zinc type, they must be removed from the set as soon as they become exhausted or serious damage may result. When trying out a set on the short-wave bands, it should be remembered that a short-wave station is surrounded by a skip area, perhaps with a radius of 500 to 1,000 miles, within which signals from that station cannot normally be received.

As has already been mentioned above, the B.B.C. World Service is broadcast on more than 20 frequencies, but not on all of them at the same time. The frequencies used by the station, and the choice of the most suitable one by the receiver, depend to some extent on the time of day, the geographical area at which the programme is directed, and the position of the receiving ship. Since the radius of the skip area is proportional to frequency, when on approaching a station the signals fade, they may be restored by switching to a lower frequency. Darkness and winter time favour the lower frequencies; daylight and summer time the higher ones; generally the best reception is had when the whole radio path is in darkness, but I have often noticed that signals are strongest when the ship is on or near the meridian of the station, and that a station which is not audible at any other time may come in clearly during twilight. A list of the frequencies on which the B.B.C. World Service is broadcast, together with their times of use, and a schedule giving the weekly pattern of the month's programmes, is published a month in advance in a pamphlet with the title *London Calling*; this is obtainable free on application to the B.B.C., and sometimes can be had from British consulates. It does not list the times when time signals are broadcast; these often precede the news. However, the most convenient and reliable means of obtaining time signals, which are so desirable for accurate navigation, is from station WWV.

This is situated at Fort Collins, Colorado, U.S.A.; it broadcasts continuously day and night (except for a 4-minute silent period commencing at 45 min. 15 sec. after each hour) on 2·5, 5, 10, 15, 20, and 25 mc/s, and sends out a pulse, similar to the tick of a clock, at 1-second intervals. In addition to the pulse a steady whistling tone is sent out for the first 2 minutes in every 5, starting precisely at the hour, at 5 minutes past, 10 minutes past, and so on. At the end of the second minute in each group of five, the whistling tone stops and in its place is broadcast throughout the third minute a signal which does not concern us here. Through the whole of the fourth minute and half of the fifth minute the pulse continues without any other sound. In the final half of the fifth minute the Greenwich mean time of the *following minute* is given in morse, and this is followed by a voice announcement of G.M.T. Thus, if we listen in just after 1115 G.M.T. we shall hear the 1-second pulse together with the whistling tone. At 1117 the pulse and the tone are replaced by the signal which is not our concern. At 1118 the pulse alone will be heard again. At 1119$\frac{1}{2}$ the announcement '1120' will be made in morse, and this will be followed by the voice saying:

'WWV Fort Collins, Colorado. Next tone begins at eleven hours twenty minutes Greenwich mean time.'

A second or two later the whistling tone will start, and the instant it does so will be the beginning of the minute announced, accurate to $\frac{1}{10}$ second. It is unlikely that the hour will be of much interest to us unless every time-piece on board has stopped, and if we are sure of the minute and require only to check the rate of the chronometer in seconds, we can listen in to WWV at almost any time, and will not have to wait for the tone to return at the end of the five-minute cycle, because the 59th pulse in each minute is omitted and is followed by a double pulse; we will therefore have to listen only for this.

Information about time signals will be found in the chapter Standard Frequency Service in *Admiralty List of Radio Signals*, Vol. V, but the following list should serve most purposes; note, however, that their schedules may differ from that of WWV and that most have no voice announcement.

Country	Place	Call sign	Frequencies, mc/s
Argentina	Buenos Aires	LOL	5, 10, 15
China	Peking	BPV	5, 10, 15
France	Bagneux	FFH	2·5, 5, 10, 15
Hawaii	Maui	WWVH*	5, 10, 15
India	New Delhi	ATA	10
Italy	Turin	IBF	5
,,	Rome	IAM	5
Japan	Tokyo	JJY	2·5, 5, 10, 15
S. Africa	Johannesburg	ZUO	5, 10
U.K.	Rugby	MSF	2·5, 5, 10
U.S.A.	Fort Collins	WWV	2·5, 5, 10, 15, 20, 25
U.S.S.R	Moscow	RWM	5, 10, 15

* WWVH covers the Pacific, and is similar in most respects to WWV, but the 4-minute silent period in each hour commences at 15 minutes after each hour.

During *Wanderer III*'s voyages we managed always to get time signals when required, no matter where we might be. In the North and South Atlantic, when out of range of local stations, we used either the B.B.C. or WWV, and in the Pacific WWVH. In the Indian Ocean we were rarely able to receive either WWV or WWVH, and only picked up the B.B.C. as we neared Africa, but we did manage to get time signals from Singapore and Colombo. Everywhere it is possible to hear some English-speaking station, but to obtain time signals from it may entail hours of profitless listening unless one happens to have local knowledge; even then it may be found that programme changes are made, and the Sunday programme is almost certain to differ from that of weekdays.

Each year more and more yachts are fitted with radio telephones, and although such equipment might be the means of saving life, it should not be allowed to take the place of constant vigilance and good seamanship. In many instances if the money spent on buying and installing the set, and the time taken learning to use it, had been devoted instead to the careful preparation of the yacht and her gear, the need to radio for help would never have arisen. The transmitting range of Simplex or Duplex radio telephone equipment broadcasting on 2,000 kc/s will of course depend, other things being equal, on the power of the set. For example, a 25-watt transmitter with a thoroughly efficient aerial and earth system (this above all else is absolutely essential) might be expected to have a range of 100 miles in average conditions, and perhaps twice that distance in good conditions with a skilful operator. Most radio telephone equipment can be used also as domestic receivers, and many of them can be used as direction-finding sets by the addition of a rotating loop or ferrite aerial. Thus some of the original high cost may be offset. With all radio sets headphones are desirable, not only for excluding ship noises when signals are weak, but so as to avoid disturbing other members of the crew.

Automatic helmsmen

Aboard a short-handed yacht which will not readily steer herself, or one which is too large to be steered by vane gear, an electric automatic helmsman will be of great value, for although this will probably

---

PLATE 21

The 46-foot cutter *Waltzing Matilda* was designed and built in Tasmania for Philip Daven-port. In 1968 while crossing the North Atlantic by the trade wind route I met her, then under new ownership and 9 days out from Tenerife, running slowly before the light wind under twin sails. A few months later she was lost at sea.

not be kept in use all the time because of the drain it puts on the yacht's batteries, a watchkeeper can more comfortably take a long watch if he can hand over to the automatic gear for a little while whenever he wants to leave the helm to get a cup of coffee or attend to navigation.

The general principle on which the automatic helmsman works is that a magnetic north-seeking device, which in some instances may serve also as the steering compass, is employed to detect (usually by means of a photo-electric cell or cells) the smallest deviation from the set course, and the information derived from this reference unit is used to activate a relay system which controls a reversible electric motor. This motor is chain-coupled to the steering wheel or some other part of the steering gear, and applies the required amount of helm to bring the yacht back on to the set course. A clutch enables the gear to be engaged or disengaged at will.

Beyond was fitted with a Sperry automatic helmsman for her world voyage, and although her owner reported favourably on it, he had some difficulty in providing enough electricity, for its consumption was 40 watts when switched on and 160 watts when actually turning the wheel. Since then improved design and the use of transistors have reduced the amount of current required; for example, the average consumption of the Pinta (made by Marine Automatic Pilots Ltd., of Hove, Sussex) is only 25 watts—about 1 amp. on a 24-volt supply.

To reduce consumption even more, Vancil took a completely different approach when devising and constructing the automatic helmsman for his lovely Alden-designed ketch *Rena* (pages 118–22), and with the generous thought that others might like to use it, suggested that I mention it here. I should, however, add that in my opinion only a man with a sound knowledge of electronics and practical ability as an electrician and mechanical engineer could possibly undertake such a task. In *Rena* the free-spinning propeller drives a small hydraulic pump to furnish oil, on demand, at a variable pressure, and the helm is moved by an hydraulic ram driven from this supply. Only a very small amount of electricity is used (10 watts) to activate the sensing device and the hydraulic ram control, and it does not matter how fast or how slowly the propeller turns. The low figure of 10 watts

H

consumed in the control circuitry is achieved by the use of distinctive pilot-operated hydraulic valves and a transistorized amplifier.

The chief disadvantage of the automatic helmsman is that it is related not to the wind but to the compass. Whereas with vane gear a change in wind direction will instantly produce a change of course, the automatic helmsman will continue to steer the set course no matter what the wind may do; therefore it cannot efficiently steer a vessel to windward, and is not safe to use when running dead before the wind as a shift of wind might then cause a gybe.

6

SOME NOTABLE
OCEAN-GOING YACHTS

*Beyond—Kochab—Little Bear—Omoo—Rena—Saoirse—
Trekka—Waltzing Matilda—Wanderer III*

I N this chapter will be found the plans and some description of
a selection of yachts which have made successful long ocean
voyages. Most of them are famous, all but one of them were still
in commission in 1969, and they range from old *Saoirse*, built on
fishing-boat lines in 1920, to some of the most modern and lovely
creations of the yacht-designer's art. I may be accused of partisan-
ship because I have included in so small a selection three designs from
the board of one architect; but these three are of widely different
types, and I believe that more small yachts built to plans by the late
Jack Laurent Giles have crossed oceans than have those from the
board of any other British yacht designer. The selection will serve to
show that it is not so much the size or type of vessel that contributes
to the success of a long voyage as the seamanship, ability, and staying
power of the people who sail her.

It will be noticed that the lines of all the yachts are not included.
The reason for this is that there are a few unscrupulous people who
will steal the plans of a yacht from such reproductions, usually by
photographing up, so as to avoid paying the royalty that is the
designer's due for each vessel built to his plans. That such methods do
not produce perfect full-size yachts is beside the point, but naturally
designers are aware of the risk of plagiarism, and are chary of permitting lines plans to be published. Perspective drawings of *Beyond*
and *Trekka* are therefore reproduced here instead of the lines, and
they provide a very good likeness; indeed they may even give a better
idea of the general shape of the hull to those who are not accustomed
to the reading of lines, because of their three-dimensional quality. Can
you resist taking a second look at that fascinating view of *Beyond* as
seen by a fish low down on the starboard quarter (see page 101), to
enjoy again the harmonious curves that speak so eloquently of sea-
kindliness, buoyancy, and power, and visualize her forging her way
across some wide ocean with the trade wind in her sails?

Beyond (Plans on pages 101-4 and Plates 18 and 20)

An outstanding small-yacht voyage of the early 1950s was that of the 43-foot auxiliary cutter *Beyond*. In her the late Tom Worth and his wife Ann made an efficient and seamanlike voyage round the world in 2 years and 2 months. For the first half, by way of Panama to New Zealand, they were accompanied by Peter Taylor and his wife; but the Taylors decided to settle in New Zealand, and the Worths were alone for the arduous homeward voyage by way of the Great Barrier Reef, Torres Strait, and Suez. This received little publicity, which is, perhaps, as the Worths would wish, but for it they were awarded the premier British cruising trophy, the challenge cup of the Royal Cruising Club, which had previously been awarded to Muhlhauser and O'Brien for their circumnavigations. Everything went according to plan, there were practically no incidents, the ship equalled every expectation, and it is obvious to anyone who has spoken with this modest couple, or who has read the account which was published in the 1953 and 1954 issues of the *Royal Cruising Club Journal*, that they obtained the utmost pleasure and satisfaction from the whole great enterprise.

Beyond was designed to incorporate her owners' special requirements by Laurent Giles & Partners of Lymington, as a fast, medium-displacement type of vessel, safety and sea-kindliness being the first considerations, and she was built by the Sussex Shipbuilding Company at Shoreham, the hull entirely of Birmabright light alloy, and the main deck and internal joinery work of Burma teak. She was insulated below deck with an asbestos spray.

At the time she was built it was most unusual for a sailing yacht to have a central cockpit; but this position was partly dictated by the fact that she was to have a diesel motor and big-capacity tanks for fuel and water, and it permitted these heavy weights to be placed in the best of all positions, amidships. It certainly splits the accommodation, and as can be seen on the plans, the owners' after cabin can only be reached from the saloon by way of the cockpit. Some might consider this a disadvantage, but no doubt it has its merits when two couples are to make a long voyage together. One might suppose the central position to be wetter than the aft position when the wind is forward of the beam, but here the deckhouse (Plate 20, *top*), in which are the compass, automatic helmsman, and chart table, gives good protection to the helmsman on that point of sailing, and I believe the only time that water was shipped in the cockpit was when running before a gale

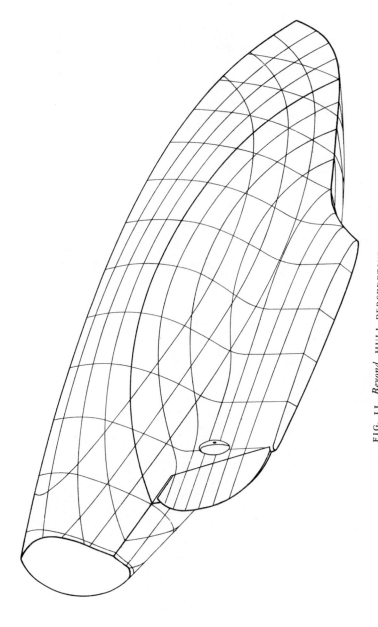

FIG. 11. *Beyond*, HULL PERSPECTIVE DRAWING

Designed for Tom Worth by Laurent Giles & Partners, her dimensions are: l.o.a. 43 ft.; l.w.l. 32 ft.; beam 10·7 ft.; draught 7 ft.

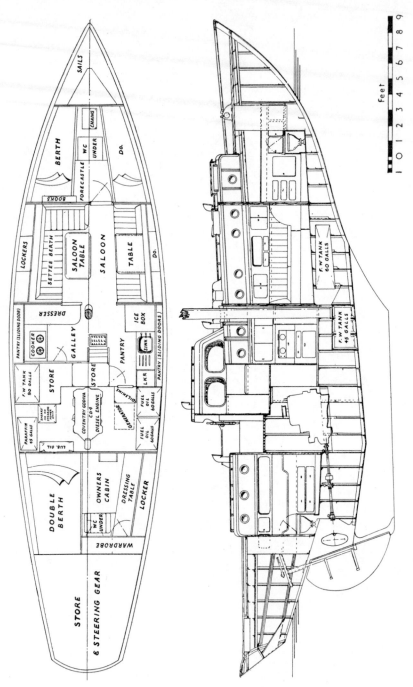

Feet
0 1 2 3 4 5 6 7 8 9

FIG. 12. *Beyond*, GENERAL ARRANGEMENT

off Portland Bill at the very end of the circumnavigation. Worth was an engineer, so naturally his ship had a good share of gadgets and mechanical devices, but the principle was that these were for added comfort, ease, or convenience, and should they all go wrong he would still have a normal, well-found ship. He said he was more than pleased

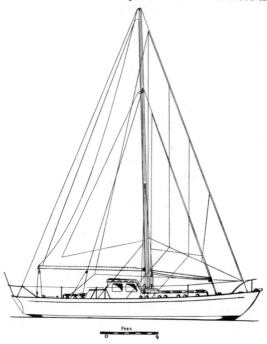

FIG. 13. *Beyond*, SAIL PLAN

Sail areas: Mainsail, 350 sq. ft. Staysail, 125 sq. ft. No. 1 jib, 275 sq. ft. No. 2 jib, 175 sq. ft. No. 3 jib, 88 sq. ft. Storm jib, 40 sq. ft. Genoa, 520 sq. ft. Trysail, 180 sq. ft. Twin running sails, combined area 750 sq. ft.

with the result, although, naturally, improvements could still have been made. The hull performance, in particular, was everything he had hoped for except that the yacht did not too easily steer herself.

The plans show how well the accommodation was arranged for a complement of four to live in comfortably in the tropics or in home waters, with its large stowage space for all the stores needed for a long voyage, together with tanks to hold 150 gallons of fresh water, 120 gallons of diesel oil—giving the yacht with her 22 h.p. Coventry Godiva engine a range of 1,400 miles at an economical cruising speed of 4·5 knots—45 gallons of paraffin and 20 gallons of lubricating oil, as well as $2\frac{1}{2}$ cubic feet of refrigerated space. And all this was done on the comparatively moderate displacement of $11\frac{3}{4}$ tons, thanks to the powerful lines of the hull and the light-alloy construction.

The all-inboard, jib-headed cutter rig with mainsail, staysail, and No. 1 jib has an area of 750 square feet, and there is no sail larger than can conveniently be handled by one man or woman. For running in the trade winds she carried twin sails with a combined area of 750 square feet (see pages 74-5 and Plates 17 and 18).

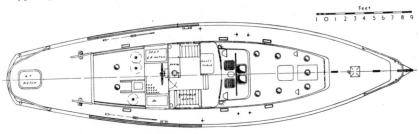

FIG. 14. *Beyond*, DECK PLAN

In this fine little ship there are many brilliantly conceived and care-fully worked-out details, such, for example, as the combined boom gallows, mainsheet horse, and after bollards; the projecting fairlead forward to keep the anchor chain clear of the stem; the closing lifeboat-type fairleads for warps; the double coachroof and wheelhouse tops with air spaces to keep the accommodation cool; the motor-cycle saddle for the helmsman; these and many others have now been widely adopted.

Kochab (Plans on pages 106-8)

It is not uncommon for yachts to be specially designed and built for men who intend to make ocean passages, as indeed were most of those described in this chapter, but it is rare for a yacht to be designed and built for a man who has already gained wide experience in blue-water sailing and who intends to make further voyages. When that does happen and he enlists the services of a leading yacht designer, the resulting plans and the yacht that materializes from them are of parti-cular interest and value.

In 1949-50 Dr. I. J. Franklen-Evans with one companion sailed the 33-foot yawl *Stortebecker III* from England to New Zealand. Two years later, and with two companions, he sailed the same yacht from New Zealand east across the Pacific to British Columbia. By then he had come to the conclusion that *Stortebecker* was too small for a complement of three to live in comfortably, and that passage-making with only two people aboard in a yacht which would not readily self-steer, except under twin running sails (this was before vane steering gear had been perfected), was too exhausting. So he decided to have a larger yacht built, not only to permit a crew of three to live aboard and cruise in comfort, but also to enable passages to be made more quickly, and to carry the stores needed without being overloaded. The length of the new yacht was not to exceed 40 feet, and she was to be capable of being sailed single-handed if necessary; she was to have a comfortable motion without inside ballast, except for a small amount for trimming, and was to be able to steer herself. Additional requirements were that she should be absolutely watertight when battened down, have a freshwater capacity of 100 gallons, and full headroom, and be pleasing to the eye; it was also important that she should go well to windward on a moderate draught, so as to be able to beat through narrow passes.

For the design he went to Arthur Robb, a New Zealander who came to England in 1936, and until his death in 1969 was one of the foremost yacht designers. The yacht was built by Herbert Woods at Potter Heigham, Norfolk, in 1956, was christened *Kochab* (a star of the constellation Ursa Minor), and in the autumn of that year set out straight from her builder's yard on a voyage to New Zealand. She took part in the Honolulu race on the way, and after some fast passages (Appendix I) arrived safely at her destination.

The accompanying plans show how the designer interpreted the owner's requirements, with a handsome, characterful yacht, moderate in all respects, as a result, and I understand that the only alterations he would like to make would be to increase the draught and the length of the overhangs a little, and this he has done in a near-sister ship.

As electrolytic action had been experienced with the muntz metal sheathing of *Stortebecker III*, *Kochab* is planked with teak below the waterline, and relies on paint instead of sheathing to protect her against worm. A feature of her construction, which was uncommon then but widely used today, is that the coachroof and doghouse tops are sheathed with glass fibre so that they require no upkeep, and their undersides are covered with Formica, the air space between acting as insulation. Two sling bolts are fitted to the keel so that the yacht can be lifted by crane or ship's derrick for scrubbing and painting in places where there is insufficient range of tide and no slip is available, the two parts of the sling passing out through the skylights.

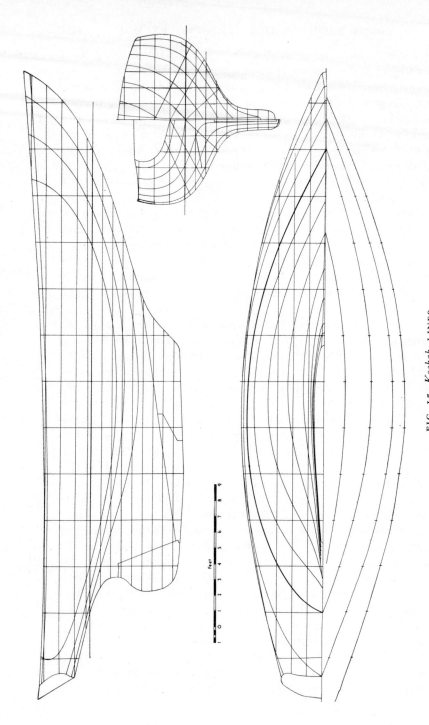

FIG. 15. *Kochab*, LINES

Designed by Arthur C. Robb for Dr. I. J. Franklen-Evans, her dimensions are: l.o.a. 39·4 ft.; l.w.l. 29 ft.; beam 10·7 ft.; draught 6 ft.

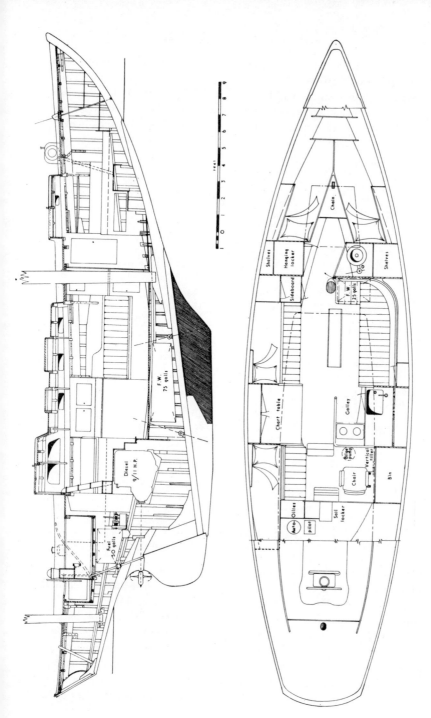

FIG. 16. *Kochab*, GENERAL ARRANGEMENT

At the owner's request the yacht was originally provided with no fewer than five methods of steering: (1) By wheel; (2) by tiller, which could be shipped on the rudder head in the event of trouble developing with the wheel steering; (3) by a vertical tiller in the doghouse, worked from an aircraft-type chair which tilted as the yacht heeled;

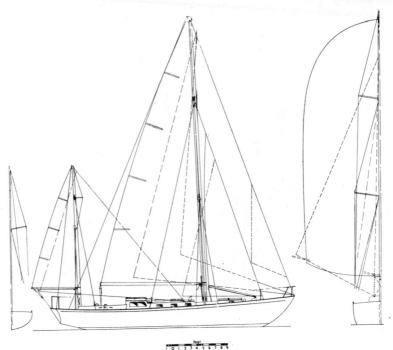

FIG. 17. *Kochab*, SAIL PLAN

Sail areas: Mainsail, 371 sq. ft. Mizzen, 80 sq. ft. Staysail, 105 sq. ft. Reaching staysail, 138 sq. ft. Mizzen staysail, 216 sq. ft. Balloon jib, 308 sq. ft. Genoa, 435 sq. ft. Storm jib, 30 sq. ft. Spinnaker, 1,000 sq. ft. Twin running sails, combined area 450 sq. ft.

but this did not work satisfactorily, and it and the chair were removed on arrival in the West Indies; (4) by twin sails with their braces led to the wheel or tiller; (5) by automatic helmsman; this got wet, became corroded, and was scrapped at Los Angeles. The 12-volt battery is charged by the main engine, a Coventry Victor 9/11 h.p. diesel, which gives the yacht a speed of 6 knots, and with a fuel capacity of 50 gallons a range of about 500 miles. In addition a generator was arranged to be driven by the two-bladed, variable-pitch propeller, but this was not a success because the propeller, as usually happened before the alternator was adapted for use in yachts, failed to drive it at sufficient speed, and on more than one occasion during the voyage to New Zealand, the battery was run down and it became impossible to start the engine. It would appear that Franklen-Evans was unfortunate with much of his ship's mechanical equipment, but some of the failures may well have been due to his hurried departure from England without proper trials, a course which was forced on him by late delivery and the season of the year.

To keep the mainsail of a manageable size for single-handed sailing, and to permit the use of a mizzen staysail, yawl rig was chosen. It will be noticed that the mizzen mast, unlike so many of its kind, is not pushed right aft but is in a position where it can be properly stayed, and there is sufficient clearance between it and the end of the main boom for it to be stayed independently of the mainmast. Both main and mizzen backstays are set up by levers close to the helmsman. With a light wind on the beam, when working sails are inclined to flog and do no work, they can be taken in and the yacht sailed under genoa and mizzen staysail only, a combined area of more than 500 square feet of light sailcloth. On a reach a total sail area of 1,330 square feet can be set, and for running there is a masthead spinnaker of 1,000 square feet. In addition there is a pair of twin running sails with a total area of 450 square feet. The booms for these sails are shipped in the ends of a crossbar $2\frac{1}{2}$ feet forward of the mast, this being braced to the deck and to the mast by metal tubes, the whole fitting being unshipped when the twin sails are not in use. It was found that the yacht would not self-steer under that rig until the tacks of the sails had been moved farther forward, when she did so with the wind up to two points out on either quarter, which is about normal with this rig.

It is always interesting to learn in which directions a new yacht comes up to her owner's expectations, and in which she has disappointed him. Arthur Robb kindly let me have for inclusion here the more important comments of the owner together with his remarks on them.

Owner. With a following sea she is lively and responds very quickly to the helm, but is always easy to steer with one hand. With a long, straight keel she might be less tiring to steer, but I think if we really got down to trimming sails she could be got to steer herself very well.

Designer. This quickness on the helm is surprising, as the profile is not all that much cut away, but, of course, the beamier the yacht the more will she wander. In my view, cutting down unnecessary wetted surface is important for speedy ocean passage-making, particularly in terms of bottom fouling. It is interesting that the cutter *Uomie*, of my design, which is said to be extremely good at steering herself, has a greatly cut-away underwater profile, and I firmly believe that more experiment with trim, particularly of the mizzen sheet, would make *Kochab* steer herself.

Owner. The yacht tends to go down a little by the head when running, and this was particularly noticeable towards the end of the Honolulu race, when for a moment it looked as though she was going to put her bows under. In a squall she was down to an inch or so of freeboard with a 6-foot wave at each side by the shrouds.

Designer. I do not believe this to be very serious, and certainly no more noticeable than it is in any other short-ended yacht; it has happened to me in a variety of types, and I guess that at the time *Kochab* was going very fast. Her lines could not well be more full forward than they already are, and her run is extremely clean. It should also be borne in mind that the lines were planned to avoid any effect on performance, steering, or roll period, regardless of immersion. *Kochab* is the owner's only home, and practically all his possessions are carried aboard on long passages. Indeed, the initial loose gear put aboard filled a half-ton truck four times, and with subsequent additions put the yacht down from 2 inches above her datum waterline to $3\frac{1}{2}$ inches below it.

Owner. Her motion is never violent, and I would say she is dry, but close-hauled in a seaway plenty of spray comes aboard, and with a quartering sea we get an occasional crest that strikes just aft of amidships and sometimes fills the cockpit. In bad weather she has always been well behaved, and I think we could keep her sailing in most conditions. She heaves-to in moderate winds, force 3–4, quite well, but in strong winds we have always taken all sail down. She lies a-hull in force 6–7 with the wind abeam, or with her head slightly down wind; in force 7–9 she is reasonably comfortable and lies with her head slightly up into the wind. We have had a staysail set in these conditions but it was of no real advantage. The yacht herself is admirable and is just the right size for two to handle; with a good, steady wind she will easily log 160 miles a day, and is capable of 180.

Little Bear (Plans on pages 111–13)

At Balboa in the Panama Canal Zone, in 1953, my wife and I met Buzz and June Champion, and experienced much kindness at the hands of this charming American couple. They were living aboard their 37-foot ketch *Little Bear* while they made preparations for a cruise, but we did not know then that their intention was to sail right across the Pacific and back again.

With great skill and industry, and without any professional assistance, they had built the ketch themselves in their spare time at San Francisco, taking seven years over the job. The design is a standard

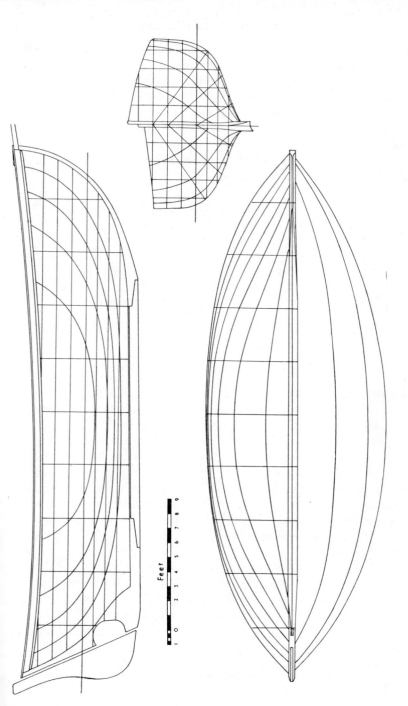

Feet

FIG. 18. *Little Bear*, LINES

Designed by the late John G. Hanna as the standard Carol Cruiser, this 37-foot ketch was built at San Francisco by her owners, Buzz and June Champion. Her dimensions are: l.o.a. 36·7 ft.; l.w.l. 32·9 ft.; beam 12 ft.; designed draught 4 ft., but this was increased to 4·5 ft. when building.

one from the board of the well-known American designer, the late John G. Hanna, and is known as the Carol Cruiser. Several yachts built to this and similar plans have made fine voyages; *Tropic Seas*, for example, in which the Caldwells and their young family made a two-year crossing of the Pacific. But Hanna is probably best remembered for his similar, but smaller, Tahiti Ketch design. Many of these, mostly American, have made great voyages, notable among them being *Adios*, 32 feet l.o.a., which Tom Steele has twice sailed round the world.

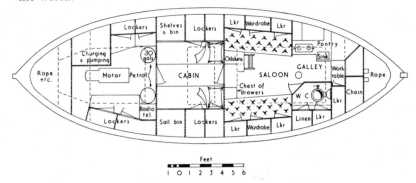

FIG. 19. *Little Bear*, GENERAL ARRANGEMENT

By British standards *Little Bear* is very beamy and shallow, but because of her great beam she has an immense amount of space below decks. The only alterations the Champions made to the lines as shown on page 111 were to increase the depth of the keel to 12 inches for better windward sailing, making the draught 4½ feet, and to heighten the bulwarks a little. Keel, stem, and sternpost are of Douglas fir, the scarfs in the stem being backed with ironbark, which timber is also used on the after side of the sternpost; frames and floors are of Indiana white oak, and the planking, of rift-sawn Douglas fir, is iron-fastened; the deck is of Douglas fir and the spars of Sitka spruce. She is not copper sheathed.

The designer's layout showed a central cockpit as well as one aft; but as the Champions intended to use *Little Bear* for commercial albacore fishing off the southern Californian and Mexican coasts, they built a shorter forward coachroof, and in place of the central cockpit

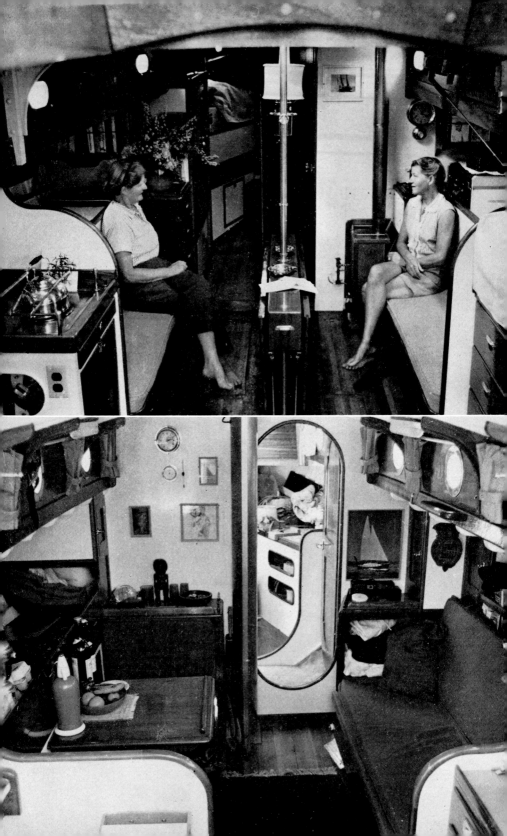

arranged a fish-well with a large hatch, which is shown in dotted lines on the owner's rough sketch plan reproduced on page 112. They fished successfully for two seasons, using lines similar to those of a tunnyman, and going up to 200 miles offshore; they fished all day and

FIG. 20. *Little Bear*, SAIL PLAN

Sail areas: Mainsail, 370 sq. ft. Mizzen, 162 sq. ft. Jib, 168 sq. ft.

slept at night, and their best catch was a ton in one day. When they had done fishing and were preparing for the long cruise, they converted the fish hold into a two-berth cabin and arranged huge lockers at the sides for provisions, and tanks for their 110 gallons of fresh water under the bunks.

Most of the beamy American ketches with which I am familiar have powerful auxiliary engines and a plentiful supply of fuel for them, and *Little Bear* is no exception. She has a 50 h.p. Universal petrol engine driving a 22-inch by 13-inch two-bladed propeller on the centre line through a 2½-to-1 reduction gear; this gives an economical speed of 5½ knots in smooth water, and, with a fuel capacity of 115 gallons, a range of about 450 miles.

PLATE 24

Top: The saloon in *Rena*, the Vancils' floating home, has a pilot berth each side outboard of the settees, and the forepeak is used as a workshop, being fitted with a lathe and drill press. *Bottom*: The single-hander David Guthrie prefers the dinette arrangement and one pilot berth aboard his 29½-foot sloop *Widgee*. Chart table and galley face one another at the after end.

I

The total sail area of only 700 square feet is very snug for a yacht of this size; but in strong winds some good runs have been made under it; for example, the 610-mile passage from Pago Pago, American Samoa, to Suva, Fiji, was made at an average speed of 155 miles a day, which is fast sailing for this type of vessel. Alterations to the rigging, which are not shown on the sail plan, are few: they comprise a back-stay for the mainmast to take some of the strain off the upper shrouds, and a jumper stay on the fore side of the mizzen to stop the mizzen masthead from whipping. Champion's comments after making the two-way crossing of the Pacific are interesting, for although he is justly proud of his ship he is not blind to her few shortcomings.

I feel she is about as comfortable and dry a boat as one could find of her length, and she can carry an unbelievably large quantity of stores. It's pretty hard to get into trouble with her low and simple rig. We have never been in really bad weather—nothing more than force 8, I am sure—but with the wind at force 8 we have never felt any anxiety when properly shortened down. If I might scrounge a few rubs on Aladdin's lamp, I would put the galley aft where it belongs, arrange things for a doghouse, put in a diesel motor with a bountiful fuel storage, install a wheel-steerer and auto-pilot, and may-be try Bermuda rig on the main. She is a beast at anchor in a swell and works round broadside (unless there is a wind to hold her otherwise) and rolls maddeningly. Not the perfect boat, you know, but one could do worse, and she does not tire us out at sea.

Omoo (Plans on pages 116-18)

This handsome 45-foot ketch was designed jointly by Louis Van de Wiele, her owner, and the late F. Mulder. She is built of 5-mm. steel plates riveted; this was not because her owner preferred steel to wood, although he was well aware of its merits, but because the yacht was built in Belgium shortly after the war when timber was exceedingly difficult to obtain, and Belgium, like Holland, is a country where the building of small vessels in steel is well understood and is common practice. The deck is of kambala wood on steel beams, and the coamings, hatches, and interior woodwork of African mahogany. The ballast is a compound of cement and rivet heads weighing 5 tons and cast in the keel, and there is 13 cwt. of inside ballast; the dis-placement is 18 tons. The intention was not to over-concentrate the ballast, or to stress its importance at the expense of the weight of superstructure and rigging, so that the rolling motion should be slow and easy, making the yacht comfortable to live in when running, which is the most common point of sailing for an ocean-voyager. Speed was a secondary consideration, but the lines are harmonious and pleasing, and *Omoo* is not a slow ship; her owner has told me that, in his opinion, she is as near perfection as an ocean-voyager and a self-steerer as could well be.

Ketch rig was chosen because of the ease with which it can be handled by a small crew, the mainsail being only 380 square feet, and gaff was preferred to the jib-headed rig because most of the passages were to be made down wind. Twin running sails with a total area of approximately 440 square feet were hanked to stays set up to the deck 5 feet forward of the mast, their booms being pivoted on the mast. It was found that there was no need to take their braces to the wheel, for the ship steered herself very well without, and the rig continued to work with the wind as far as four points out on the quarter, which is exceptional, but a ship of this size, especially with the attention that was given to her ballasting, will have a slower roll than a smaller one, and the wind will therefore not be thrown out of the sails so readily.

The general arrangement plan shows the good use to which the available space was put, the separate engine-room, with access from the deck only, being a particularly good feature in keeping heat and smell from the rest of the accommodation. The 27 h.p. Kermath diesel engine drives a three-bladed 20-inch diameter propeller through a 2-to-1 reduction gear, and there are tanks holding 80 gallons of fuel in the engine room. A 12-volt battery with its own charging plant was used for lighting until the wiring gave trouble in the tropics, when it was abandoned in favour of paraffin. Although there was a coal-burning stove in the galley, cooking—of an elaborate continental style —was mostly done on a paraffin-burning pressure stove; 10 gallons of paraffin was found to be an ample supply for cooking and lighting for three months. Originally a w.c. was installed in the compartment on the port side just forward of the mast, but unlike most of these machines, which are generally reliable, it gave so much trouble that it was scrapped at an early date, a bucket being used instead, and the space was given over to the stowage of sails. One ton of fresh water is carried in tanks beneath the cabin sole.

Louis Van de Wiele and his wife Annie, with one companion and a dog, sailed *Omoo* (in Marquesan dialect the word signifies a native who wanders from one island to another) west-about round the world in two years, returning to Zeebrugge in August 1953. In achieving their ambition, sight-seeing was a secondary consideration, their chief aim being to complete the circumnavigation without accident or incident; they took the longer route by way of the Cape of Good Hope, instead

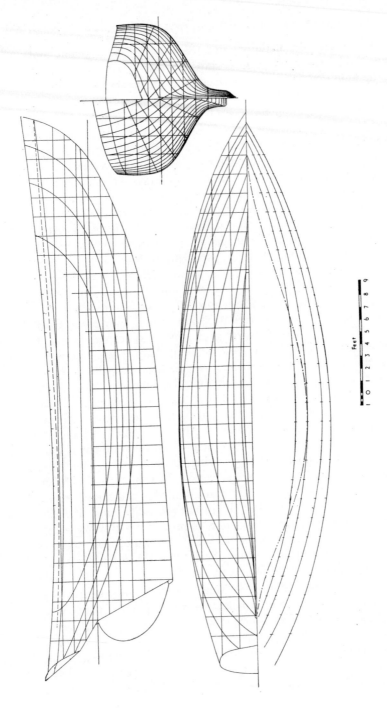

Feet

10 1 2 3 4 5 6 7 8 9

FIG. 21. *Omoo*, LINES

This 45-foot ketch was designed by Louis Van de Wiele and the late F. Mulder, and was built of steel in Belgium. Dimensions: l.o.a. 45·3 ft.; l.w.l. 37 ft.; beam 12·1 ft.; draught 6·2 ft.; her displacement is 18 tons.

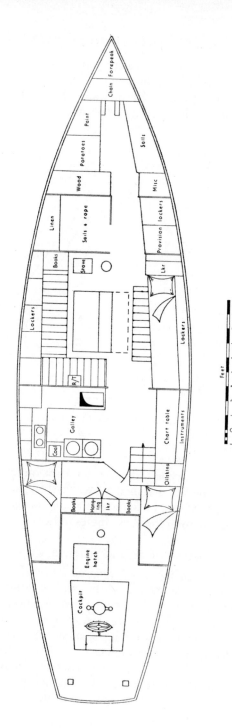

FIG. 22. *Omoo,* GENERAL ARRANGEMENT

of passing through the Suez canal, and found that the space of two years, to which they were limited, was not long enough for such a venture. They had to hurry all the way to keep to their schedule, and at the conclusion of the voyage Van de Wiele was so exhausted that he

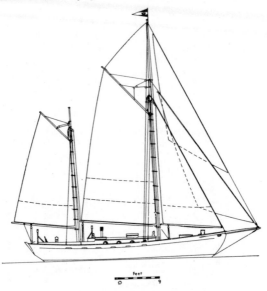

FIG. 23. *Omoo*, SAIL PLAN

Sail areas: Mainsail, 380 sq. ft. Mizzen, 210 sq. ft. Staysail, 150 sq. ft. Jib, 136 sq. ft. Topsail, 100 sq. ft. Jib-topsail, 136 sq. ft. Twin running sails, combined area 440 sq. ft.

gave up sailing for a time and became a white hunter in Kenya. Since then, however, he has designed a smaller steel yacht, the gaff cutter *Hierro* (Plate 7, *top*), and I had the pleasure of meeting him and his wife making a round trip to the West Indies in her.

Rena (Plans on pages 119-21, and Plates 22, 23, and 24)

I first saw this 46-foot ketch moored fore-and-aft in the Pontinha at Madeira, rolling very gently in the swell; but the next time we met she was lying absolutely still above her own reflection in a tiny tree-shaded Florida lagoon, with a gangway rigged to the near-by lawn; on our third meeting she was lying at anchor in the lee of a low, scrub-covered Bahama island. On each occasion I considered her to be one of the most beautiful yachts I had ever seen. As beauty lies in the eye of the beholder, I know that everyone will not agree with my perhaps old-fashioned opinion, but to me that clipper bow with its carved

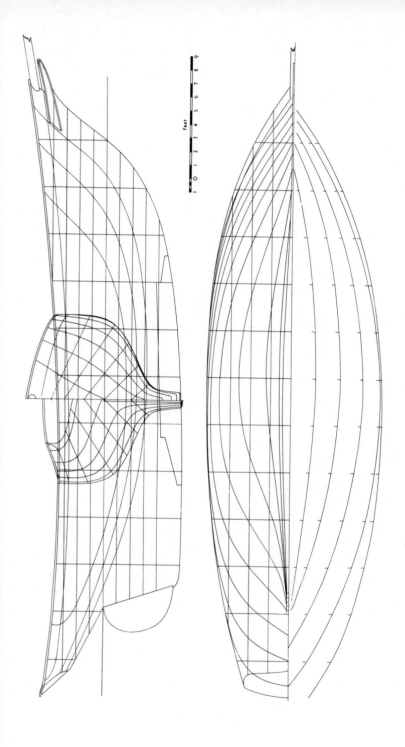

Feet

FIG. 24. *Rena*, LINES

The clipper bow ketch which the Vancils built to designs by John G. Alden of Boston, Mass.
Dimensions: l.o.a. 45·6 ft.; l.w.l. 34·8 ft.; beam 12·7 ft.; draught 5·9 ft.; displacement 18 tons.

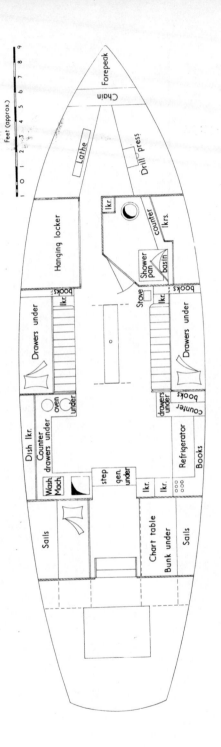

Feet (approx.)
0 1 2 3 4 5 6 7 8 9

Forepeak

Chain

Lathe

Drill press

Hanging locker

lkr.

Shower pan

basin

counter

lkrs.

books

Drawers under

Stove

lkr.

books

Drawers under

Dish lkr.

Counter
drawers under

oven

under

drawers
under

books

Counter

Wash
Mach.

step

gen.
under

Refrigerator

Books

lkr.

lkr.

Sails

Chart table

Bunk under

Sails

FIG. 25. *Rena*, GENERAL ARRANGEMENT

trailboards and upthrust bowsprit, the bold sheer, and the neatly tucked-up counter stern, are quite perfect in their sea-kindly grace and harmony. I noticed, too, the smaller details of rig, gear, and hull: the rake of the masts, the snug-fitting sailcoats, the immaculate deck and sides, the golden shine of the brightwork; all combined to make what the American press might call a 'million-dollar luxury yacht'.

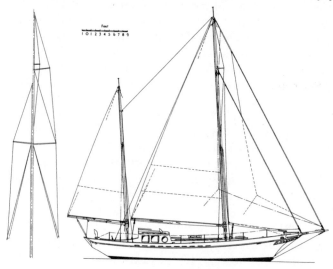

FIG. 26. *Rena*, SAIL PLAN

Sail areas: Mainsail 423 sq. ft. Mizzen 175 sq. ft. Staysail 262 sq. ft. Jib 200 sq. ft.

But *Rena* is nothing of the sort, she is the practical, floating, cruising home of Commander Vancil and his wife, and the whole of her hull and accommodation, all her sails, and most of her fittings, including even the barometer and washing machine, and the automatic helmsman (see page 97), were fashioned by the Vancils with their own hands and no outside help. After Vancil, a pilot and engineer, had retired from the U.S. Navy he and his wife Jo vanished for a time from social life; they rented a field beside the Intracoastal Waterway at Great Bridge, Virginia, threw up a shed, and for four years, seven days a week from 6 a.m. to 10 p.m., they worked to make their dream come true. They named the yacht after their daughter Rena, and on her completion in 1962 set out to prove her worth on a voyage to Europe and back.

The design comes from John G. Alden & Co. of Boston, Mass., a firm which understands so well the need for a cruising yacht to possess some character, and I am grateful to them for allowing me to reproduce here the sweet, sea-kindly lines of this beautiful little ship. Here we may see how a clipper bow, which in Britain is almost forgotten, but in America shows signs of increasing popularity, should be drawn. It makes not only an attractive but a good, practical end for a cruising yacht of this size, gives extra space on the foredeck and keeps the foredeck dry, and with its bowsprit permits a fine spread of sail to be carried without unduly tall masts or excessive overlap; how well the stern matches it, and how firm are the sections to enable the yacht to stand up to her canvas with a draught of less than 6 feet.

After the hull had been completed it was sheathed all over, including deck and coachroof, with glass reinforced plastics to reduce maintenance, and the masts and booms were treated in a like manner. A handsome curved bipod was built instead of the single spar bowsprit shown on the sail plan. The Vancils worked out their own general arrangement plan, and some may criticize this because of the large area of 'wasted' space at the after end of the saloon; but often something which may look wrong on paper proves to be correct in reality, and so it is here, and a wonderfully generous impression of space and comfort has been achieved. The engine is a Sheppard 50 h.p. diesel.

Saoirse (Plans on pages 123–5 and Plates 25 and 26)

In 1922 the late Conor O'Brien designed the 42-foot ketch *Saoirse*, had her built at Baltimore in Eire, and the following year set out on his famous 31,000 mile voyage round the world, at about the same time as the late George Muhlhauser was completing his circumnavigation in the yawl *Amaryllis*, and took exactly two years over it. O'Brien's voyage was unique in that, instead of sailing west-about in the trade winds, which is the usual route for a yacht, and an easier and more pleasant one, he chose to sail eastwards, and, in proper sailing-ship fashion, ran his easting down in the roaring forties and came home by way of Cape Horn. His voyage was also unusual because he called at only twelve ports; but his book *Across Three Oceans* makes fascinating reading, not only because of the simplicity and directness with which the story is told, but because he was always ready to experiment with his rig and with the use of the ocean weather systems. His account is, however, so modest that it gives the reader little idea of the good seamanship, vigilance, and endurance required to drive such a vessel, her uncoppered bottom growing ever fouler, at speeds of from 150 to 170 miles a day, or of the magnitude of the seas that must sometimes have been met with. The ship's complement varied between two and four, the crew usually consisting of paid hands.

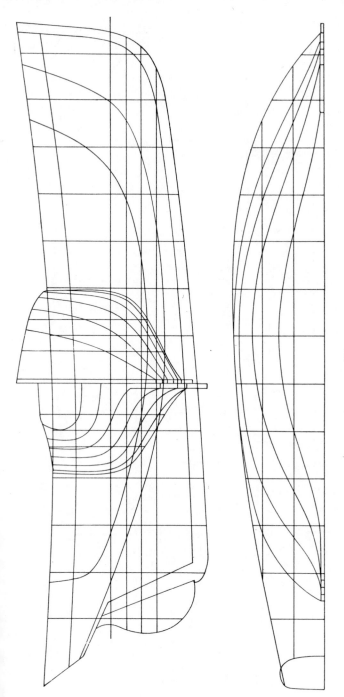

Feet

FIG. 27. *Saoirse*, LINES

The ketch which the late Conor O'Brien designed for his voyage round the world by way of Cape Horn. Dimensions: l.o.a. 42 ft.; l.w.l. 37 ft.; beam 12·2 ft.; draught 6·8 ft.; her displacement in seagoing trim is about 21 tons.

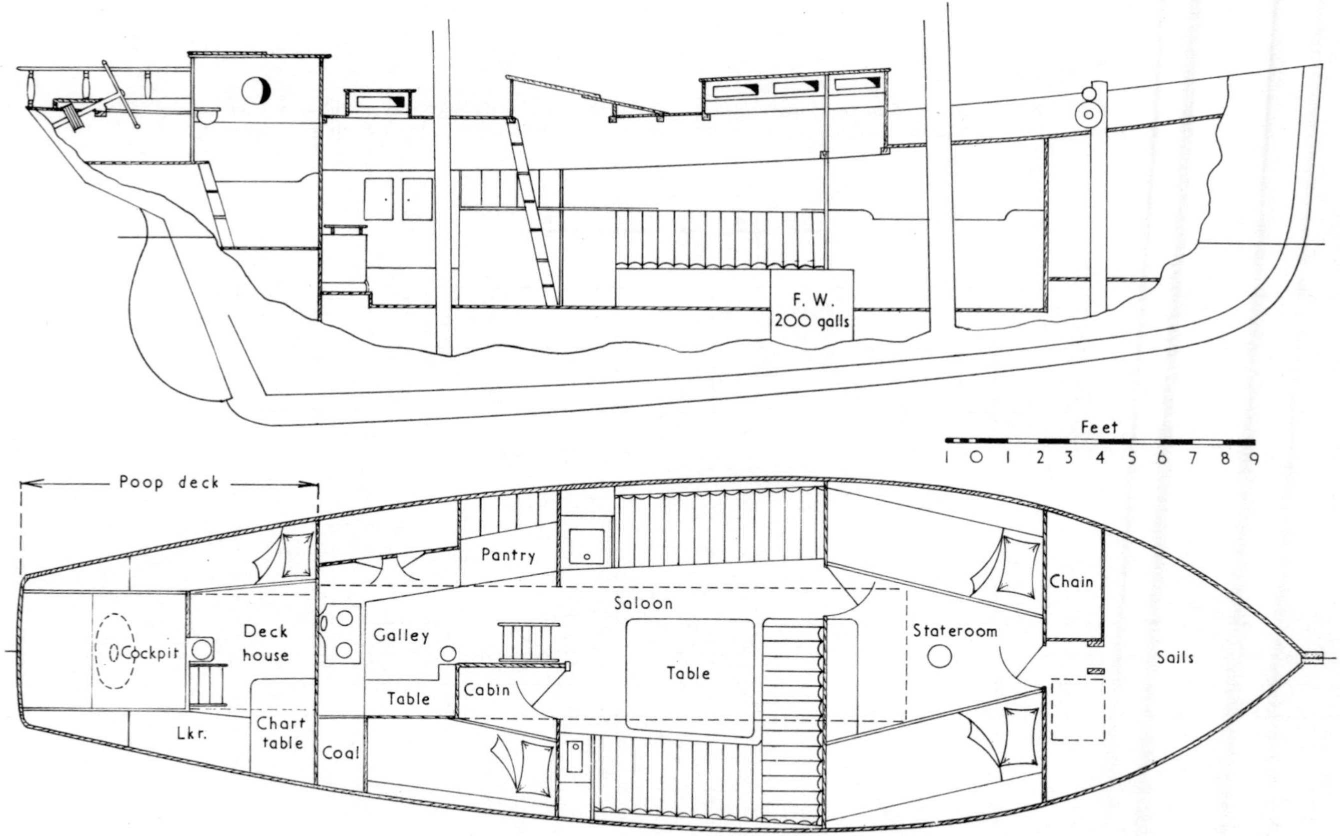

Feet
0 1 2 3 4 5 6 7 8 9

F. W.
200 galls

Sails

Chain

Stateroom

Saloon

Table

Pantry

Cabin

Galley

Table

Coal

Deck
house

Chart
table

Lkr.

Cockpit

Poop deck

FIG. 28. *Saoirse*, GENERAL ARRANGEMENT

Saoirse is of fishing-boat type and build, with a long, straight wood keel, a very stiff midship section—almost too stiff until she was given a heavier mast at Cape Town—and with about 8 tons of scrap-iron ballast inside. She carries 200 gallons of fresh water, and during her great voyage had no auxiliary engine. She is planked with pitchpine, iron-fastened to oak frames. One of her most interesting features is

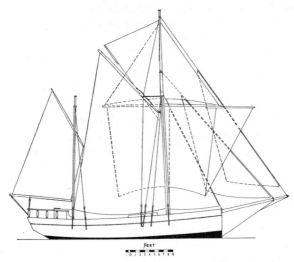

FIG. 29. *Saoirse*, ORIGINAL SAIL PLAN

Sail areas: Mainsail, 450 sq. ft. Mizzen, 210 sq. ft. Staysail, 110 sq. ft. Jib, 140 sq. ft. Topsail, 140 sq. ft. Flying jib, 160 sq. ft. Total area of square canvas, 500 sq. ft.

the raised poop deck, guarded by a high balustrade. The poop has a small cockpit and a sunk deckhouse, in which are the compass, chart table, and a bunk, and so close is it to the wheel that it is possible for the helmsman to sit inside and steer in bad weather. The poop provides a great feeling of security, and gives a comfortable impression of height above the sea.

For her world voyage *Saoirse* was rigged as a ketch, and to allow the big saloon to be unobstructed the mainmast was placed too far forward where the sails on it tend to depress the bow. The jib was hanked to a stay, the mainsail had no boom, but its gaff was controlled by a vang leading to the mizzen masthead, and the mizzen was a lugsail. The bowsprit extended 5 feet beyond the jib-stay, and on it was set a flying jib, but most frequently this was hoisted right up the topmast stay as a jib-topsail. O'Brien considered that *Saoirse*'s ability to steer herself with the wind on the quarter was largely due to the sails set on this long bowsprit. The total fore-and-aft sail area was 1,050 square feet. In addition there was a squaresail, and above it two raffees, one of which could be set upside down as a stunsail when required. The area of square canvas was 500 square feet. It is interesting to note that the ship's passages improved consistently as the possibilities of the square rig were realized and made use of, and O'Brien became a staunch advocate of square rig in ocean-going yachts, and subsequently made a number of alterations to *Saoirse*'s rig; at one time she was a topsail schooner and at another a brigantine. This fine old ship, often to be seen in the English Channel, is now rigged once again as a ketch, and carries no square canvas, but as may be seen in Plate 25 her mainsail has a boom and the mizzen is jib-headed. Her present owner, Eric Ruck, keeps her in splendid order, and when I complimented him on her appearance, he said: 'I regard her as an ancient monument, and treat her with the respect that is her due.'

Trekka (Plans on pages 127–9)

In the early 1950s a number of very small yachts were built to take part in coastal races. A few of these have made notable voyages; the Laurent Giles-designed 20-foot cutter *Sopranino*, for example, crossed the North Atlantic in 1951 by the southern route, sailing from the Canary Islands to Barbados, a distance of 2,700 miles, in 28 days. But it would not be correct to assume that only in this decade have very small craft made such long passages; between 1877 and 1903, for instance, the North Atlantic was crossed from west to east by at least ten craft with an overall length of 20 feet or less. The 21-foot ketch *Trekka* is of the same type as *Sopranino*, and in hull form is little more than a large sailing dinghy with a fin keel added after construction; she is a particularly interesting example of what one might presume to be the smallest practical ocean cruiser; indeed, many people might consider her to be a lot too small. She also was designed by Laurent Giles, and was built at Vancouver, B.C., by her owner, John Guzzwell.

Although she is very little longer than *Sopranino*, her displacement is just double that of the earlier boat; this is largely because she was given thicker planking to withstand possible collision with floating logs in Canadian waters. An essential feature of both craft, and one to which their designer attached great importance, is that they are arranged so far as is possible to be unsinkable; this is achieved by having watertight compartments at bow and stern capable of floating the little vessels even if their central portions should become flooded.

Reverse sheer, which can sometimes be hideous, and in a larger yacht often has little to commend it, is the only practicable method of making habitable so small a vessel as *Trekka* without resorting to excessive freeboard at stem and stern, but, as may be seen here, when reverse sheer is drawn by an artist it can be pleasing to the eye. An unusual feature is that the upper few inches of the topsides are canted inboard to minimize a boxy appearance, but at the same time, of course, reducing the deck area a little.

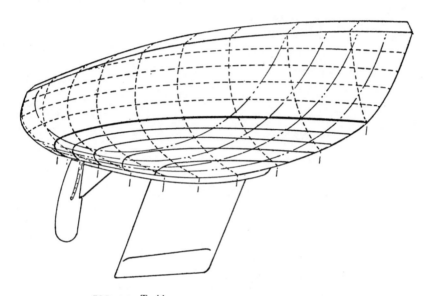

FIG. 30. *Trekka*, HULL PERSPECTIVE DRAWING

Designed by Laurent Giles & Partners for John Guzzwell, her dimensions are: l.o.a. 20·8 ft.; l.w.l. 18·5 ft.; beam 6·5 ft.; draught 4·5 ft. (Drawing by courtesy of the editor of *Yachts and Yachting*.)

In such a craft it is of the first importance to keep the weight down to the minimum, and here the construction has been cleverly devised to that end. Permanent plywood bulkheads act as moulds and are tied together with the bunks and other internal parts to make a light, rigid, and strong framework. The laminated keel and the steam-bent timbers are of oak, and the skin planking is of $\frac{9}{16}$-inch red cedar, edge-glued. The hull and plywood deck are covered with g.r.p. The fin keel and rudder skeg are of $\frac{3}{8}$-inch steel plate through-bolted to two lengths of angle-iron bolted to the wood keel; the angle irons run from the forward end of the fin to the after end of the rudder skeg, and no doubt help to stiffen the yacht longitudinally. The fin with its bulb weighs 1,300 lb., and can easily be removed without touching any bolts through the hull if the yacht is to be transported overland. *Trekka* is rigged as a ketch to make her more readily steer herself in the days before vane gear was available. The working area is 184 square feet, but with the masthead genoa and mizzen staysail set, the area can be increased to 340 square feet.

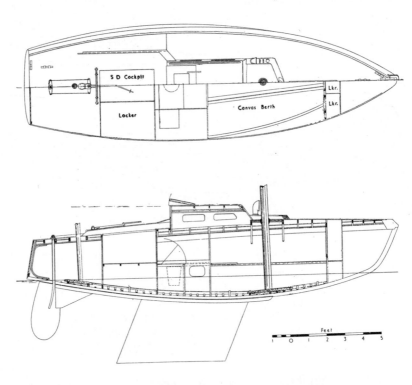

FIG. 31. *Trekka*, GENERAL ARRANGEMENT

    In September 1955 this miniature cruiser, deeply laden with provisions for 60 days and 24 gallons of water in plastic bottles, left Vancouver with her owner/builder as sole crew, and made a perfectly executed and almost trouble-free voyage around the world by way of the Cape and Panama, and was the smallest vessel ever to have done so. Since then, under new ownership, she has made a second circumnavigation.

In 1965 Guzzwell, again entirely with his own hands, built the 45-foot cutter *Treasure*, and with his wife and twin sons has sailed her from England to New Zealand.

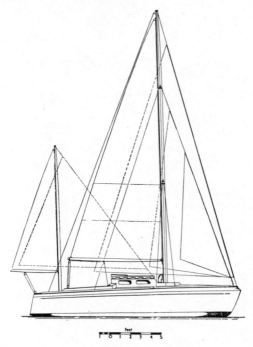

FIG. 32. *Trekka*, SAIL PLAN

Sail areas: Mainsail, 100 sq. ft. Mizzen, 26 sq. ft. No. 1 staysail, 58 sq. ft. No. 2 staysail, 25 sq. ft. Mizzen staysail, 66 sq. ft. Genoa, 148 sq. ft.

Waltzing Matilda (Plans on pages 130-2 and Plate 21)

During a visit to Sydney in 1954 I had the good fortune to meet Philip Davenport and his wife Rosetta. He, an Australian air-line pilot, told me in his quiet and modest manner something about the voyage they had made to England by way of Magellan Strait in their 46-foot cutter *Waltzing Matilda*. For fifteen years he had worked and planned towards that end, and by 1949 had saved sufficient money to be able to ask J. Muir, of Hobart in Tasmania, to design and build for him the sort of ship he had set his heart on. The folk-song *Waltzing Matilda* tells the story of a swagman, one of the tramps who lead an independent life wandering in the Australian outback; the name is appropriate, for it signifies independence and freedom.

K

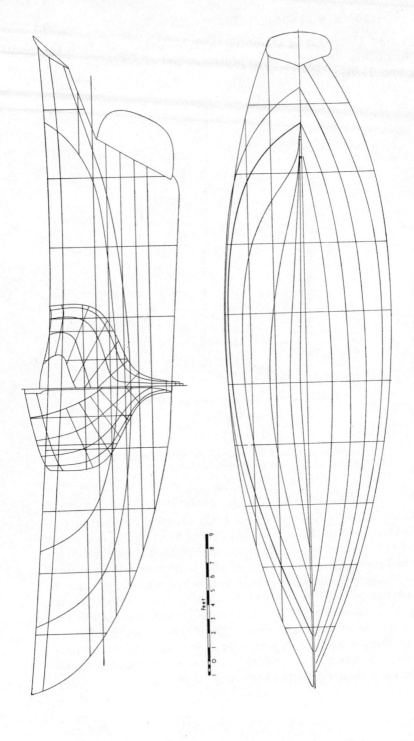

FIG. 33. *Waltzing Matilda*, LINES

Designed and built in Tasmania by J. Muir for Philip Davenport, her dimensions are: l.o.a. 46·4 ft.; l.w.l. 36 ft.; beam 12 ft.; draught 6·5 ft.; her displacement is 15 tons.

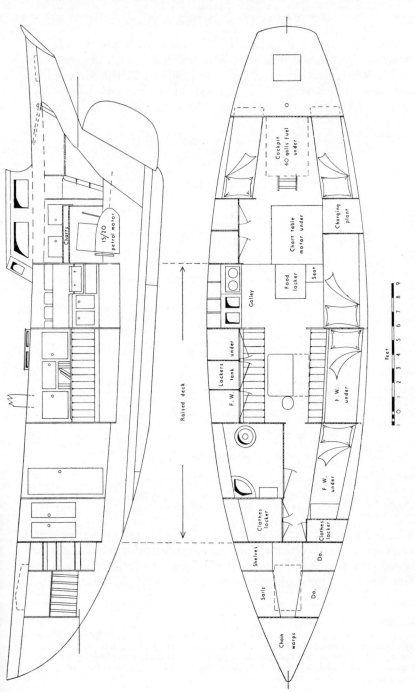

FIG. 34. *Waltzing Matilda*, GENERAL ARRANGEMENT

Charts

15/20 petrol motor

Raised deck

Cockpit
40 galls fuel
under

Charging
plant

Chart table
motor under

Food
locker

Seat

Galley

F. W.
under

Lockers
F. W. tank under

F. W.
under

Clothes
locker

Clothes
locker

Shelves

Do.

Sails

Do.

Chain
warps

Feet
0 1 2 3 4 5 6 7 8 9

The lines show a long-keeled hull with a beam one-third of the waterline length, and with waterlines which are much finer forward and fuller aft than are usually given to modern British yachts. It is also unusual for a vessel of this size to have the deck raised for 20 feet of its length amidships; but this enables a coachroof to be dispensed

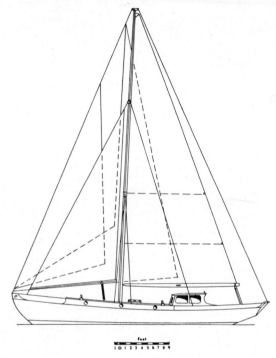

FIG. 35. *Waltzing Matilda*, SAIL PLAN

Sail areas: Mainsail, 480 sq. ft. Staysail, 234 sq. ft. Jib, 260 sq. ft.
Yankee jib, 300 sq. ft. Genoa, 650 sq. ft.

with and a simpler and stronger form of construction to be employed. Davenport remarked that the feeling of spaciousness below under the raised deck is probably more important than the actual gain in cubic feet. The top of the doghouse overhangs the cockpit by 2 feet, and this

---

PLATE 25

Saoirse (see pages 122–6), the 42-foot ketch designed by the late Conor O'Brien, and sailed by him east-about round the world by way of Cape Horn. The photograph is a recent one, but the present fore-and-aft rig differs from the original only in having a boomed instead of a boom-less mainsail, and a jib-headed instead of a sliding-gunter mizzen; the main gaff is controlled by a vang taken to the mizzen masthead.

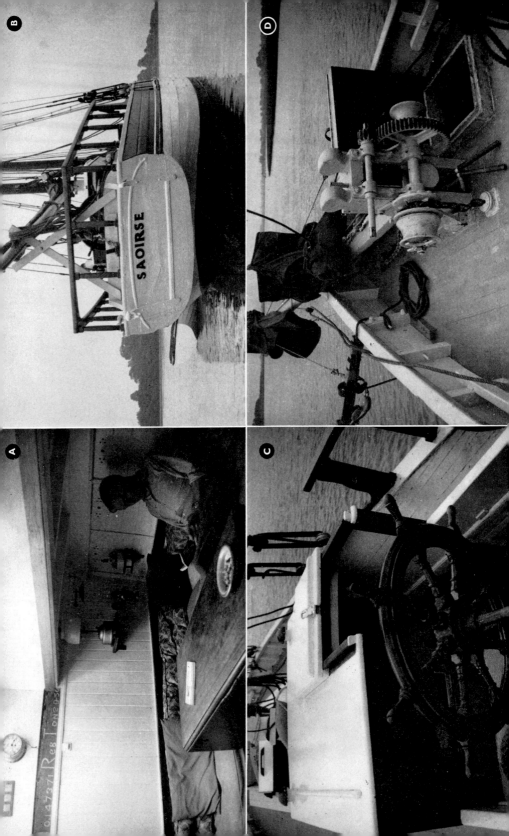

'back veranda', as it was called, proved to be of great value in protecting the cockpit, the companionway, and the helmsman. The general arrangement was planned with a long cruise in mind. Settees, which are not used as berths, are always available, the chart table is large and permanent and is not required for meals, and there is space enough for each member of the crew to go about his or her business without interfering with the others. Although separate sleeping and day accommodation is arranged, the impression of space has not been spoilt by dividing the interior into too many small compartments. Fresh water is in four tanks, one each side of the main cabin, one under the galley bench, and one beneath the forward bunk, the weight being well distributed; the total capacity is 240 gallons. In high latitudes damp was the cause of great discomfort, particularly in the bunks close above the water tanks where condensation made the bedding wet; no doubt a cabin stove would have been a comfort. The only alterations the owner told me he would like to make would be to devote more space amidships to clothes lockers that could be kept ventilated by the natural forward-moving circulation of air, and to shift the chart table to the side of the doghouse so that the bulkhead between doghouse and cabin could be done away with. The 15/20 Universal petrol engine, driving a two-bladed propeller under the port quarter, gave a speed of 3 knots only in calm conditions, and was later replaced by a more powerful diesel engine. Cooking was done on a two-burner gas stove, a cylinder containing 65 lb. of gas lasting about 12 weeks when cooking was done for four people.

Having crossed the Tasman Sea, the Davenports with two male companions set out from New Zealand at the end of 1950 on the 5,000 mile passage across the South Pacific to the Golfo de Peñas, Chile, where they entered the Patagonian Channels; they made their way through Magellan Strait, and after calling at ports on the east side of South America arrived in England in August 1951. It was a fine voyage in the best tradition along a route rarely taken by yachts, and although there were exciting moments, there were none of those

---

PLATE 26

A: *Saoirse*'s saloon is 12 feet wide and is lit by a big skylight. On the coaming between clock and barometer are three small silver plaques, recording the fact that for his voyage round the world O'Brien was awarded the challenge cup of the Royal Cruising Club for three consecutive years. *B*: Old-fashioned though it may seem, *Saoirse*'s poop, guarded by a high balustrade, is a practical and pleasing feature; *C*, it has a small cockpit and a sunk deckhouse in which are the compass, chart table, and a bunk, and in bad weather the helmsman can reach the wheel from its shelter. *D*: Massive bitts, high bulwarks, and a cathead for securing the anchor are big-ship features of this fine old yacht.

incidents which reflect bad planning, inefficient seamanship, or poor navigation. Apart from a weakness in the rudder head, which showed up when the ship was lying a-hull in the very violent sea which was experienced when the centre of a disturbance passed over her, she stood up splendidly to all that was asked of her, and there were times when she was driven hard. No special running or self-steering sails were carried on the voyage because the crew was large enough for watches to be stood at all times. The final 3,400 miles of the South Pacific were sailed under trysail and small spinnaker only at an average of 136 miles a day, and the best day's run was 181 miles. No doubt this might have been bettered had a larger area of sail been set on the quieter days, but the mainsail and large spinnaker were preserved for use in the areas of light and variable winds in the Atlantic, and for the Fastnet race, in which *Waltzing Matilda* was to take part on her arrival in British waters. Under new ownership the yacht was lost off St. Lucia in 1969.

Wanderer III (Plans on pages 136–8 and Plate 27)

When my wife and I decided to have a yacht built in which to attempt a voyage round the world, we went for plans to Laurent Giles & Partners, the designers of our previous yacht, *Wanderer II*, and of the 25-foot Vertue class, which I regard as one of the finest of small cruisers. A number of outstanding blue-water voyages have been made in these little yachts, notable among them being Humphrey Barton's crossing of the Atlantic by the northern route in *Vertue XXXV* (Plate 1, *right*), Hamilton's voyage from Hong Kong to England by way of the Cape of Good Hope in *Speedwell of Hong Kong*, John Goodwin's voyage in the same little vessel from England to the West Indies and Cape Town, and Bill Nance's wonderful single-handed voyage in *Cardinal Vertue* round the world east-about by way of Cape Horn. We wanted an overall length of 30 feet, and it so happened that Jack Giles had already drawn the plans of a yacht of that size which we all agreed would most nearly suit our purpose. So, after some modifications and a completely different general arrangement plan had been made, we got William King of Burnham-on-Crouch, Essex, to build to that design.

The hull is of heavy displacement type (7 tons) with all the ballast (3 tons of lead) on the keel. Apart from size, the most obvious differences between it and that of the Vertue class are that the latter has a greater proportion of beam to waterline length, exactly one-third, and greater rake to the sternpost. It was hoped that *Wanderer*'s more vertical sternpost might assist her to steer herself, but I now doubt whether this has any bearing on the matter, while it has the disadvantage of increasing the wetted surface.

As the new yacht would have to carry a considerable weight of stores, water, and gear, the freeboard of the original design was increased, and this proved to be a fortunate precaution, for when she was launched she was 6 inches down on her marks, due partly to the fact that the available West African mahogany—a light timber which had been specified for planking—was of such poor quality that iroko, a heavy timber, was used instead. Fortunately the increased displacement has no effect on performance beyond making her a little slow, and in light winds she could do with more sail than the maximum of 600 square feet, which is all that she can set with her jib-headed sloop rig. I chose that rig because I already had wide cruising experience with the gaff cutter, and wanted to find out for myself whether or not the modern rig is the best for blue-water sailing, a matter which has already been mentioned in Chapter 3.

The coachroof provides sufficient headroom in the galley/chartroom, so that there was no need for a doghouse, and it is therefore possible to carry a $7\frac{1}{2}$-foot dinghy in the best of all places, amidships abaft the mast, where, as there is no skylight, it steals neither light nor air from the accommodation. The cockpit is self-draining, and the only water to get below enters through the hinged lid of the lee cockpit locker; these lockers must be made to open for easy access to the gear stowed in them; no doubt the lids should be held down on rubber gaskets, but in a small cockpit the wing nuts or clamps required for that purpose are dangerous.

Below deck the arrangement is unusual in having only two berths; there is a large amount of locker space, and the galley/chartroom, where the most important work is done when at sea, is of considerable size. The Taylors' two-burner paraffin-cooker is slung on fore-and-aft pivots; the working bench is 4 feet long and is covered with stainless-steel sheet, and there are self-stowing arrangements for crockery, pots, etc. It is not obvious from the plan that the chart table extends right out to the ship's side beneath lockers which are arranged above it, or that there is stowage space beneath it in drawers and lockers for 400 Admiralty charts. Forward of the sideboards over the feet of the settee-berths, lockers extend out to the side of the ship and up to the deck; they are fitted with shelves for clothes, typewriter, photographic equipment, etc., and ventilation holes enable them to

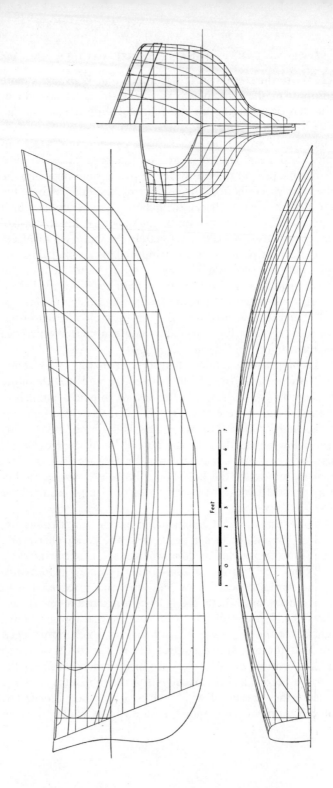

FIG. 36. *Wanderer III*, LINES

Designed by Laurent Giles & Partners, her dimensions are: l.o.a. 30·3 ft.; l.w.l. 26·3 ft.; beam 8·4 ft.; draught 5 ft.

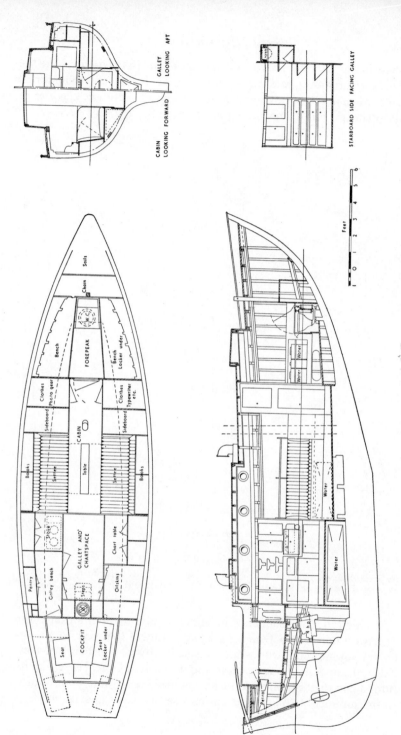

CABIN
LOOKING FORWARD

GALLEY
LOOKING AFT

STARBOARD SIDE FACING GALLEY

Sails

Chain

W.C.

Bench

FOREPEAK

Bench
Locker under

Clothes

Sideboard

Photo gear

Clothes
Sideboard
Typewriter
etc.

Books

Settee

CABIN

Table

Settee

Books

Stove

Pantry

Galley bench

GALLEY AND
CHARTSPACE

Chart table

Steps

Oilskins

Seat

COCKPIT

Seat
Locker under

W.C.

Water   Water

Water

Water

Water

Petrol

Feet

FIG. 37. *Wanderer III*, GENERAL ARRANGEMENT

share in the natural forward-moving flow of air. In the forepeak are the heads, chain locker, rack for water cans, a bin locker, and room for all sails; when in port it can be converted into a darkroom: see Chapter 12 and Plate 48. For details of engine see page 83, and sails page 52.

FIG. 38. *Wanderer III*, SAIL PLAN

Sail areas: Mainsail, 280 sq. ft. No. 1 staysail, 144 sq. ft. No. 2 staysail, 92 sq. ft. No. 3 staysail, 43 sq. ft. Trysail, 75 sq. ft. Genoa, 322 sq. ft. Twin running sails, combined area 250 sq. ft.

Wanderer III was launched in March 1952, and, after a trial cruise in Irish waters, my wife and I set out in July of that year on a west-bound voyage round the world. We sailed by way of Panama, New Zealand, Australia, Torres Strait, and the Cape of Good Hope, and re-turned home within three years, having made good some 32,000 miles. In the summer of 1959 we set out on a second circumnavigation by a route which differed from the earlier one in many respects, and again took three years over it. In 1965-7 we made a two-year trip to the East Coast of America, bringing *Wanderer*'s total sea mileage up to about 110,000. She is a fine small cruiser, and never caused us any anxiety, but I formed the opinion that greater beam and more spread-out ballast might have made her motion less violent. Apparently Laurent Giles & Partners thought so too, for they have since produced such a design with a beam of 9·3 feet on a 24·5-foot waterline, and several yachts, known as the Wanderer class, have been built to this, including David Guthrie's *Widgee* (Plates 1, *left*, 9A, and 24, *bottom*), in which he made a single-handed double crossing of the Atlantic.

Top: The 30-foot sloop *Wanderer III*, in which my wife and I twice circled the globe, is an orthodox, heavy displacement type of cruiser. Her best day's run was 169 miles, and once she made good 440 miles in three consecutive days although she was 6 inches or so below her designed waterline. *Bottom*: Her saloon by night, the photograph being taken by the light of the pressure lamp on the table only. At the top at each side can be seen the drip-catching handrails, and at their forward ends part of the two steel beams which strengthen the coachroof in way of the mast.

PART II

VOYAGING

∿∿

7

PLANNING THE VOYAGE

Wind systems of the world—Seasonal winds—Tropical revolving storms—Currents—Sailing directions—Pilot charts—Navigational charts—Time and route planning

CONTRARY to a common belief among laymen, a successful ocean voyage in a sailing yacht is not a haphazard undertaking, commencing at any moment convenient to owner and crew and leading wherever the wind may blow. Certainly a number of voyages have been started in such a casual manner, but few of them were carried through to a successful conclusion, and the majority were abandoned at an early stage.

Like any other worthwhile undertaking a voyage needs planning, and it is probable that the greater the care with which this is done the greater will be the success of the voyage and the pleasure and satisfaction to be had from it. The two chief considerations are to avoid dangerously bad weather, notably the great storms known variously as hurricanes, cyclones, or typhoons, to which certain areas of the oceans are subject at certain times of year; and to make the greatest possible use of fair winds, in which it will sometimes be found that the longer way round is the quicker in the end, as well as being kinder to ship and crew. An example of the kind of thing that can happen if this matter is not given proper consideration, was provided by the 63-foot auxiliary ketch *Ululani*. In 1966 she set out from the Marquesas Islands bound for the Galapagos, but instead of going south

---

PLATE 28

Top: Waiting for wind in the doldrums, the region of calms, light airs, heavy rains, and thunderstorms, which lies between the trade wind belts. *Bottom*: With dorsal and tail fin breaking surface, a shark swims close alongside as the yacht lies without steerage way; the little striped pilot fish, *bottom left*, which usually accompanies it, sometimes changes over to a yacht, and may then remain in company with her for many hundreds of miles.

and running her easting down in the westerlies, she attempted to make the 3,000-mile voyage direct against the trade wind. In this she failed, and after being at sea for 113 days (having received urgently needed supplies of food from a Japanese fishing vessel) returned to her port of departure.

The basis of planning is to be in the right place at the right time, but first one must decide how long can be allowed before one must be home again, or at some specified place, and even the fortunate few who can start on a voyage with no fixed time of return must lay their plans with care, especially if the voyage is to extend through tropical areas.

Usually those who intend to make a voyage have had the idea in mind for some time, or at least have taken an intelligent interest in the subject, and have read some of the published accounts of other small-ship voyages (see Appendix III). Such reading is a great help, not only in providing a realistic picture of the conditions likely to be encountered and the difficulties that will have to be overcome, but in deciding which ports or places are worth visiting, and which are best avoided. Such reading may form the basis of the plan; but alone it is not enough, for some understanding of the wind systems of the oceans is needed, together with the detailed information and advice which is to be found only in certain textbooks and charts.

Wind systems of the world

A full description of the winds and weather of all oceans will be found in *Ocean Passages for the World* and the appropriate *Pilots* (see page 157-8); but the following brief outline should serve as a sufficient introduction to the subject to enable the reader to understand the general principles and make the best use of the pilot charts (see page 159 and endpapers), which provide in the simplest and clearest manner the essential information needed when an ocean voyage is being planned. Some of the matter in the following sections is based on information obtained from *Ocean Passages* and the *Admiralty Navigation Manual*, Vol. I, with permission of the Controller of H.M. Stationery Office.

If all parts of the earth's surface had the same temperature, the atmospheric pressure at the surface would everywhere be the same, and there would be no wind. The primary cause of wind is a difference of temperature, which in turn is responsible for differences of atmospheric pressure. Warm air rises and cooler air flows to take its place;

also air tends to flow from an area of high pressure to an area of lower pressure.

Between the latitudes of 20° and 40°, both north and south, there are belts of relatively high pressure over the oceans; on each side of these belts the pressure is relatively low. The belt of nearly uniform low pressure in the neighbourhood of the equator is known as the doldrums, a region of calms, light variable winds, intermittent heavy rains, and thunderstorms; its average width varies between 200 and 300 miles. If there were no other wind-creating factors, and if the

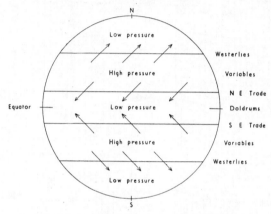

FIG. 39. PRESSURE AND WIND BELTS OF
THE WORLD

earth was stationary, wind would blow on the surface from the high-pressure belts to the low-pressure belts, the winds being north and south, and the displaced air returning high up in the contrary direction. But the earth is rotating on its axis in an easterly direction, and this movement causes any mass of air which is being drawn towards a low-pressure centre to be deflected to the right in the northern hemisphere, so that a circulation is set up in an anti-clockwise direction about the centre. Around a centre of high pressure a clockwise circulation is set up in the northern hemisphere. In the southern hemisphere these directions are reversed. Therefore the winds on the equatorial sides of the high-pressure belts, blowing towards the doldrums, do so from a north-east and south-east direction, Fig. 39. It was because of the steady assistance these winds gave to the trade of sailing ships that they became known as trade winds, and so far as is possible a sailing yacht's route should be planned to make the greatest use of them, even though this may add considerably to the distance.

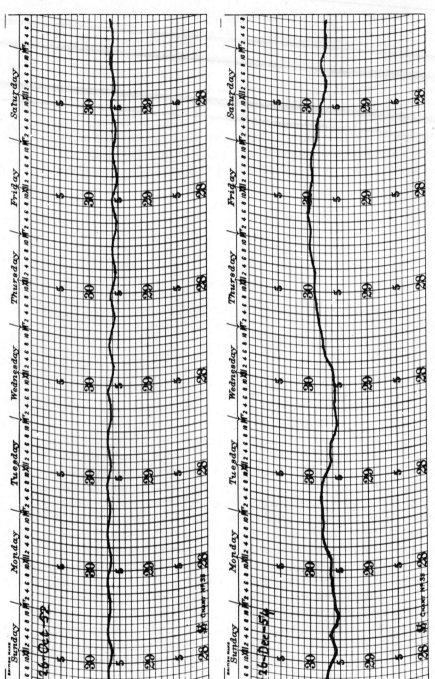

FIG. 40. BAROGRAPH RECORDS

Top: Traced in the tropics between latitudes 17° N. and 14° N., it clearly shows the diurnal rise and fall. *Bottom*: Traced outside the tropics, the diurnal movement is faint and is masked to some extent by greater changes of pressure.

Normally the trade winds do not blow with gale force, and the weather within their belts is mostly fine and clear with a pleasantly warm temperature, while the barometer remains steady, except for the diurnal rise and fall, which results from the atmospheric pressure waves with a period of 12 hours which sweep regularly round the earth from east to west. These waves are at their maximum in the tropics, and as may be seen in the top part of Fig. 40—which is a re-production of a week's barograph record traced during a crossing of the Atlantic, while in the tropics and between latitudes 17° N. and 14° N.—the pressure rises from approximately 4 a.m. to 10 a.m., then falls until 4 p.m., after which it rises again until 10 p.m. If this diurnal rise and fall ceases, or if there is a marked upward or downward trend of the glass, the presence of a tropical disturbance may be suspected. The lower part of Fig. 40 shows a barograph record traced in the neigh-bourhood of the Cape of Good Hope; but as the Cape is outside the tropics, the diurnal movement, though it may still be seen, is not so large or so clear as it is in the upper record, and is masked to some ex-tent by the greater barometer movements caused by weather systems moving across from the west.

The trade-wind belts of both hemispheres, and therefore the zone of doldrums that lies between them, move bodily north and south during the year (see Figs. 41-4), following in a small measure the declinational path of the sun, which ranges from 23° 27′ N. (Tropic of Cancer) to 23° 27′ S. (Tropic of Capricorn). Their average limits in the different seasons are as follows:

NE. trade wind of the Atlantic Ocean,	January	2° N. to 25° N.
	July	10° N. to 30° N.
SE. trade wind of the Atlantic Ocean,	January	0° to 30° S.
	July	5° N. to 25° S.
NE. trade wind of the Pacific Ocean,	January	4° N. to 25° N.
	July	12° N. to 30° N.
SE. trade wind of the Pacific Ocean,	January	4° N. to 30° S.
	July	8° N. to 25° S.
SE. trade wind of the Indian Ocean,	January	15° S. to 30° S.
	July	0° to 25° S.

The south-east trade wind of the Pacific is not so constant or steady as are the trade winds of the other oceans. Only during the months of June, July, and August is there a continuous belt of south-east wind blowing steadily across the ocean. During the rest of the year there is a space 600 miles wide and extending diagonally across the belt in

L

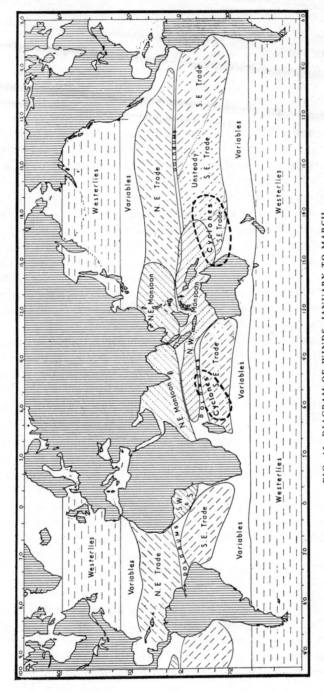

FIG. 41. DIAGRAM OF WINDS, JANUARY TO MARCH

Off the coast of north-west Australia willy-willies blow during these months and until April.

(Figs. 41–4 are based on the former Admiralty Diagram of Oceanic Winds with the permission of the Controller of H.M. Stationery Office and the Hydrographer of the Navy.)

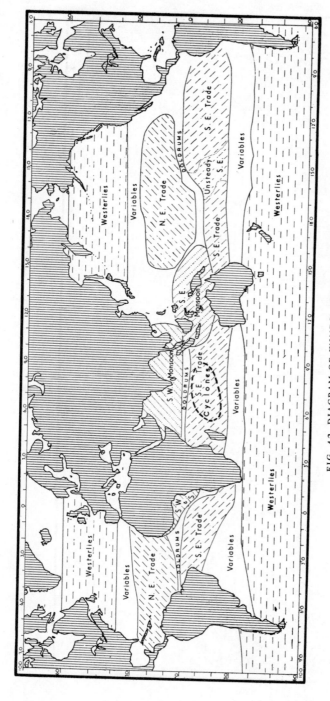

FIG. 42. DIAGRAM OF WINDS, APRIL TO JUNE

The cyclone season in the western part of the South Pacific finishes in April. The south-west monsoon starts in May.

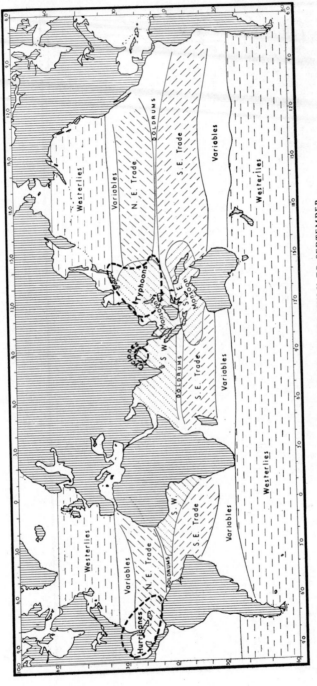

FIG. 43. DIAGRAM OF WINDS, JULY TO SEPTEMBER

Pamperos blow in the River Plate during these months.

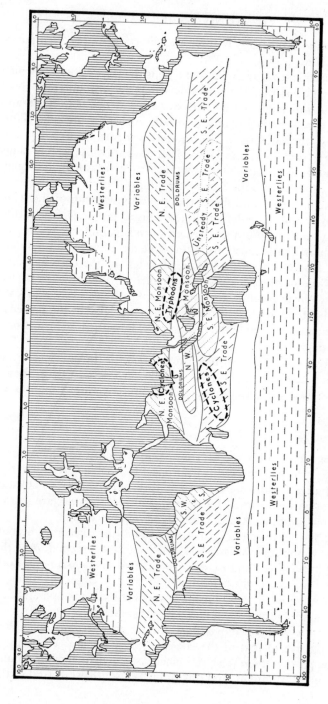

FIG. 44. DIAGRAM OF WINDS, OCTOBER TO DECEMBER

West Indian hurricanes may be experienced until mid-November. The north-west monsoon starts in November. The cyclone season in the western South Pacific, and willy-willies off the north-west coast of Australia, start in December.

which the wind is unsteady in force and direction, and is often punctuated by calms.

On the polar sides of the trade-wind belts there are regions in which calms and light or moderate variable winds are encountered; these extend approximately between latitudes 25° and 35°, both north and south, and correspond more or less with the high-pressure belts. They are known as the variables of Cancer and Capricorn respectively, and are often referred to as the Horse Latitudes, because sailing-ships becalmed for long periods there sometimes had to jettison their cargoes of livestock to conserve the freshwater supply. On the polar sides of latitude 35° in each hemisphere, the westerly direction of the wind becomes more and more definite as the latitude increases. The forties of south latitude, where these westerly winds commonly reach gale force and blow continuously from the same direction for several days at a time, are known as the Roaring Forties. In the corresponding latitude of the northern hemisphere, namely, in the more restricted areas of the North Atlantic and North Pacific, the westerly winds are generally lighter and more variable in direction, due to the presence of large land masses, and to the succession of low-pressure formations, known as depressions.

The above main wind systems, together with the seasonal variations which are described in the following section, are shown diagrammatically in Figs. 41-4, each figure covering 3 months of the year.

Seasonal winds

In certain parts of some oceans seasonal winds are experienced. Such winds, which blow steadily from one direction for several months, and then, after a short pause, become reversed and blow as steadily from the opposite direction for the remainder of the year, are known as monsoons, from an Arabic word signifying 'seasons'.

During the summer the great land mass of Asia becomes heated, and in the winter months is cooled, relative to the temperature of the sea along its southern and eastern coasts. Sea water does not absorb or release heat as rapidly as land, and in spite of large alterations in the temperature of the air, remains at about the same temperature all the year round. During the summer months, therefore, a great region of heated air is established over Asia, and the area becomes one of low pressure relative to that of the cooler air over the sea. While these conditions prevail, there is an anti-clockwise circulation of air around this part of the continent, and this is felt as a south-west wind in the northern part of the Indian Ocean and in the China Sea. But during the winter months the conditions are reversed, and there is then a large high pressure area over the cooler land, and low pressure over the warmer sea, producing a clockwise movement of air, which is felt as a north-east wind over the same waters. These two winds are known as the south-west and north-east monsoons.

Monsoons also occur in northern Australian waters, and extend in a comparatively narrow band westward from Australia across the Indian Ocean, filling the space between the equator and the northern limit of the south-east trade wind. In this case a north-west or north monsoon blows during the southern summer, and a south-east wind— known either as monsoon or trade wind—blows for the rest of the year, i.e. from April to October.

The approximate seasons of the monsoons are:

> SW. monsoon—May to September.
> NE. monsoon—October to March.
> NW. monsoon—November to March.

Over the full length of the Red Sea the prevailing wind is NNW. from June to September; but from October to April a SSE. wind prevails in the southern part while the NNW. wind continues in the northern part. Between them lies a calm belt.

On the south-east coast of South America, south of about 15° S., the SE. trade wind does not reach the coast, and during the southern summer a NE. monsoon blows along the coast as far as the River Plate; during the winter the winds there are SW. and SE. Violent SW. gales, known as pamperos, are experienced in the vicinity of the River Plate from July to September.

The SE. trade wind does not extend into the Gulf of Guinea on the west coast of Africa; there the prevailing wind throughout the year is south or south-west. Between June and September this wind extends as a monsoon in a narrow belt westward from the African coast towards South America, and occupies the region between the equator and the southern limit of the NE. trade wind as far as longitude 32° W.

There are some other seasonal winds, but these are not usually given the name monsoon. A description of them will be found in *Ocean Passages*, and of local winds for all coasts in the *Pilots*.

In the vicinity of Cape Horn the prevailing wind is westerly, and gales from between south and north-west occur at all seasons. Off the Cape of Good Hope south-east winds are common in the summer, and westerly in the winter. Gales from between north-west and south-west are most common in June and July, but they may occur at any time of year.

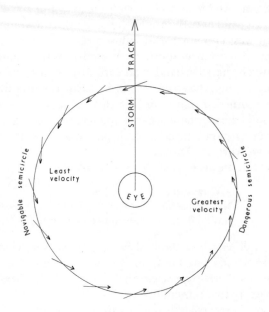

FIG. 45. TROPICAL REVOLVING STORM, NORTHERN HEMISPHERE

Tropical revolving storms

The strongest winds at sea are encountered in tropical revolving storms, when they may attain a speed of 125 miles per hour or more, and raise a huge sea which is likely to overwhelm a small vessel, particularly if the centre, or eye, of the storm passes over her. These storms are so named because they originate in the tropics, and the wind blows round an area in which the lowest pressure is in the centre. The direction of rotation is anti-clockwise in the northern hemisphere and clockwise in the southern hemisphere. The wind does not revolve round the low-pressure centre in concentric circles, but has a spiral movement in towards the centre of the disturbance; see Fig. 45.

These storms occur for the most part at the western sides of the oceans, though they are also experienced in the Bay of Bengal, and off the north-west coast of Australia, while some Atlantic hurricanes have started near the Cape Verde Islands; they are unknown in the South Atlantic. Usually they form between latitudes 10° and 20°

either side of the equator, and in addition to their circular motion they have a forward movement. At first this is in a westerly direction, but frequently the storms recurve, turning to the north or north-east in the northern hemisphere, and to the south or south-east in the southern hemisphere; they lessen in intensity and increase in area and speed as they leave the tropics, and may then travel for long distances, bringing destructive gales of great strength to places far removed from their origin. It is, for example, not uncommon for hurricanes originating near the American coast to travel across the Atlantic to the coast of Europe.

A vessel lying in the track of the centre, or eye, of the storm will be in particular danger, for the wind will remain constant in direction until the eye reaches her when, after a short calm, it will blow with great violence from the opposite direction and raise a wickedly steep and confused sea. In the northern hemisphere the right hand is the dangerous semicircle, for a vessel caught in this will tend to be blown towards the track of the centre, or the storm may recurve and its centre pass over her; also she will encounter a greater wind velocity than would be found in the left-hand, or navigable, semicircle, for the storm may be travelling at anything between 5 and 50 knots, and where the wind is blowing in the same general direction as the eye is travelling the two velocities combine. In the left, or navigable, semicircle the velocities are subtractive.

One of the most obvious and important considerations when planning a voyage will be to ensure that the vessel will not find herself in an area subject to tropical storms during the danger months, or if she has to pass through an area which is never entirely free from them, that she does so during the months of reduced risk.

Tropical revolving storms are most frequent during the late summer or early autumn of their hemisphere, being comparatively rare in the northern hemisphere from mid-November to mid-June, and in the southern hemisphere from mid-May to mid-November. But in the Arabian Sea they are most likely to occur in October–November and May–June, i.e. at the change of the monsoon. Out-of-season storms occur from time to time, and in the Western North Pacific no month is entirely safe. The following table shows the average number of severe tropical revolving storms recorded per annum for the various areas, together with the names that are given to them according to the part of the world in which they occur; but some of the figures may be under-estimates, since in the less-frequented parts of the world some storms may have escaped notice.

Western North Atlantic (hurricanes)   5
Western North Pacific (typhoons)   25
Western South Pacific (cyclones)   3
Bay of Bengal (cyclones)   2
Arabian Sea (cyclones)   1
North-west Australia (willy-willies)   1
Southern Indian Ocean (cyclones)   6

The storms occur in different parts of the world as follows:

Western North Atlantic. Hurricanes may blow here from June to mid-November. They are most frequent during August, September, and October when the trade wind is feeble, and they generally curve to the northward by the Bahamas and between the American coast and Bermuda. Some pass to the Gulf of Mexico, others curve well out into the Atlantic east of Bermuda.

Western North Pacific. Typhoons may blow here during all months of the year, and are most frequent from July to October; in the China Sea October is the worst month. They generally move in a west or north-west direction at first, then curve north or north-east.

Western South Pacific. Cyclones blow from December to April, but are most common from January to March.

Bay of Bengal. Cyclones blow from May to December, but are most frequent during September.

Arabian Sea. Cyclones blow from April to January and are most frequent during June, October, and November. They rarely blow in August.

North-West Australia. Willy-willies blow from December to April.

Southern Indian Ocean. Cyclones blow from October to June and are most frequent from December to April.

More detailed information regarding tropical revolving storms in any area, together with the rules for avoiding them, will be found in the relevant *Pilot*.

Currents

An ocean current is a permanent or semi-permanent movement of water; it should not be confused with a tidal stream, which depends on the movements of celestial bodies and is subject to hourly changes.

Currents may be caused by differences of specific gravity, such as result from variations in temperature and salinity; but the major cause is wind. When wind blows over water, friction causes the surface particles of the water to move in the same direction as the wind, and these particles act on those below them so that a current is created. Such a current, while it is still under the influence of the wind that created it, is known as a drift current. Permanent winds, such as the trade winds, create drift currents, but these do not follow the exact direction of the wind because the rotation of the earth has a similar effect on them to its effect on wind; in the northern hemisphere this causes a current to flow to the right of the direction towards which the wind is blowing, and in the southern hemisphere to the left. Having regard to the general wind systems of the world, it will be realized that ocean currents north of the equator tend to flow in a clockwise direction, and those south of the equator in an anti-clockwise direction. Ocean currents not under the influence of the parent wind may be caused by the following: (1) The momentum of a drift current which carries water into regions where it is no longer under the influence of the wind that created it. (2) The deflection of a drift current by land or shallow water, so that it sweeps on under its own momentum in a new direction. (3) The replacement, in the form of a counter or compensating current, of water carried away by the primary current.

The currents of the North Atlantic (Fig. 46) may be taken as a fair example of ocean currents in general. The north-east trade wind, blowing almost continually, creates the west-flowing North Equatorial Current, which has a rate of from 10 to 40 miles a day. Some of the surface water displaced by this current, together with some from the South Equatorial Current (created by the south-east trade wind), returns to the east along the narrow doldrum belt where there is no wind to hinder it or set up a drift current; this is known as the Equatorial Counter Current, and has a rate of from 10 to 30 miles a day. But the bulk of the North Equatorial Current presses on along the north coast of South America, and passes through the openings between the West Indies to pile up with a rise of level in the Caribbean Sea. The only escape for this water is between the Yucatan Peninsula and Cuba, and thence through the gully separating Florida from the shallow Bahama Banks; the current taking that route is known as the Gulf Stream, but that is a misnomer, for it does not originate in the Gulf of Mexico, indeed, no water goes into that Gulf as there is no exit from it. The Gulf Stream, a band of warm and very salt water some 40 miles wide, flowing at a rate of up to 100 miles a day, and in narrow bands within the Stream sometimes at a greatly increased rate—it is said that a speed of 9 knots has been recorded—sweeps along the eastern coast of North America to the Newfoundland Banks, where it meets the cold Labrador Current and is deflected to the east.

No doubt it would lose its impetus there were it not for the westerly winds which continue the easterly drift; this is sometimes called the Gulf Stream, but is more commonly known as the North Atlantic Drift; it has a rate of from 10 to 25 miles a day. In the neighbourhood of the Azores it divides, one part setting north-east to the arctic regions, the other turning down the eastern side of the ocean, where

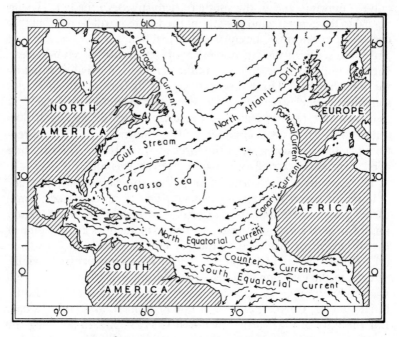

FIG. 46. CURRENTS OF THE NORTH ATLANTIC

(Based on Admiralty chart No. 5310 with the permission of the Controller of H.M. Stationery Office and the Hydrographer of the Navy.)

inshore it is known first as the Portugal Current, and then as the Canary Current. Eventually it swings to the south-west and west as the North Equatorial Current, thus completing the cycle.

This great clockwise circulation of the surface waters of the North Atlantic revolves round a large area where there is little or no current; flotsam comes to rest there, together with a quantity of Sargassum weed, from which the area got the name Sargasso Sea, a place said to be so stagnant that it gave rise to legends of ships held fast until they rotted.

The other oceans, except the northern part of the Indian Ocean where the currents are affected by the monsoons, have similar circular current systems, clockwise in the northern hemisphere, anti-clockwise in the southern hemisphere. They also have their named currents, as, for example, the warm Agulhas Current of the Indian Ocean, which flows south-west along the south-east coast of Africa; and the Benguella and Humboldt Currents of the South Atlantic and South Pacific, both of which are cold and flow up the west coasts of South Africa and South America respectively.

There is much yet to be learnt about ocean currents, and the navigator should remember that they are variable both in strength and direction. Even when a trade wind has been blowing steadily for many days, the surface drift may be in any direction, and may vary as much as sixteen points from one day to the next; indeed, I have sometimes found the least current when the trade wind is blowing with strength and constancy. Although the more powerful currents, such as the Gulf Stream and the Agulhas Current, keep a more constant direction, considerable variations in strength occur; Irving Johnson told me that during the six voyages he had made round the Cape of Good Hope at the time of our meeting, he had found the Agulhas Current to run with greatest strength when the wind was opposed to it. Current arrows on charts should therefore be regarded only as an approximate indication of the general trend, and no reliance should be placed on them or the rate written against them when a landfall has to be made under difficult conditions.

Sailing directions

Ocean Passages for the World, a large volume compiled by the Hydrographic Department of the Admiralty,[1] and obtainable from chart agents, contains a clear description of the winds, currents, and weather for all the oceans. It also gives details of the recommended routes between major points of departure and arrival for all oceans at different seasons for power-driven vessels and sailing-ships. A pocket at the back of the book contains current and wind diagrams for the world, together with a chart showing the recommended sailing-ship routes for the world, and separate charts showing the recommended routes for power-driven vessels in the Atlantic, Pacific, and Indian Ocean. Although this book cannot be regarded as essential, it is of

[1] The former Admiralty, including the Hydrographic Department, is now incorporated in the Ministry of Defence, but the terms 'Admiralty Chart' and the like officially continue.

considerable value when one is planning a voyage, because it contains between one pair of covers most of the relevant information.

Books of sailing directions, known as *Pilots*, are also compiled by the Hydrographic Department, and can be obtained from chart agents. There are more than 70 volumes, they cover all coasts and islands, and an index map in the *Catalogue of Admiralty Charts* shows the area covered by each. Each volume contains some general remarks on navigation, including the rules for avoiding tropical revolving storms, and an index chart showing how its area is covered by Admiralty charts. This is followed by remarks on the countries or states dealt with, their peoples, forms of government, currency, etc., and the meteorology, currents, signals, buoyage system, etc., to be encountered there. The body of the book gives detailed descriptions of the coasts, ports and adjacent waters, often illustrated, together with sailing directions for them. When visiting any coast with which one is not familiar the relevant *Pilot* is essential, and will be found of considerable assistance when planning the trip, in indicating which are safe or easy landfalls to make and which are good anchorages; also, should the navigator lack a chart of some place he wishes to call at, he may be able to draw one from the information given in the *Pilot*. An example of this is given by Smeeton in his book *Sunrise to Windward*:

We were navigating on a tracing of the general chart of the Nicobars, but I had drawn a chart of the harbour from the *Pilot* on a plotting diagram, using a blown-up scale. . . . I felt I knew it all well, almost as if I was a native, as a result of the laboured drawings that I had made during the passage. Lights go out and beacons decay, buoys change colour and are missing, but the land remains the same and the sailing directions are usually given with reference to a recognizable landmark.

A yearly supplement of corrections and additions is published for each *Pilot* (available free from chart agents), and when this grows inconveniently large, the volume is withdrawn and a new edition is published.

The Coast and Geodetic Survey of the U.S.A. also publishes *Pilots* covering the U.S. and foreign coasts. In some instances these are of more value to the yachtsman than their British counterparts as they contain special information of use to small craft. E.g., *U.S. Coast Pilot 4* (Cape Henry to Key West) lists all bridges on the Intracoastal Waterway with their clearances and times of opening, and gives for every marina the number of berths, depth in the approach and alongside, and states what facilities are available.

Pilot charts

Pilot charts are fascinating documents, showing as they do practically everything that it is necessary for the mariner to know when planning an ocean voyage. By a simple system of colours and symbols they indicate the average strength and direction of the winds and currents, and the percentages of calms and gales that have been encountered in all navigable parts of each ocean at different times of year. A set of such charts for the area in which the voyage is to be made is the most essential requisite when planning that voyage.

The originator of the pilot charts was Matthew Fontaine Maury, a lieutenant in the U.S. Navy, who started collecting material for them in the early part of the nineteenth century by persuading mariners of all nations to fill in the special log-books with which he provided them. Some idea of the immensity of the task may be gained from the fact that the meteorological data of the April pilot chart for the North Atlantic are based on more than 4 million observations, and that is only one chart of a set of twelve for that ocean.

Pilot charts measure 38 in. by 26 in., are published by the U.S. Navy Oceanographic Office, and can be obtained through British chart agents if advance notice is given. They are published as follows: Monthly for the North Atlantic; quarterly (December–February, etc.) for the South Atlantic; monthly for the North Pacific; quarterly for the South Pacific; monthly for the Indian Ocean. They are continually being amended as fresh information is obtained. A full-size reproduction of a portion of one of the North Atlantic charts will be found on the front and back endpapers.

Four colours are used for the symbols; blue, green, red, and orange. The chart is divided by pecked and solid black lines into 5° rectangles. In the centre of each rectangle is a small blue circle. The figure in that circle indicates the percentage of calms and light airs that has been experienced within the rectangle during the month for which the chart is valid by the observers who have supplied data. The blue arrows flying towards the circle fly with the wind, and indicate the direction of the wind (concentrated on eight points of the compass) which has been experienced within the rectangle. The length of éach wind arrow, measured against a scale in a corner of the chart, expresses the percentage of wind from that particular point. When, as sometimes happens in trade-wind areas, the percentage of wind from one point is so great that the arrow would be inconveniently long, the arrow is broken and the percentage in figures is inserted, as may be seen in the lower part of the endpaper. The number of feathers on each arrow indicates the average force of that wind expressed by the Beaufort scale. The average percentages of all available ships' observations in which gales (winds of force 8 and above) were reported in each 5° rectangle are shown in red figures on a small inset chart. The direction of currents is shown by green arrows, and the figures printed in green against some of these arrows express the speed of the currents in knots and tenths.

The tracks of some representative storms which occurred during the month (or quarter) in the past are shown in solid red lines, with the dates of their positions at Greenwich noon, and reference numbers in roman figures marked on them; see top left of endpaper. The maximum and minimum limits of ice, and the extreme limit of icebergs, are also shown in red, and the percentage of fog in dotted blue lines.

With these essentials before him, the navigator can plan his ocean passages to the best advantage, seeing at a glance where winds and currents are likely to be most favourable and where they are adverse, and what the risk may be of encountering a great storm, gales, fog, or ice. But when he puts his plans into practice and finds himself out in the wide ocean, he must not be disappointed if he encounters conditions different from those shown on the chart, but should remind himself that the symbols can only represent the average of many observations taken in the past, and that no efforts of man to reduce nature to a formula are ever entirely successful. This fact was realized as long ago as 1740, when Anson set out on his famous voyage. In his book, *A Voyage Round the World*, he comments that during the passage from Madeira to St. Catherine's Island, Brazil, the trade winds did not blow as expected or as had been experienced by earlier navigators; but he then goes on to say:

I mention not these particulars with a view to cavilling at the received accounts of these trade winds, which I doubt not are in general sufficiently accurate; but I thought it a matter worthy of public notice that such deviations from the established rules do sometimes take place.

In addition to the information which has already been mentioned, the pilot charts contain certain other matter that is of value and interest to the navigator. Lines of equal magnetic variation (at intervals of one degree) for the year stated are shown by orange lines, and

PLATE 29

Perfect sailing weather. With mainsail and spinnaker pulling hard, and the brilliant sunshine casting clear-cut, ever-moving shadows on the slanting deck, *Wanderer II*, homeward bound from the Azores, hurries on her way.

the rate of annual change in the variation is shown, also in orange, in a small inset chart. The normal atmospheric temperatures are shown by dotted red isotherms, the positions of ocean station vessels are shown by bold red circles, and the normal atmospheric pressure is shown in blue isobars on an inset chart. The dashed red lines, and the large letters *B* and *D* in the reproduction, are connected with the division of the chart into hydrolant areas for the purposes of correction, and therefore do not concern us.

Finally, pilot charts show a number of regular shipping routes with ·distances, but one is naturally sorry to discover that on up-to-date charts the sailing-ship routes, which from the yachtsman's point of view formed such an interesting and valuable feature of earlier charts, have been omitted. On the reverse side are printed articles on a variety of nautical subjects; some titles are: 'Fog', 'Tropical revolving storms', 'The Gulf Stream', 'National flags and ensigns'.

Her Majesty's Stationery Office publishes a set of monthly climatological charts covering the North Atlantic. These do not appear to be based on so many observations as the U.S. pilot charts, but they provide similar information though with a different and not so easily read set of symbols, and may be obtained from chart agents.

Navigational charts

Great feats of seamanship, perseverance, and endurance were performed in the making of many of the early charts, for the work was done in the days of sail, and when there were no sounding instruments except the hand lead and line, no motor launches to do the inshore work, no radio signals with which to check the chronometers, and no refrigeration to keep the ships' companies supplied with fresh, health-giving food. Little is heard today of the men who performed this valuable service, and to most of us who use charts based on their surveys the only reminder is the credit line sometimes engraved beneath the title of the chart: 'Surveyed by Captain Fitzroy, R.N., and the officers of H.M.S. *Beagle*, 1836.' 'Compiled from the discoveries and surveys of Cook, Kotze, Bellinghausen, Wilkes, and other navigators.' 'From the surveys of Captain H. M. Denham, R.N., F.R.S., and the officers of H.M.S. *Herald* and *Torch*, 1854–56.'

PLATE 30

The American 36-foot, steel-built, centreboard cutter *Altair* (Charles and Chris Grey), leaving San Juan, Puerto Rico, in a steep swell which, *middle*, often hid her hull from my view and, *bottom*, broke angrily on the ramparts of El Morro beside the narrow harbour mouth.

M

Today, unfortunately for those who love charts and admire the men who made them, more and more of the credit lines are being dropped as new editions are printed, and their place taken by the unromantic statement: 'Compiled from the latest information in the Hydrographic Department.'

Although a small-scale chart of the ocean or oceans in which one intends to voyage is of value when planning, it will not be necessary at that stage to buy all the coastal charts and harbour plans that one may think will be needed during the voyage. Indeed, if cost is a major consideration, it will be better to plan the route in some detail before buying them so as to avoid spending money on charts of coasts which one may discover, on studying the *Pilots* and pilot charts, are best avoided. And while it is, of course, important to have all the necessary charts some time before sailing day so that they may be studied and arranged in a suitable order, the longer their purchase can be delayed the less correcting will have to be done, for agents supply charts corrected in most respects to the date of dispatch. It is as well to refrain from ordering very large-scale harbour plans until the general and approach sheets have been studied, for it often happens that the latter give sufficient detail to enable ports to be made by a small vessel in safety. But I would advise the navigator not to be too parsimonious, particularly in buying charts which cover coasts or islands near his proposed route, for he may well find it desirable to make some changes to his plans during the voyage, and the probability then will be that the only way of obtaining missing charts will be to have them sent out from England or the U.S.A.

Although there are in many ports abroad agents for the sale of Admiralty charts, usually their stock is small and of only local coverage. I therefore prefer to deal direct with one of the several class A agents in Britain, which maintain complete stocks of charts covering the world, and correct them to the date of dispatch; for many years I have dealt with J. D. Potter Ltd. of 145 Minories, London, E.C. 3, and have found their service prompt and efficient. On receipt of a roll of charts I open it as soon as possible and flatten each chart by rolling it the opposite way, because charts kept rolled are awkward to use and take a lot of stowage space. I then fold each along its original crease or creases, and with the help of the index chart in the appropriate *Pilot* (this is much more convenient to use than the unwieldy *Catalogue of Admiralty Charts*, which when open requires a space 3 feet 9 inches by 1 foot 5 inches) arrange the charts in geographical order, ticking off on the index chart the number of each before stowing them away flat.

If the short title printed on the back edge is not sufficiently descriptive, I add to it; e.g. '868, Five Fathom Hole, The Narrows and Murrays Anchorage.' To this I would add 'Bermuda'.

Since the British government has decided to change to the metric system, the use of fathoms on Admiralty charts will be discontinued as new editions of charts are published, and the navigator will have to accustom himself to working in metres; during the change-over, which presumably will take many years, he will have to be on his guard against confusion, particularly when he is tired or is reading a chart in a poor light.

For cruising in American waters American charts are essential. They can be obtained by mail, prepaid, from Coast and Geodetic Survey, Environmental Science Services Administration, Rockville, Maryland 20852. Neat little catalogues with index charts may be had gratis from the same address. For some other parts of the world also I have found U.S. charts preferable to British. The Galapagos Islands west of Panama, for example, extend over 2° of latitude and 2½° of longitude, yet the only British chart is one half-size sheet with a scale of 10 miles to an inch; this group is covered by a dozen large-scale American charts. American charts, which are in colour, are a little cheaper (most cost $1.00) and are more boldly printed than their British equivalents, and sometimes provide more information, thus saving the need to consult other sources. However, they come in a variety of shapes and sizes and may need folding several times before they will fit the space of 2 feet 6 inches by 1 foot 9 inches required by British charts, and are therefore more bulky. Their titles are not printed on their backs, so these need to be written on. Some of the abbreviations and symbols, notably those indicating buoys (diamonds) and beacons (triangles) will be unfamiliar to the British cruising man, but chart No. 1 (a small booklet) lists all these and costs only 25 cents. U.S. charts on which the number is followed by the letters S–C contain additional information for the use of small craft, and often come in a folder on which are printed tide tables for the year, times of weather forecasts, etc. Such charts are not hand-corrected, but new editions are published each year.

The Canadian Hydrographic Service (Chart Division, Department of Mines and Technical Surveys, 615 Booth Street, Ottawa, Ontario) will supply free on request an Information Bulletin showing the chart coverage available for any specified area in Canada, together with a list of marine publications, and the authorized agents from whom they may be obtained.

Time and route planning

As has already been mentioned, the essentials when planning a voyage are to make the greatest use of fair winds and avoid dangerously bad weather by choosing a suitable course and being in the right place at the right time. Therefore, the destination having been decided on, one will study *Ocean Passages* and/or the pilot charts, to obtain as much information as possible about the winds and weather between the point of departure and the destination, and make brief notes. Possibly an area in which tropical revolving storms occur will be the most serious consideration round which the plan will begin to grow, and from the safe months for that area one will work backwards and forwards.

To illustrate this, let us take a simple example and suppose that we wish to make a round trip from the English Channel to the West Indies and back by way of Bermuda, a voyage which is often made by British yachts.

From a previous section in this chapter we have learnt that the West Indies hurricane season extends from June to November, but it is generally considered that the last half of November is safe. Although navigation among the islands is possible during the hurricane months, though it is not pleasant because of the squalls and the lightness of the trade wind during that season—indeed, a number of yachts have safely made the West Indies during the summer months—the risk is one which a prudent seaman will do his best to avoid. So here is the first time factor: we should not arrive at the West Indies area before mid-November, and we ought not to remain there after May.

From the chart of the world showing sailing-ship routes, or from the text of *Ocean Passages*, it is obvious that the shortest route (great circle course, see page 190) from the English Channel to the West Indies is not practicable for a sailing vessel, because she would almost certainly have headwinds for a great part of the way; a glance at the pilot charts of the North Atlantic will confirm this. The passage recommended is by the southern route, i.e. down the eastern side of the ocean until the north-east trade wind is picked up, and then curving away to the westward. This route passes close to the Madeiras and Canaries, either or both of which will make convenient stopping places for refreshment. By measuring on the Atlantic chart, we find that the distance by the recommended route from the Canaries to Barbados, the nearest of the West Indies, is about 2,700 miles. How long will that passage take? I have usually planned on the assumption that my vessel will average

100 miles a day in the trade winds and have not often found that to be far wrong; but as readers will probably find it interesting and encouraging to know what times have been taken by yachts of various sizes at different times of year, some typical examples of the times taken on a variety of passages in each of the three oceans have been grouped in Appendix I. From this it will be seen that the crossing of the Atlantic from the Canaries to Barbados has taken between 21 and 44 days, but of course much depends on the size and type of yacht, the number of her crew, and the manner in which she is sailed, apart from the constancy or otherwise of the wind. Perhaps we shall do this trip in 25 days, and as we do not wish to arrive at the West Indies before 15 November, we should not leave the Canaries before 21 October.

The Canaries lie about 1,500 miles from Falmouth, and if we decide to sail there direct, the passage might take us something between 12 and 25 days, depending largely on the weather we encounter while crossing the Bay of Biscay. If we take the larger and very pessimistic figure, and allow ourselves a week at the Canaries, we should leave Falmouth not later than 19 September. But if we look at the pilot charts for September and October, we shall find that the north or north-east wind, known as the Portuguese trade wind, which predominates during the summer from Cape Finisterre to the southward, is becoming variable, and that the proportion of south-west winds is increasing; also we shall find that the percentage of gales is rising. If time permits, it will therefore be better to leave the Channel earlier, say in August, and put the time to good advantage at Madeira and the Canaries.

As we have already seen, we may cruise safely among the West Indies until May, but by June we should be on passage to Bermuda. We may leave earlier if we wish, but up to March the pilot charts show a high proportion of gales in the Bermuda area. The passage to Bermuda, about 800 miles from Nassau, will be in the variables, but the charts show that in May and June winds with a southerly component predominate.

From Bermuda to the English Channel we should get a large proportion of fair winds if we follow the great circle course; but in deciding on the course to steer, we should bear in mind that in the Atlantic the farther north one goes the stronger are the winds likely to be; also that the great circle course passes within the line marking the extreme limit of ice off Newfoundland, and into an area where 10 to 15 per cent of fog may be expected. The majority of cruising yachts therefore take a more southerly course to keep in a pleasanter weather area, and some call for refreshment at the Azores; but in the neighbourhood of those islands in the summer there is a considerable chance of falling in with north-east winds, which will compel one to stand away to the north or north-north-west to seek the westerlies of higher latitudes. June, July, and August are good months for making the crossing, the smallest percentage of gales being in July. The crossing may take 30 to 40 days, so if we leave Bermuda on 1 July, we should be in the Channel during the first half of August.

The planning of a voyage which is to extend across more than one ocean is naturally a more complicated and lengthy business. The method employed by Tom Worth when planning *Beyond*'s voyage round the world was as follows. He mounted the world chart showing sailing-ship routes on a piece of soft wallboard, and glued the names of the months (cut from a calendar) to the heads of drawing pins. Having roughly pencilled in the proposed route, he read *Ocean Passages* and stuck the drawing pins of the best months for each area into the chart. Then he consulted the pilot charts, measured distances, and computed the possible time it would take to sail from one place to another, and the pins were gradually moved. He spent about 300 hours at it, and at the end simply had a list of ports and approximate dates; but it was a valuable list, and much had been learnt in the making of it.

It is important to allow oneself a reasonable length of time in a port or on a coast at the end of a stage, otherwise an unexpectedly long passage could upset the time plan for the following passages. If a passage takes the expected time, one has the full allotted time in port; but if the passage is a slow one, the time in port will be reduced to permit departure on the planned date.

8

SEAMANSHIP

Self-dependence—Vigilance—Management in heavy weather— At anchor

THE art of good seamanship may be defined in the broadest terms as the safe and efficient management of a vessel at all times; the preparation of her and her gear, her safe conduct from one place to another, selection of a safe anchorage and attention to her welfare there, are all a part of it.

In *Cruising Under Sail* an attempt was made to cover the practical side of this wide and fascinating subject, though the point was made, and I must repeat it here, that seamanship cannot be learnt from reading books or listening to lectures, though both are of some assistance. One of the attractions of the study of seamanship is that it has no finality, conditions never repeat themselves identically, no two vessels behave quite alike, and always there is something fresh to be learnt. Although the principles remain the same whether a short day sail is being made from one familiar port to another close by, or a long ocean voyage to a coast of which one knows little, there are certain matters which merit special consideration from the ocean-voyager's point of view.

Self-dependence

As there is no longer any commerce done in sail, many of the ocean routes likely to be followed by a sailing yacht are empty of shipping. These routes are, of course, crossed in places by power-vessel routes, which are clearly marked on the charts supplied with *Ocean Passages for the World*, on Admiralty charts 5305, 5306, and 5307, and some are shown on pilot charts. But although there are a lot of these routes criss-crossing the oceans, one should bear in mind that on some of them there may be no more than a ship every month or so, and that as a power-vessel will be visible in clear weather up to a distance of between 5 and 10 miles only, one may see nothing when crossing a route. Nevertheless, a proper lookout should be kept on such occasions and a radar reflector should be rigged, for even out in the wide oceans collisions do occur from time to time, more commonly since the advent of radar, for it would appear that aboard many merchant ships today the only lookout kept at sea is by radar, and without a reflector (preferably arranged as high as possible) a wooden yacht may not be visible on the screen.

During *Saoirse*'s voyage round the world, and while she was sailing east-about from Recife to Recife (Pernambuco) not a single vessel was sighted when land could not be seen at the same time; throughout *Omoo*'s crossing of three oceans only two ships were sighted; and in the first 24,000 miles of *Wanderer III*'s first circumnavigation only three were seen when out of sight of land. Clearly therefore no reliance should be placed on receiving any assistance, and when one sets out it must be with the clear understanding that no matter what may happen—failure of gear, injury or illness, lack of stores or water, or the urgent need for a check on the position—the yacht's company must depend entirely on their own resources. At the start of a long voyage this is a sobering thought, but it has to be faced, and if the skipper has any serious doubts as to the ability of himself, his companions, or his ship to see it through, he would do well to abandon the venture before it is too late. But in making his decision to go ahead or withdraw, he should try to analyse his feelings, and decide whether the dread from which he is suffering is based on the knowledge that something is amiss with the ship or her people, or his own efficiency and skill, or whether it is simply the baseless feeling of apprehension from which, I believe, most people suffer before the start of any venture in the successful conclusion of which they have to depend on themselves; and which is usually aggravated by the knowledge that they will be out of touch with the rest of mankind for a time. This feeling becomes less strong as more experience is gained, and it may be of some comfort to know that it usually dies soon after the land has faded out of sight astern, and does not rear its head again until heavy weather is encountered or the time for making a landfall approaches.

Some mental preparation is therefore desirable, but this will be of little use if the vessel herself is wanting in any respect, or does not carry sufficient provisions, water, or materials with which repairs or renewals can be made. So it may be said that good seamanship begins in port with the most perfect and detailed preparations that it is possible to make, with the checking of the strength and efficiency of every single item, and careful compiling of the commissioning lists to ensure that nothing essential for the well-being of yacht or crew is lacking. If these things are conscientiously done, if the yacht has already proved her seaworthiness, and the skipper his ability as a navigator and seaman, he will more easily be able to reassure himself that his feeling of apprehension is baseless, and will obtain the enjoyment that he should from the voyage once it has begun.

Vigilance

Good seamanship implies constant vigilance. In a time of stress, such as during bad weather, or when executing a difficult manœuvre or piece of navigation, most people are keyed up to their best endeavours, are alert and quick to react in an emergency; but when all is going well, particularly when difficulties have recently been overcome, or when a passage is nearing its end, there is a risk that vigilance may be relaxed or that one will become over-confident—that is a dangerous moment, and it is then that the vessel is most likely to get into trouble. The perfect seaman will not let that happen, but we are not all perfect, no matter how experienced we may be, as the following examples will show.

I almost wrecked *Wanderer III* on a rocky Panamanian shore where a big sea, driven by the north-east trade wind, was breaking heavily, simply because I was lacking in vigilance. The 10-day crossing of the Caribbean from Antigua had been a rough one and my wife and I were glad when the lights of Cristobal/Colon lifted ahead, and we could see bright among them a powerful flashing light. Tired and relaxed, the passage nearly completed and port just ahead, I omitted to check the period of that light with the *List of Lights*, and steered for it, taking it for granted that it was the 18-mile light on the end of the breakwater. It was only my wife's alert lookout and common sense that averted disaster, for it was she who realized, just in time, that the light was an air-beacon some distance inland. I learnt from this how important it is when I am tired to place no trust in my judgement and to check every item, no matter how obvious it may appear.

My shame at that unseamanlike approach to Central America was reduced a little when I learnt that even a fine seaman, such as Tom Worth, can also be guilty of negligence, as is shown by an incident which occurred when he and his wife in *Beyond* had crossed the Indian Ocean and were sailing up the African coast bound for Aden. They were never certain of their position off that almost featureless coast except at noon, but one evening at dusk they knew that Hafum Peninsula—which sticks out about 15 miles and is joined to the shore by a low beach—lay fairly close ahead. There is a light on its end, the only one for hundreds of miles. Worth saw a headland which could be Hafum, and set a course to pass outside it, thinking it could be positively identified by its light later on, and that in any case the brilliant moon would be up in a couple of hours. Nearly 2 hours later he noticed that the swell was confused, and then he suddenly saw breakers very close ahead. He immediately altered course to starboard, and *Beyond* just—but only just—sailed clear of the lee shore.

What had happened then became obvious. Before dusk, unsuspected mist had obscured Hafum; the headland sighted was not Hafum but a part of the mainland, and the ship had almost been wrecked on the low connecting beach. Describing this incident in the *Royal Cruising Club Journal*, Worth wrote: 'The lesson is, of course, either to keep out to sea and rely on celestial navigation, or keep inshore and pay *very* close attention to business all the time. I hope I never let such a thing happen again as I would not deserve to be so lucky a second time.'

Another great sailing man, Peter Pye, nearly lost his 29-foot converted fishing boat *Moonraker* during a West Indies cruise. When sailing among the Bahamas one day, Pye had just completed a difficult piece of pilotage and manœuvring when passing through a narrow cut against a tide race, and, pleased with the success of this, he was, to quote from his book *Red Mains'l*, 'relaxed and off guard', as *Moonraker* sailed in towards the anchorage at Royal Island Harbour at about high water. From the chart he knew that two rocks lay there, one large, the other small, and he supposed both of them to be above water. Two rocks were sighted, and although these looked rather closer together than they should be, he was not very concerned, as there was no need to go close to them. The headsails had been handed preparatory to anchoring, but the old ship was still sailing fast when she struck, and with such force that her stern rose 2 feet.

Pye then realized that the 'two' rocks he had seen were in reality one, its pinnacles separated by a few feet of water at high tide, while the second rock, which his ship had struck, was covered at high water, a fact which a careful study of the chart should have revealed. Fortunately there were two motor boats in the harbour; with one of them towing from ahead and the other pulling from abeam with a rope made fast to *Moonraker*'s mast so as to heel her farther and thus reduce her draught, she was got safely off.

The above three incidents will suffice to show the danger that is attendant on relaxation or over-confidence in conditions believed to be straightforward and easy. But there is another danger, also due to the human element, which must be guarded against. This arises when land is close, and the ship's company, mentally and physically exhausted by a long-drawn-out voyage, or a continuance of bad weather, are hungry for a night of undisturbed sleep in a quiet harbour, and an end to their present wretchedness, and their one thought, to the exclusion of all else, is bent to that end. Against their better judgement they carry on running towards the land in worsening weather and poor visibility, from a position which may be open to doubt, when good seamanship would counsel that the only prudent action is to keep a good offing and remain at sea until conditions have improved, or at least until such time as the position has been verified. This is highly dangerous and may even lead to complete loss of control of the situation, for the sea will grow more confused as land is approached, and even though the danger may at last have been realized, it will probably be too late then to round up and heave-to, or attempt to regain an offing.

The 45-foot ketch *Omoo*, when nearing the end of her circumnavigation, during which there had been no bad landfalls or narrow escapes, nor any suggestion of anything but the highest order of seamanship, found herself in such a situation. She had left Ascension on 9 April bound for the Azores, had crossed the equator in about 20° W. longitude, and soon after found the north-east trade wind through which she punched close-hauled. But the trade died 600 miles before it should have done, and thereafter calms, light headwinds, and squalls were experienced, and the passage grew more lengthy than had been expected by Van de Wiele, who had said when at the Cape: 'The Atlantic is before us again, and from now on it's a piece of cake.' The squalls were the cause of much sail-drill, for the cotton sails had become rotten and could not be risked in strong winds; the more palatable stores had been consumed, though there was no real shortage of food or water, and the ship's company were suffering from boils, enteritis, and cramp in the legs. When eventually a fair wind made, it soon hardened into a gale from west-south-west, and rain reduced visibility. The chronometer had received a knock, and its rate, no longer certain, could not be checked because the radio battery was exhausted, so the ship's position was uncertain.

In the circumstances a choice of two obvious actions lay before *Omoo*'s skipper: to heave-to and wait for better weather before closing with the land, or to abandon the idea of calling at the Azores, and alter course accordingly. But he took neither of these, for he and his companions were tired with the long passage and the frequent sail-shifting, the uninteresting diet, the cold and wet after their time in the tropics, and their one overpowering desire was to have done with it all and reach the comforts of the land; so they held on under reduced sail, hoping to make the harbour of Horta on the island of Fayal. The gale increased, rising to force 10, and the seas, which were estimated at 30 feet in height, grew steeper and became confused, while visibility was as little as a quarter of a mile at times. Apparently the gravity of the situation did not sufficiently impress itself on Van de Wiele until it was too late to do anything about it; indeed, control of the situation had almost been lost—lifebelts had been got ready and the ship's papers put in a watertight bag—when soon after dawn on the fortieth day at sea, land was sighted less than a mile ahead. This proved to be the south-west corner of Pico, an island to the eastward of Fayal and one which has no harbour. Fortunately the wind was drawing along the island's southern shore, and *Omoo* ran with it, thus narrowly escaping shipwreck.

It is not in any derogatory sense that I have referred to this incident here, for I have the highest regard for Van de Wiele. But that so experienced a seaman could make so obvious an error because his judgement was warped by mental and physical fatigue surely emphasizes the grave danger attending that state, and the need for all of us to be aware of it so that we may combat it with all our power.

Management in heavy weather

Although the good seaman will take all reasonable precautions to avoid dangerously bad weather by careful time and route planning, once he has started on an offshore passage he will have to take the weather as he finds it; and as he is certain to encounter bad weather sooner or later, the management of his vessel in such conditions is a matter of practical rather than academic importance to him.

The fear of bad weather is no longer such a bogy as it used to be. This is probably due to the fact that in recent years the number of small yachts has increased enormously, they make more ambitious voyages, and as very few of them get into serious trouble it is reasonable to assume that the dangers of bad weather have at times in the past been over-stressed, and that a small sailing yacht can be just as safe, though not so comfortable, as a larger one. In the event of bad weather approaching when the yacht is at sea clear of all shoals and tides, and if she is well-found in all respects, and has a strong and experienced crew, she will no doubt be kept sailing fast as long as her skipper considers that to be prudent; but if she is short-handed or the crew is inexperienced, it will be as well to shorten sail before that becomes really necessary; but how much reduction, and at what strength of wind, will depend on a variety of circumstances. If, when the weather worsens, the yacht is sailing with the wind abeam or forward of the beam, spray will drive over her, and her lee deck may be buried now and again; later, heavier water will start coming aboard over the weather side, and conditions will be so unpleasant that sail will almost certainly be shortened before carrying on could possibly become in any way dangerous. But if the yacht is running at the time, more experience and riper judgement will be needed to decide how long it is safe to continue to run under the sail then set, for the motion will be easier, the decks drier, and the wind seemingly not so strong because its apparent speed has been reduced by the speed of the yacht.

It is, however, not the wind of itself which constitutes a danger, but the sea created by the wind. The relationship between the height of seas in feet and the wind velocity in miles per hour is given by the U.S. Hydrographic Office as approximately 1 to 2; a wind of 50 m.p.h. should therefore raise a 25-foot sea. The length, measured from crest to crest, is given as approximately twenty times the height; the length of seas 25 feet high would therefore be 500 feet. These figures apply only in open water where the fetch, the distance from the weather shore or the point where the wind commenced to blow, is a thousand miles or more. Although the speed of seas depends largely on the velocity of the wind which causes them, 25 knots would appear to be the maximum recorded. As the wind drops, so the seas will lose height and speed, but they will retain their length and continue travelling perhaps for many hundreds of miles as a long and slowly moderating swell. There is always some swell in the open sea, though it may be so easy as scarcely to be noticed by a small vessel; but a heavy swell is often the forerunner of bad weather. If a sea encounters a swell running in some other direction, or a contrary current, a shoal, or a change of wind direction, it will become shorter and steeper and therefore more dangerous.

It has often been observed by seamen that at intervals there is a succession of higher, or shorter and steeper seas. These may constitute a danger in heavy weather, as may the freak sea very occasionally caused by one sea being superimposed on another, for this could overwhelm a small vessel in certain circumstances; but the chance of encountering one of these, except perhaps in high southern latitudes, is slight.

It is difficult to estimate the height of a sea from aboard a small vessel; the only practical method is for someone to climb the rigging to a point from which the crest is in line with the horizon when the vessel is in the trough, and then to measure the height of that point above the vessel's waterline. But at a time when the sea is high enough to be interesting, one will probably be too ocupied or too anxious to do this. Since ocean weather ships have been fitted with wave-recorders, some data about the height of seas in the North Atlantic has become available. It is now known that heights of between 40 and 50 feet are not uncommon in heavy gales, but in soundings, that is where the depth is less than 100 fathoms, the height is less. However, the height of a sea is of little importance to a small vessel, except that her sails may be becalmed or even taken aback when she is in the trough; it is the steepness and the size of the crests that matter to her.

As was mentioned in Chapter 1, the displacement waves created by a vessel's movement through the water may interfere with the over-taking seas, causing them to build up astern and on the quarters more steeply than elsewhere. If running under a press of sail in a rising sea is persisted in, this interference could conceivably cause an overtaking sea to break; the movement of the water in that sea would then no longer be a harmless oscillation but a forward movement, and it could poop the vessel and sweep her deck, or even overwhelm or smash her. The other danger attendant on running too fast in a heavy sea is that the vessel may get out of control, and despite the efforts of the helms-man, broach-to, that is, swing round at right angles to her course, when there may be a risk of the next sea breaking aboard the full length of her, or of its crest, exploding under the bilge, flinging her to leeward on her beam-ends. It should, however, be noted that the more easily driven type of hull, such as is possessed by the majority of modern sailing yachts, is less likely to disturb the overtaking seas than is the more boxy, heavier-displacement type, and one that is well balanced and easily steered is less likely to broach-to. Today it is almost un-heard of for an ocean racing yacht to stop because of bad weather, and most are kept sailing at speeds which would have been considered suicidal a few years ago; but that caution should be exercised in other types of vessel was emphasized by the capsizing of Erling Tambs's *Sandefjord*, a 47-foot Norwegian ex-lifeboat (Plate 32, *bottom*), a type of vessel often regarded as the perfect ocean cruiser, short-ended and very beamy. She was running at the time under the double-reefed mainsail only; the two men at the helm had already complained that they could not hold her, when a steep sea lifted her stern, her bow dived deeply, and she turned end over end, losing one man and her mizzen mast.

In my own yachts of from 24 to 30 feet I usually decided that when the seas astern and on the quarters were noticeably higher or steeper, or had angrier-looking crests than those elsewhere, it was time to reduce speed by shortening sail; when that had been done an immediate improvement in the appearance of the seas was noticeable, and steering became easier. But I must add that I usually took such action to ease the physical strain of steering rather than because I thought it necessary for the safety of the yacht.

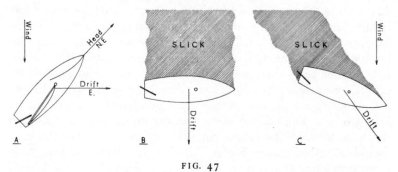

FIG. 47

A: Hove-to on the port tack and making a square drift. *B*: Lying a-hull, beam on to wind and sea, and leaving a slick dead to windward. *C*: Lying a-hull with wind on the quarter and making some headway; the slick does not give complete protection.

The management of yachts in winds of 40 knots and upwards is often a matter of controversy. The alternatives are: heaving-to under reduced sail; lying a-hull, i.e. under bare poles with wind and sea more or less abeam; lying to a sea-anchor; running before the wind, possibly towing ropes or other forms of drag.

If the gale is from ahead, one will naturally be reluctant to lose more ground than necessary, while if a danger lies to leeward every effort will, of course, be made to prevent the yacht from being driven on to it. In these circumstances heaving-to under reduced sail may be the best procedure, for when hove-to the average cruising yacht will make a square drift, that is, she will lie with the wind about 4 points on the bow and make good a course about 8 points (90 degrees) to the wind, so that she will hold her own, or nearly so. For example, if the wind is north and she is hove-to on the port tack, she will head north-east and drift in an easterly direction (Fig. 47A). The steadiness with which she will lie will depend to some extent on the strength of the wind and size of sea, but more on her size, shape of hull, and position of mast or masts, and it is questionable whether many yachts will continue to lie-to well when the wind rises above force 8. In theory the backed headsail counteracts the forward drive of the reefed mainsail or trysail, but it may be found that even with the headsail backed as much as possible, the yacht will forge ahead (fore-reach) too fast, bringing heavy water aboard, and that if the helm is lashed down far enough to prevent that, the sail set abaft the mast will shake. That must be stopped, for a sail shaking in a strong wind is likely to damage itself. In some craft the remedy is to hand the headsail and let her lie under reefed main or trysail alone; but only experiment with indivi-dual craft in the conditions prevailing will show what is the best pro-cedure. If the wind continues to freshen, almost any yacht, though it happens sooner with a small than with a large one, will fore-reach too fast for safety, and heavy water will start to come aboard. If no further reduction of sail is then possible, she will have to be stripped of her canvas, when it may be found, again depending on her type, that she will lie safely a-hull, that is, take up a position with wind and sea abeam or a little abaft the beam, making no headway and drifting slowly to leeward. To reduce the pressure on the rudder the helm should be lashed down. But if, when lying a-hull, she persists in surging first ahead and then astern, like a piece of flat wood being dragged sideways through the water, as did *Waltzing Matilda* when she was riding out a gale in the roaring forties, a great strain will be imposed on the rudder, and unless provision has been made to support the rudder blade with tackles, it will not be safe to lie like that.

At first sight it might appear unwise to expose the yacht's broad-side to the advancing seas, but if her draught is moderate and therefore does not offer too much lateral resistance, she should drift to leeward enough to leave a slick of semi-smooth water to windward (Fig. 47B); this has something of the same effect as oil and tends to smooth the sea, sometimes to a remarkable degree. But if, in spite of the helm being lashed down, the yacht persists in lying with her quarter to the wind and makes headway, the slick, no longer being left dead to wind-ward of her, will not give her full protection (Fig. 47C). On several occasions *Wanderer III* has lain a-hull safely in winds of force 8-9, drifting to leeward at half a knot or so, but I would not consider it wise to continue lying a-hull in stronger winds, for in a severe gale a mass of broken water may pour down the advancing face of a sea and, with

---

PLATE 31

A and *B*: A conical type sea-anchor with its metal hoop hinged for convenient stowage. *C*: A pyramid type sea-anchor, with its mouth held open by oak crossbars. *D*: Ropes from the sea-anchor brought together round a large thimble to form a bridle. *E*: To steady the sea-anchor there should be a small hole at the apex of the bag to allow an escape of water. *F*: When there is no windlass, a pawl fitted over the stemhead roller will hold each link of anchor chain as it is hove in, and so make weighing easier.

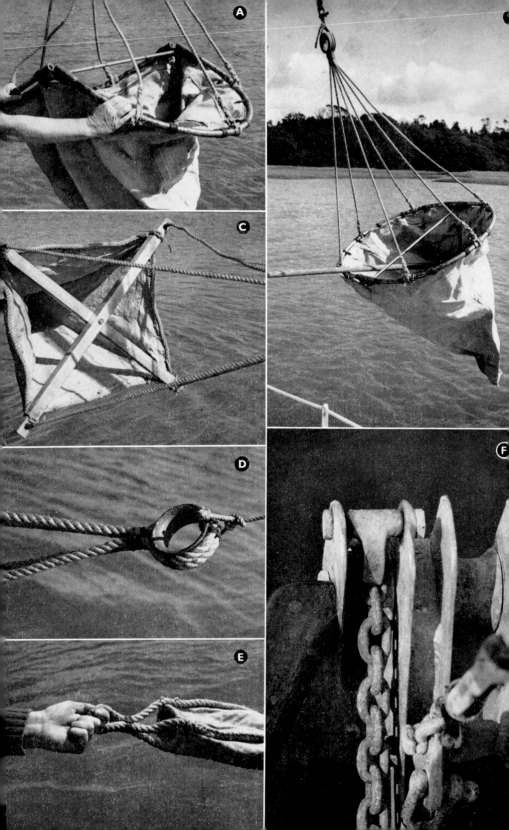

the yacht's keel held in the still water beneath, could throw her on to her beam-ends or even roll her right over. I believe this is what happened to Chichester's *Gipsy Moth IV* in the Tasman Sea, and that she did not lie a-hull again during the voyage to Plymouth. I am convinced that the only safe procedure in such weather is to get the yacht end on to the seas and keep her so, thus presenting the smallest target to the broken water.

In the past many abortive attempts have been made to get yachts to lie head to wind and sea by streaming sea-anchors from their bows— I have made such an attempt myself—and even today, when there has been some clearer thinking on this subject, that is still sometimes thought to be the best means of survival in a great gale.

This is largely due to the writings of Voss, who used a sea-anchor successfully in that manner during his long voyages in the canoe *Tilikum*; but she, like a ship's lifeboat, with which a sea-anchor is equally effective, had about the same windage and the same draught forward as aft, and could therefore be made to lie bows on to wind and sea with the help of a small riding sail set aft. But a normal yacht, drawing more water aft than she does forward, and having greater windage forward than she has aft, will not lie like that. No matter how large the sea-anchor, she is bound to make sternway; her bow, having less grip than her stern on the still, deep water, is more affected by the wind, breaking crests and surface drift, so that it falls off to leeward; the hull pivots on its heel, and eventually takes up a position more or less beam on to wind and sea (Fig. 48A), just as it will when lying a-hull. If a riding sail is set aft and sheeted flat, the position may be improved (Fig. 48B), but even then the yacht will not lie head to wind, though she may come up occasionally and fall off on the other tack, the sail flogging dreadfully at times, and the strain on the rudder caused by sternway being great. By attempting to make the yacht lie bows-on one is fighting against her natural inclinations and against the elements, and achieving no good result. It is therefore much better to work with her in her natural desire to head down wind, and to stream the sea-anchor—if it must be streamed (see below)—from the stern (Fig. 48C).

PLATE 32

Top: For his beautiful 70-footer *Varua*, here lying at Papeete, her Tahitian home port, W. A. Robinson modernized the brigantine rig so that his ship could be handled by two men only, if necessary. She is of composite construction, and was planned jointly by her owner and the famous American designer Starling Burgess. *Bottom*: The 46-foot ketch *Sandefjord* is typical of the beamy, short-ended Norwegian ex-lifeboats, which are often thought to make perfect ocean cruisers. While in the ownership of Erling Tambs, she was driven too hard and capsized, losing one man and her mizzen mast, but righted herself and reached port. She later made a circumnavigation in the hands of the Cullen brothers.

N

Like that she will lie with her stern or quarter to the sea and present a smaller target; as she is moving ahead there will be no abnormal strain on the rudder, and within limits she should answer the helm, which may be of great importance if there is some danger to leeward.

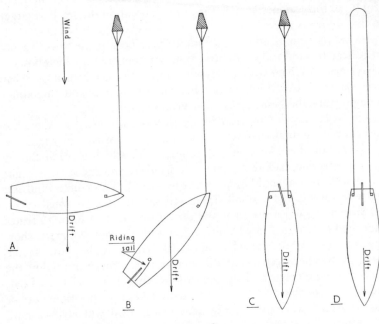

FIG. 48

A: With a sea-anchor streamed from the bow, a normal type of yacht will lie beam on to wind and sea. *B*: A riding sail set aft will improve her position a little, but will not hold her head to wind. *C*: With the sea-anchor streamed from the stern, she will lie stern on to wind and sea and will answer to her helm. *D*: The bight of a large rope streamed over the stern, with its ends made fast on each quarter, will act as a drag and will tend to quieten the sea.

On one occasion when near the Tonga Islands in the South Pacific, *Wanderer III* lay for 4 days to a sea-anchor streamed from the stern; her helm was lashed down 10 degrees so that she steered a little across the wind and so went clear of an active volcano and a reef on to which she would otherwise probably have drifted, for the gale was so strong and the sea so rough that I feared if she ran with it on the quarter under bare poles and without some form of drag she might ship dangerously heavy water. Also no helmsman was needed, and this is an important consideration in such conditions when mental and physical fatigue are most likely to occur, and when a wind-vane steering gear may fail to function properly. In those 4 days riding to the sea-anchor, *Wanderer* drifted 120 miles at an average speed of 1¼ knots.

If there is plenty of sea-room to leeward, particularly if the desired course lies in that direction, and there is a good helmsman available, undoubtedly the best action is to run before the gale, reducing sail as required, but keeping a small headsail set and sheeted flat. If wind and sea continue to increase, the skipper may decide to stream a sea-anchor over the stern. Many people, however, now maintain that this will hold the yacht too firmly against the overtaking seas, preventing her from giving way to them, and that she may suffer damage as a result. There are indeed several instances of yachts riding to sea-anchors being severely damaged in great gales, as happened to *Vertue XXXV* during her famous east-to-west crossing of the North Atlantic. She, in a hurricane-force wind created by an intense depression near Bermuda, had a 21-inch diameter sea-anchor streamed on a 20-fathom warp from the starboard quarter. The helm was lashed 20 degrees up, and there was nobody on deck when she was boarded by a great sea which smashed her doghouse windows, split one of the coachroof coamings for almost its full length, and partly flooded her below. From then on, with a helmsman keeping her exactly stern on to the seas, and still with the sea-anchor working until, some hours later, its warp parted, she suffered no further damage. It may well be that if she had been steered in a like manner at the time when the freak sea struck her, and had been kept stern on to it, she would have escaped damage, for it was probably the fact that the sea struck her on the quarter that caused the damage, and not, as has been suggested— though not by her skipper Humphrey Barton—that the sea-anchor held her too firmly.

However, it is only with a danger close to leeward that I would now seriously consider the use of a sea-anchor, and I question whether this piece of equipment is worth carrying for that single purpose. But those who consider that it is may find the following remarks of some use.

The commonest type of sea-anchor consists of a conical canvas bag, with its mouth held open by a metal hoop, which is usually hinged for convenience of stowage (Plate 31A and B). Ropes from the hoop, which are usually also sewn to the canvas, are brought together round a thimble to form a bridle (Plate 31D), and the riding warp, which should have an eye-splice fitted with a large thimble at each end, is shackled to that thimble; but it is a good plan to insert a swivel at that point as the anchor may tend to spin while it is being streamed. The metal hoop will probably provide sufficient weight to keep the

anchor submerged, and a small hole at the apex (Plate 31E), by allowing a small escape of water, will tend to steady it. Another type
(Plate 31C) consists of a pyramid-shaped bag, its square mouth held
open by wooden crossbars. This will stow in a slightly smaller space
than the other kind, but there is some risk of the four parts of the bridle
being chafed where they pass through the holes at the ends of the
crossbars; a piece of lead should be fastened to one of the crossbars to
prevent the anchor floating too near the surface. Neither tripping line
nor buoy is necessary, and such complications, if fitted, will almost
certainly get in a snarl and be a cause of chafe. A third type of sea-
anchor is favoured by some New Zealand yachtsmen, but of this I
have had a bad report from John Letcher, who tried it with his 20-foot
Island Girl. (I think it was not weighted sufficiently.) This consists of
two planks bolted to 90-degree angle-irons, with a two-inch gap between their inner edges. The angle-irons project a little beyond the
outer edges of the planks, and four metal rods to form the bridle are
bolted to them. Such a sea-anchor is immensely strong, is proof against
chafe, and when dismantled will stow compactly.

The strain on a sea-anchor and its warp is great; the whole affair
cannot be too strongly made, and every care must be taken to prevent
chafe of the warp. Parcelling at the fairlead does not last long, and will
probably need to be renewed every two or three hours; so it may be
better to shackle the inboard end of the warp to a short length of chain,
or to use the end of the anchor chain for the purpose.

But I consider it is preferable to slow a yacht down by towing ropes
rather than a sea-anchor, for her speed can be regulated by their length
and number, and the bight of a large rope streamed over the stern
with its ends made fast on each quarter (Fig. 48D) may have a smooth-
ing effect on the sea, though some people, including David Lewis in
Cardinal Vertue, did not find that so. Lewis reported that with the
wind at 43 knots and when he had no sail set, he streamed a 20-fathom
warp in a bight astern, and attempted to steer before the wind. The
yacht became unmanageable; she persistently ran across the seas, and
repeatedly broached-to. The warp hindered steering, and each large
sea swept it alongside.

PLATE 33

Top: The forerunner of a great gale. This huge swell in the South Pacific was quickly whipped
into a heavy sea when the wind increased to 60 knots, and during the 4 days that the yacht lay
by the stern to a sea-anchor she drifted 120 miles. *Bottom*: With startling whiteness beneath the
low grey sky, a crest breaks harmlessly astern while the yacht lies hove-to near the Azores in a
gale of force 8.

However, the 31-foot sloop *Samuel Pepys* encountered the same storm as damaged *Vertue XXXV*. Under bare poles she ran too fast, so 40 fathoms of 3½-inch nylon rope was towed in a bight from the stern, and two separate 3-inch hemp ropes of 20 fathoms each were also streamed. This checked her speed to an estimated 3 knots, and cut a smoother track through the seas which gave her protection so long as she was kept exactly stern on, but an attempt to move out of the path of the storm by bringing the wind on the quarter at once resulted in heavy water coming aboard. The seas were estimated to exceed 30 feet in height with a length of 400 feet. The yacht was steered with little physical effort but with considerable nervous strain throughout the storm, and she suffered no damage.

Robinson in *Varua* (Plate 32, *top*), when he encountered the ultimate storm of a lifetime in 47° S., 103° W., also ran before it towing ropes. It took a 75-fathom 6¼-inch rope towed in a bight, plus four 12-fathom ropes of the same size, each dragging a large eye-splice, and about 100 fathoms of assorted ropes of smaller sizes, to slow her down from the 6–7 knots she was doing under bare poles without drags to what Robinson considered a safe 3 knots. One of the advantages of running before a gale at slow speed is that oil can be used effectively if required; Robinson used oil from bags each side of the ship and from the forward heads, but how much good it did he found it difficult to decide. This experience certainly shows that running before a gale with drags out aft can be just as effective with a large as a small vessel.

The first *Tzu Hang* incident (page 43) occurred when the yacht was running at an estimated speed of 4 knots before a very heavy sea in the South Pacific, a little to the east and south of the position where *Varua* encountered her great storm, but not during the same year. She was under bare poles at the time with one rope towing from the stern, and was being steered dead before the seas. The sea that caused the damage was phenomenally high and steep, but was not breaking or showing signs of being about to break. *Tzu Hang* was upended on her bows until she reached a position a little beyond the vertical, and was then rolled over sideways by her ballast keel. It seems to me that she stumbled, that is, her bow was held in still water while the momentum of her forward movement, coupled with the surface drift, continued to carry her stern on and over the vertical position, just as a man would stumble if, when running, his feet caught in something immovable. It seems reasonable to wonder, therefore, whether the disaster might have been averted if the yacht's speed had been reduced further by putting out more drags; for after her masts had been swept away, thus reducing her windage so that she was moving only slowly, she was not again upended or further damaged, although the gale continued; but probably the sea was a freak from which no small vessel could have escaped without damage, for no others like it were seen before or after. John Guzzwell, who was aboard at the time, suggests that the phenomenally high and steep sea may have been caused by a shoal, and Robinson firmly believes that *Varua* did pass over a shoal during her ultimate storm. Although no shoals are shown on the charts of the area, one must accept the fact that they could well exist between the widely scattered soundings. *Tzu Hang*'s second incident occurred after she had been refitted in Chile and had reached latitude 48° S., longitude 82° W. on her second attempt to round the Horn. She was lying a-hull at the time in a full gale during which some of the seas had a great weight of broken water on their advancing faces. One of these caught *Tzu Hang* broadside on and rolled her completely over, dismasting her.

The seas in the forties and fifties of south latitude, driven unimpeded by the great westerly gales with a fetch of thousands of miles, are probably the heaviest in the world, and only a very brave man or woman—Beryl Smeeton was the driving force which launched *Tzu Hang*'s second attempt on the Horn—with the springs of adventure strong in him would venture to take a small vessel there. It is good to know that in December 1968 the Smeetons finally achieved their ambition: they succeeded in rounding the Horn—the hard way from east to west—and sailed from 50° S. to 50° S. in the remarkably fast time of 13 days.

In 1966 something happened to shake the long-accepted theory of small craft management in heavy weather. The Frenchman Bernard Moitessier and his wife sailed their 40-foot ketch *Joshua* non-stop from Moorea, Society Islands, to Alicante, Spain, a distance of 14,216 miles, in 126 days. Moitessier had made a study of the behaviour of small craft in heavy weather, and when he was overtaken by a great gale between latitudes 43° and 45° S., ran before it with 5 warps weighted with iron pigs towing astern. These apparently had no effect on the yacht's speed, but they prevented her answering her helm correctly—by then the vane gear had been disconnected and steering was being done by hand. One great breaker caught *Joshua* on the quarter, instead of dead aft as Moitessier had intended, and this gave him the idea for the unusual technique which he then adopted. In a letter to a friend he wrote:

> . . . it was this alone which saved the masts, otherwise the boat would have plunged forward and downward where the undertow was fiercest. In a flash I grasped the technique of Vito Dumas [the Argentinian who had sailed his 31-foot ketch *Lehg II* single-handed south of the three stormy capes]: to keep moving at about 5 knots, running full or nearly full [dead before it?] with as little canvas as possible [a small storm staysail], and just before the arrival of each roller to give a touch of helm and luff to take the roller at an angle of 15–20°. This trick made the boat skid, so that, as she planed, she heeled on to her beam. Thus she could not pitch-pole nor be thrown on to her beam-ends, as the sea was taken somewhat on the quarter. Directly I realised this—and none too soon—I cut away my hawsers and *Joshua* was no longer in danger.

A year after this event I took part in a *Yachting World* forum, in which four of us, including my old friend Adlard Coles—perhaps the most widely experienced British ocean-racing man of the day, and whose latest book, *Heavy Weather Sailing*, had recently been published—discussed yacht management in heavy weather. Naturally the Moitessier method cropped up, and we all found it rather startling. Coles said he had never dared to try it, but he repeated what he had already written: in his opinion the best method of survival a small yacht has in storm conditions is to run before it under as much sail as she can stand up to, even though she may be laid flat on the water at times; but he added that he shrank from recommending this because if it proved wrong it could lead to loss of life.

It is difficult in the face of so much conflicting evidence and advice to lay down any rules, especially as so much depends on the type of yacht and the conditions in which she may find herself; but I would suggest that if she is caught in a gale which does not exceed force 9— anything beyond that creates conditions over which one can have little real control—she should adopt the well-tried expedient of running before it with warps streamed astern to reduce her speed.

At anchor

The natural desire of a skipper at the end of a passage, particularly if it has been a long, difficult, or rough one, is to reach a comfortable and secure anchorage so that for a time he may relax his vigilance and rest with an easy mind. But he should exercise some caution when entering foreign commercial or naval ports by night, for although he may be using up-to-date charts and sailing directions, it is possible that changes have been made either to the buoyage or the harbour works, but of which no notice has yet reached the hydrographic offices of other countries. On my only visit to the Cape Verde Islands I nearly got wrecked on an unmarked, submerged breakwater, which was at the time under construction at Porto Grand, but of which neither chart nor *Pilot* gave any hint. Navies, including our own, are in the habit of laying moorings with enormous unlighted buoys in waters through which a yacht may wish to pass, and because most towns are today so brilliantly lighted, it may be impossible to see these dangerous obstructions or even the navigational lights against the blinding glare. There is also the risk of some illuminated advertisement sign ashore being mistaken for a navigational light. E.g., Port Elizabeth in South Africa used to have, and for all I know may still have, a red sign in the town with the same characteristic as, and a greater range than, the breakwater light; this confused me for a time. Local pilots of course know about such things, but the stranger cannot be expected to know, yet he scarcely wants the bother and expense of engaging a pilot. It is for such reasons as these that I prefer to find my way in the dark into some natural and unlighted anchorage rather than risk the dangerous trap of a man-made or man-improved one.

The choice of an anchorage is an important matter, and for this the chart or the *Pilot*, or the account of some trustworthy cruising man who has been there, will be the surest guide. It would be unwise to ignore local advice which might warn one of some unsuspected danger, such as poor holding ground, or the risk of theft, or of damage from badly handled local craft; but an anchorage recommended contrary to the advice given in the *Pilot*, or by one's own common sense, should be avoided. This may appear obvious, but the temptation to act on it may be great if it offers certain advantages, such as cleanliness, access to a good landing place, or conveniently shallow water.

The schooner *Lang Syne*, having sailed without mishap across the Pacific, through the labyrinth of the Great Barrier Reef, among the Indonesian Islands and across the Indian Ocean, was wrecked at Zanzibar because for once her competent owners, Bill and Phyllis Crowe, acted on local advice when choosing an anchorage. Before following this advice they questioned the lack of protection from an onshore wind, but were assured that there was no need to worry as there would be no such thing at that time of year; it was not until later that they read the sailing directions and learnt that although the southwest monsoon, bringing heavy rains and winds, is said to set in about March, the seasons are so uncertain and subject to such variation that no reliance can be placed on them. One evening a few days after *Lang Syne*'s arrival, the wind did come suddenly onshore, the engine failed to start, and before sail could be made she was driven on to the sand and coral beach and suffered severe damage. She was refloated by her owners unaided, and they displayed fine seamanship and perseverance at a time when a tug which was sent to their assistance considered it too dangerous to help; they repaired the schooner themselves and completed their circumnavigation.

But in some places the *only* anchorages are exposed and deep. The lagoons of many of the Pacific islands, for example, are large enough to give a fetch of several miles, so that a strong wind can raise a considerable sea, and they are often between 10 and 20 fathoms deep, while the anchorages at some other islands which have no lagoons are frequently much deeper. It is therefore imperative for the visiting yacht to have adequate ground tackle, and for her skipper's peace of mind it is desirable that this should be a little heavier than is normally specified for the use of yachts in home waters.

Wanderer III carried only two anchors during her voyages; each was of the plough type (genuine C.Q.R.) and weighed 35 lb. She had 45 fathoms of $\frac{5}{16}$-inch tested, short-link chain; 30 fathoms of this was shackled to the port anchor; the remaining 15 fathoms, a separate length, was held in reserve either for increasing the scope on the port anchor when the anchorage was a deep one, or for use with the starboard anchor. There should have been a third anchor as insurance against loss of one of the others. She had no windlass, but a strong chain pawl was fitted over the chain roller at the stemhead (Plate 31F) to hold each link as the chain was hove in, and no serious trouble was ever experienced in weighing, though in deep water the work was heavy. She brought up in a great many anchorages and on a variety of bottoms, but only on a few occasions did she drag. Once was in English Harbour, Antigua, where inadvertently the anchor was let go on mud so soft as to be almost liquid; another was in the roadstead under the lee of Ascension Island, where the anchor was at first dropped on a patch of volcanic cinders.

The nature of the bottom in recognized anchorages is usually indicated on the chart (for the abbreviations used see Admiralty chart No. 5011 and U.S. chart No. 1) or mentioned in the *Pilot*. The following should be avoided if possible: cinders, coral, ooze, rock, shingle, stones, and weed. Any note on the chart or in the *Pilot*, to the effect that the holding ground is poor, should be taken seriously, even though the sample obtained with one's own armed lead, or the appearance of the bottom seen through clear water, may appear to be all right; sand, for example, is sometimes only a thin covering to smooth rock, and may afford no hold for the anchor. Weed is particularly dangerous with the plough anchor, especially if the anchor is allowed to drag at all, as it may then become so clogged with weed as to be incapable of biting into the ground. With this type of anchor it is important to veer plenty of chain at once on letting go, so that the moment the strain comes on the anchor it will be in a horizontal direction; the anchor will then turn over and start to dig in instantly.

Beyond's bower anchor was of the fisherman type and her second anchor of the plough type. During the 2 days that she lay in Suvorov lagoon in the indifferent shelter of Anchorage Island, the wind blew hard from the south. She was lying in a depth of 5 fathoms to her bower anchor with 25 fathoms of chain, and the second anchor with 4 fathoms of chain and 20 fathoms of nylon rope. The hinge pin of the plough anchor broke (this is not a common failing, but *Moonraker* had a similar trouble with hers during her West Indies cruise) and the yacht snubbed badly on her bower. Eventually she lay to the bower and third anchor, using a 3-fathom nylon spring. Many dives were needed to untangle the chains from the coral heads before she could leave.

The usual method of preventing snubbing is to slide a weight part way down the anchor chain to increase the catenary and act as a spring; but in shallow water I have found a nylon spring to be more effective.

Anchoring with a rope instead of a chain is dangerous, and except in a very small yacht, where the weight of chain might be too great, has little to recommend it; but if rope must be used it should be of nylon as that material has great elasticity and strength. Chain is easier to handle, is self-stowing, occupies much less space than is required by a similar length of rope, needs less scope, and of course it is impervious to chafe. There are many instances of ropes parting and of yachts anchored with them dragging. On their world voyage in *Viking*, the Holmdahls had to anchor twice during contrary tides when approaching Darwin on the north coast of Australia. On each occasion the rope parted and they lost both their anchors. At the conclusion of her circumnavigation the yacht was bought by Joe Pachernegg, who renamed her *Sunrise*, and started off on a single-handed circumnavigation. He anchored her with a rope off the north-west coast of Santa Cruz, one of the Galapagos Islands, where she blew ashore in the night and was a total loss. Her owner endured great hardship during his 4-day walk across the island to the settlement at Academy

Bay, drinking the blood of goats he managed to shoot, as he was unable to find any water.

Nothing could be more reassuring during a rough night in a coral-fringed anchorage than to know that the vessel's anchors and chains are heavy enough and strong enough to hold her securely, and it is therefore all the more surprising that yachtsmen so often prefer to pick up a mooring and lie to it without having any knowledge of its composition or condition. The temptation is, of course, considerable when the water is deep or the bottom believed to be foul, and it must have been one of these reasons which induced that experienced voyager T. H. Carr to secure *Havfruen III* to the seaplane mooring at Bora-Bora, Society Islands. During a gust of wind the mooring parted, but fortunately he and his wife were aboard at the time and were able to let go an anchor before the ship blew ashore.

9

NAVIGATION

*Course and distance—The compass—The sextant—Taking sights—
Star recognition—Time—Astronomical navigation—Working sights—
Making a landfall—Pilotage among coral*

THERE is no mystery or any very great difficulty connected with
the business of navigating a vessel from one place to another
across a wide stretch of sea. Setting the course, keeping the
dead reckoning up to date, and fixing the position by observations
of the celestial bodies, call for nothing more than simple arithmetic,
a little geometry, and some dexterity in handling the sextant. There
are many ways of working sights, but only the simplest and most
practical methods are described in this chapter, because the navigator
of the small ocean-going yacht is usually also a watchkeeper, and therefore needs to reduce to the minimum the time spent at the chart table
without sacrificing reasonable accuracy. It is probable that if he has
stood a night watch he will not be quite so alert as usual, and the chance
of making an error then will be reduced if his calculations are short
and simple and the navigation tables easy to use; in this connection
the methods devised for aircraft navigation have much to recommend them.

Course and distance

The difficulty of drawing a chart of part of the world's curved surface on a flat sheet of paper has been overcome by the system known
as Mercator projection, after the Dutch mathematician Gerard Mercator who invented it in 1580. On such a chart the meridians (lines
joining the north and south poles) are laid down as parallel straight
lines instead of converging on the poles, as they do on the earth, so
that a degree of longitude (angular measurement east or west from the

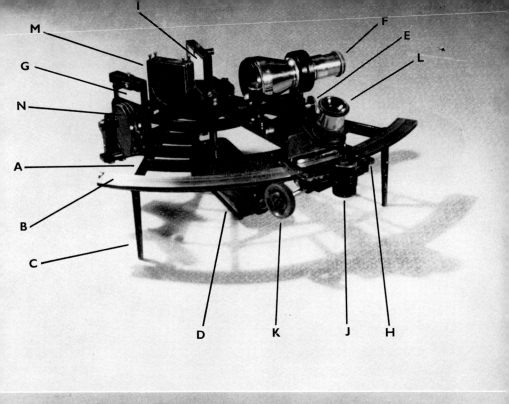

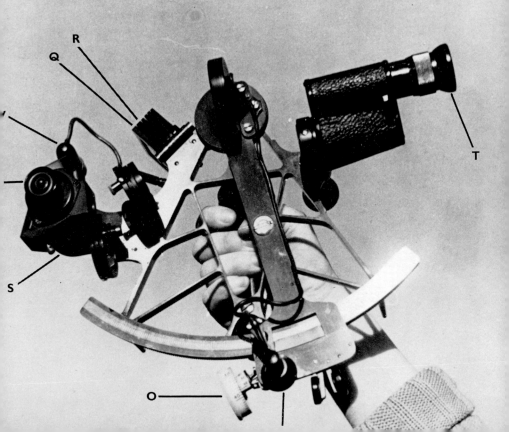

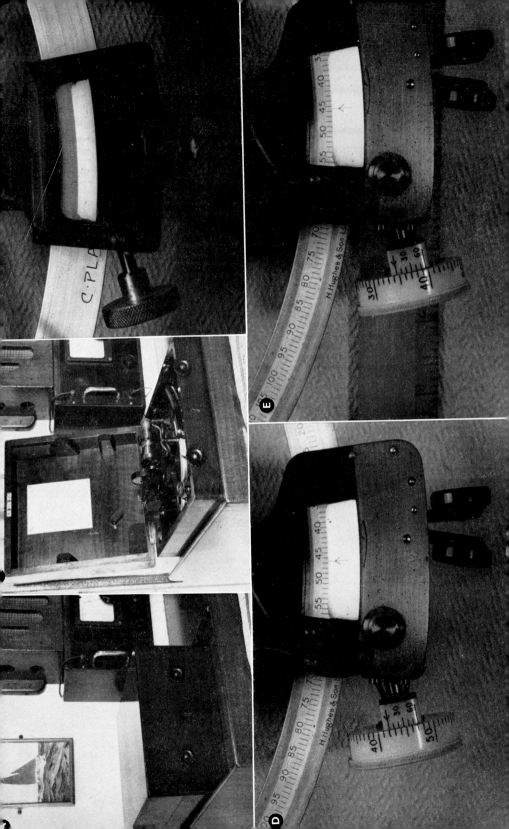

meridian of Greenwich) is kept the same size over the whole chart. This means that the chart is increasingly distorted as the pole is approached, and in order to preserve the correct proportion of length to breadth in every part of it, the parallels of latitude (lines drawn round the earth parallel to the equator) are laid down at increasing distances apart as they get nearer to the pole (Fig. 49). Latitude is the angular measurement north or south of the equator, and one minute ('), which is the 60th part of a degree (°), measured on the earth's surface, is one nautical mile of 6,080 feet. So the minutes of latitude scale on the east and west margins of the chart can be used as a scale of nautical miles; but when using it for that purpose on a chart which covers a large area, the scale in the same latitude as the distance to be measured must be used, or there will be an error due to the lengthening of the minutes of latitude as the polar edge of the chart is approached. On large-scale charts, however, the difference in length of the minutes of latitude near the north and south extremes of the chart is so slight as to be of no practical importance. The longitude scale marked on the north and south margins of the chart must not be used as a scale of distance, because nowhere, except on the equator, does one minute of longitude equal one nautical mile.

All circles which bound the maximum circumference of the earth, that is, circles which have their centres at the centre of the earth, are known as great circles. All meridians are great circles, but of the parallels of latitude the equator is the only great circle. If a flat surface is laid to touch a globe at a certain point, all great circles projected on to it from the centre of the globe will appear as straight lines, and a chart constructed on that principle is called a gnomonic chart (Fig. 50); here the meridians converge towards the pole, and all parallels of latitude, except the equator, appear as curves.

Any course represented by a straight line on a Mercator chart, except one which runs due north or south, is known as a rhumb line, and a vessel adhering to that course will reach her destination, but not by the shortest route. A rhumb line is, however, near enough for distances up to 300 or 400 miles, or in low latitudes, i.e. near the equator. The shortest distance between two points on the earth's surface is the arc of the great circle which passes through them, and it is in determining this course that the gnomonic chart is used.

Suppose, for example, you wish to determine the great-circle course from Bermuda to the Isles of Scilly. On the gnomonic chart of the North Atlantic rule a straight line from Bermuda to Scilly, as has been done in Fig. 49. Note the latitude in which this line crosses

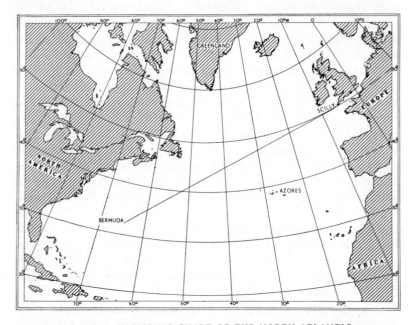

FIG. 49. A GNOMONIC CHART OF THE NORTH ATLANTIC
on which Bermuda and Scilly have been joined by a straight line for the purpose of ascertaining the great-circle course between them. (Based on Admiralty chart No. 5095 with the permission of the Controller of H.M. Stationery Office and the Hydrographer of the Navy.)

each of the six 10° meridians on the chart. (To enable this to be done easily, all the latitude and longitude lines on the chart, and there are many more of them than are shown in the figure, are marked at 10′ intervals; but for the purpose of reproduction here these marks have had to be omitted.) Transfer these positions to the Mercator chart of the North Atlantic, and a curved line swept through them will be the great-circle course. But to follow that line exactly is impossible because it would involve a slight but continuous change of course. Therefore each position is joined to the next by a straight line, as has been done in Fig. 50, and the course is altered at each position, i.e. at each 10° meridian, to keep to those lines.

It does not often happen that a sailing vessel can keep to a great-circle course, but in high-latitude sailing she should keep as closely as possible to it, weather and other circumstances permitting, because the higher the latitude the greater will be the saving in distance. But the chief use of the great-circle course is to show which is the most favourable tack in turning to windward, and in the absence of any

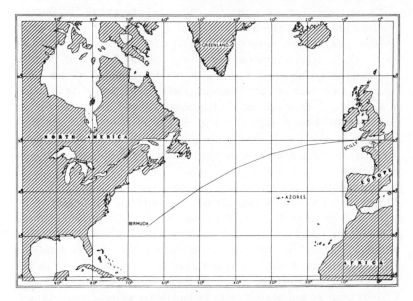

FIG. 50. A MERCATOR CHART OF THE NORTH ATLANTIC
with the great circle course from Bermuda to Scilly drawn in. (Based on Admiralty chart No. 2059 with the permission of the Controller of H.M. Stationery Office and the Hydrographer of the Navy.)

special reason to the contrary, the vessel should be put on the tack which brings her nearer to the great-circle course.

By means of the traverse table, it is possible to find the course and distance between any two points on the surface of the earth; or, given the course and distance made good from a known position, to find the new position. But that table is inconvenient to use, and I would advise the small yacht navigator to avoid it and take his courses and distances from and to mark his positions on, the chart. For that purpose he will need to have an ocean chart such that his point of departure and destination, or turning point, are both on one sheet. He will mark his noon positions on this chart and thus be able to see the situation of his vessel in relation to her surroundings at any time. But in order that a whole ocean, or a large part of it, may be included on one sheet, ocean charts have to be drawn to a very small scale, perhaps half an inch, or even less, to a degree of latitude; a pencil line may then be a mile or more in thickness, and obviously it will not be possible to do any accurate plotting or drawing on such a small-scale chart, and distances measured on it will be only approximate.

To overcome this difficulty I recommend the use of the Baker's position line chart, which is published by Imray, Laurie, Norie & Wilson, and can be obtained from that firm or through any chart agent. The chart comprises a sheet of stout paper 26 in. by 18 in. with a scale of longitude marked along its north and south margins. On folding flaps at each side are latitude scales in proportion with the scale of longitude, from 0° to 40° on the west side and 40° to 60° on the east. Such a chart is shown in Fig. 63 (pages 234-5), but for convenience in reproduction the latitude scales have been detached from the chart and placed on the facing page. An explanation of the drawing shown on this chart will be given later.

The longitude on the Baker's chart remains constant, as on a Mercator chart, but all measurements of latitude (or distance) are taken from the latitude scale, using the scale marked with the latitude in which the vessel lies, or in which the work has to be done. In other words, by using the appropriate latitude scale, one converts the Baker's chart into a blank Mercator chart for any desired latitude between 0° and 60° north or south. By using the large circle marked in degrees, courses and bearings may be laid down with the parallel rules. If a 'B' pencil is used for drawing, the work can be rubbed out easily and the chart used over and over again; indeed, when only a small-scale ocean chart is available, the Baker's chart will be in daily use, and because the work is done diagrammatically, instead of in figures (as with the traverse table), any serious error should at once be obvious.

As magnetic variation (the difference between true north and magnetic north) varies over the earth's surface, a chart covering a large area could not usefully be provided with magnetic compass roses, such as are usually printed on coastal charts, for each rose would only serve one small part of the chart. Ocean charts are therefore provided with isogonic lines, that is, lines drawn through points of equal variation. The amount of variation for the year stated on the chart is printed against each line together with the amount of the annual change in variation. The navigator can therefore convert true courses and bearings into magnetic, and vice versa, by applying the variation for that part of the ocean in which his vessel is sailing, and will make the necessary alteration each time she crosses an isogonic line. Isogonic lines are also drawn in orange on the American pilot charts (see endpapers).

The compass

As the only compass roses printed on ocean charts are true and are marked in notation 'A' (o° clockwise to 360°), and as frequent conversions of courses and bearings from magnetic to true and true to magnetic will have to be made, some trouble will be saved if the vessel's compass is marked in notation 'A', rather than in notation 'B' (from north and south to 90° at east and west, as in many old compasses) or in points and quarter points.

Many agree that steering is easier with the grid type than with the normal type of compass, and this is important when long periods have to be spent at the helm, especially at night. In such an instrument the north–south diameter of the card is boldly marked with a line, wire, or some other straight device. On top of the bowl two parallel wires are fixed to a verge ring; this is engraved with degrees and can be turned round on the bowl so that the course to be steered can be made to correspond with the lubber line. To steer that course, the helmsman has only to turn the ship until the north–south line on the card is parallel to the wires on top of the bowl, and then keep it so. Some compasses have the grid in the form of a T. The human eye is good at judging parallelism, and with grid steering a more accurate course can thus be steered than would be possible with the ordinary type of compass, where a small degree mark on the card has to be kept against the lubber line. Also, as parallax (the apparent displacement of the lubber line, which with ordinary compasses increases as the helmsman's eye moves farther off the centre line) need not be considered with grid steering, this type of compass can be placed in any position within the helmsman's view. Another advantage is that the north–south line and the wires on top of the bowl can be made luminous, so that no electricity is needed for night steering. Luminous paint tends to grow dim when submerged for long in compass fluid, but this may be overcome by having the north–south line made from a slender glass tube which is filled with luminous powder, a similar smaller tube making a cross at the north-seeking end to reduce the risk of a reciprocal course being steered. Such a compass, made for me by Henry Browne & Son of Leadenhall Street, London, mounted beneath a hinged cover of unbreakable glass in the bridge deck, is shown in Plate 37, *bottom*. The bowl is painted black to enable the luminous grid to stand out more

o

clearly, especially during twilight, and it will be noticed that the card is marked in points and quarter points; this is a personal preference, but the verge ring is, of course, marked in degrees, 0–360.

Deviation of the compass was discussed in *Cruising Under Sail*, but since the second edition of that book was published some additional causes of deviation have come to my notice. Not all stainless steels are non-magnetic; it is common for a horseshoe-shape lifebuoy to be placed close to the steering compass, sometimes even embracing it, but if this is of the light plastic type it may have been formed on a steel wire core; the steel cans in which beer and other beverages are so often sold are so thin that one might suppose they could have little effect on the compass, but often they will have been handled in the brewery or factory by magnetic means and are therefore highly magnetic; the pointer type of echo-sounder is also magnetic, as is its repeater, and this should be at least 18 inches from the compass; I am told that fluorescent lights cause deviation, but have not checked this; compass adjusters tell me that the alternator, which is now so widely used for battery-charging, also creates magnetic problems. One or more of the above have in recent years been the cause of mysterious deviation difficult to trace and sometimes impossible to correct.

The sextant

The sextant is an instrument used by seamen for measuring vertical or horizontal angles with great accuracy. It is most commonly used for taking an observation, or sight, i.e. measuring the vertical angle between the visible horizon and a celestial body—sun, moon, planet or star—and this is the use that should be kept in mind when reading the following description.

As may be seen in Plate 35, *top*, the sextant consists of a rigid frame (*A*) in the shape of a segment of a circle, with divisions engraved along its arc (*B*). On the under side of the frame are three legs (*C*), on which the sextant stands when not in use, together with the handle (*D*) by which it is held vertically when an observation is being taken. At one side of the frame is the rising piece (*E*), which can be raised or lowered within small limits; attached to the rising piece is the collar into which can be screwed one or other of the various telescopes (*F*) with which the sextant is provided. At the opposite edge of the frame is the horizon glass (*G*), which is half clear glass and half mirror. Pivoted at the apex of the frame is the index bar (*H*); this has a mirror, known as the index glass (*I*), fixed to it immediately over the pivot; movement of the index bar therefore pivots the index mirror about its own axis. At the other end of the index bar, where it swings along the arc, provision is made for reading the angle that one mirror makes with the other. At its extremity the index bar has a quick release clamp (J) so that it may be held firmly at any point along the arc; a worm, actuated by a milled wheel, engages with teeth on the under side of the arc, and is used for making the delicate final adjustment when taking a sight; this is known as the tangent screw (K).

The optical law on which the construction of the sextant is based is this: if a ray of light from an object is reflected twice at the surface of two plane mirrors, and the final reflection is brought into line with some other object, the angle between the planes of the mirrors is half the angle between the objects observed. If, therefore, we look directly at the horizon through the clear half of the horizon glass, and move the index bar until the image of a star is reflected from the index glass into the silvered half of the horizon glass so that it appears to touch the horizon, and then back to our eye at the telescope, the angle shown by the index bar on the arc would, if the latter were engraved with degrees, be half the angle between the star and the horizon. For this reason, and to avoid the need to double the angle read on the arc to obtain the true angle between star and horizon, the divisions engraved on the arc and ✓ numbered as degrees are in reality the size of half degrees.

The arcs of most sextants are engraved up to 120°, i.e. one-third of a circle, but because the degree markings are half-size, the arc is in fact only one-sixth of a circle; hence the term sextant. But many full-size sextants, especially the older ones, have their arcs engraved beyond 120°, and sometimes up to 150°; there is no practical advantage in this from the yachtsman's point of view. The arc of every sextant is engraved for 5° or 10° to the right of the zero mark. The reason for this will be explained later. Angles read there are said to be 'off the arc'.

The radius is usually 7 inches and the weight between 4 and 4½ lb.; I do not consider that a light-weight sextant has advantages, though one that much exceeds 4 lb. is tiring to hold, and unless the observation is quickly made, the observer's hand may start to shake. Neither do I consider that small size is desirable, except in a very small yacht where it may be impossible to find a convenient stowage space for a full-size instrument, the box for which will measure about 12 inches square and 5½ inches high. The box is best kept in a readily accessible position so that the sextant can be lifted quickly from it, otherwise the opportunity for taking a sight on a cloudy day when the sun shows only momentarily may be lost. The method used in *Wanderer III* was to keep it on a sideboard, Plate 36A, where it was a snug fit between the sideboard fiddle and a piece of wood screwed to the sideboard. The box had only to be pulled towards the sideboard edge, Plate 36B, its lid opened and lodged against the bulkhead, and the sextant could at once be lifted out.

All sextants work on the same principle, and all serve the same purpose, but the more expensive instruments do it with slightly greater accuracy, and possess features which make it possible to take an observation and read the angle with greater ease and speed. There are many excellent sextants to be bought cheaply on the second-hand market, and I would strongly advise a beginner to buy one of these. Later, when he has gained experience, he will be in a better position to decide what refinements he would like and which would be of little value to him, and can choose the sextant which best suits his needs and his purse. If he is as fortunate in his first choice as I was, when in 1935 I bought an excellent twelve-year-old Plath sextant for £5, he will probably treasure it for the rest of his life. This old sextant of mine, shown in Plate 35, *top*, was in almost constant daily use during *Wanderer III*'s first circumnavigation, and I had no fault to find with it except that it was not so easy to read as is a sextant fitted with a micrometer.

It is not practicable to engrave the arc with minutes of arc, for they would be impossibly close together. Yet the sextant can be read to the nearest minute, and with a degree of accuracy even greater than that, if desired.

To enable this to be done before the advent of the micrometer, a vernier was used. This is a small-scale or auxiliary arc attached to the lower part of the index bar and sliding in contact with the arc. It contains one more division than the number of divisions on a similar length of the arc, and it makes use of the fact that the human eye is able to judge with great accuracy whether a line is continuous or whether it has a break in it. Plate 36C shows the position of the vernier. The engravings on arc and vernier are too fine to be seen in the reproduction, but a simplified vernier, together with a part of the arc adjoining it, is shown in Fig. 51. From this it will be seen that the arc is engraved in degrees, every fifth degree being numbered. Each degree is divided into six equal parts with shorter lines, each of these parts therefore equals 10 minutes. To make the counting of these small divisions easier, the 30′ mark is a little longer than the 10′, 20′, 40′, and 50′ marks. To read the angle measured by the sextant, swing the microscope (L in Plate 35), which is pivoted on the index bar, over the vernier, and note where the zero mark of the vernier (often engraved with an arrow head) cuts the arc. In Fig. 51 this is to the left of the 31° mark, and between it and the first of the small 10′ marks; so the sextant reads 31° + something less than 10′. Now swing the microscope slowly to the left and note which of the numbered lines on the vernier is exactly coincident with any line on the arc. In the figure this is the line numbered 5. The vernier therefore reads 5′, and this added to the 31° read on the arc makes a total reading of 31° 5′. If, however, you look at the vernier on a sextant, you will find that each of the 1′ divisions is divided into six equal parts, enabling the sextant to be read to one-sixth of a minute, i.e. to 10 seconds (″) of arc; for the sake of clarity these 10″ divisions have not been included

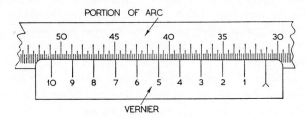

FIG. 51. A SIMPLIFIED SEXTANT VERNIER

in Fig. 51, and in fact such accuracy of reading is not of great value in the ordinary course of yacht navigation, because one cannot take an observation from a small vessel, where there is always some movement, with an accuracy greater than 1′.

If the engravings on arc and vernier are clearly defined—a touch of lamp black mixed with light oil will help to make them bolder—no great difficulty should be experienced in reading the angle measured, provided the light is good. Nevertheless it is not uncommon for mistakes in reading to be made, especially if the navigator is tired, or when the reading has to be done in artificial light.

A micrometer makes reading easier and quicker; Plate 35, *bottom*, shows a modern Husun sextant fitted with this refinement and with some accessories which will be mentioned later. The ordinary tangent screw, with which the final adjustment is made when taking a sight, is replaced by a larger wheel which is engraved with 60 equal parts, the divisions exactly filling its circumference. This is known as the micrometer head (O in Plate 35). The worm it actuates, and the teeth with which this engages on the under side of the arc, are so cut that one complete revolution of the micrometer head moves the index bar exactly 1 degree. The arc of a micrometer sextant is engraved with

whole degrees only, and an arrow, known as the index, is engraved on
the index bar in place of the vernier. In Plate 36D, which is a close-up
view of the micrometer and part of the arc of the sextant shown in
Plate 35, *bottom*, it will be seen that the index points to the left of the
46° mark, so the sextant reads 46° and something more. Adjoining
the micrometer head will be seen another index arrow; this points to
mark 42; the micrometer therefore reads 42′ and the total reading of
the sextant is 46° 42′. It will also be seen that there is a small vernier
adjoining the micrometer; there the reading can be taken to the nearest
10″ if desired. For example see Plate 36E. Here the reading on the
arc is 45° and something more. The micrometer index arrow points
between the 34 and 35 marks, and as it is a little nearer to the 35 than
the 34 we can say that for all practical purposes the sextant reads
45° 35′. But if greater accuracy is required, we look along the small
vernier and note which line there is coincident with any line on the
micrometer. It can be seen in the Plate that this is the 40″ line, so the
sextant reads 45° 34′ 40″. Part of a ruler marked with inches and
tenths is shown beneath the micrometer so that some idea of the size
and clarity of the sextant engravings may be had.

A micrometer sextant can be read with the naked eye even in a poor
light, and such sextants usually have a small electric light, *P* in Plate 35,
on a swinging arm, so that when star sights are being taken the observer
can read the sextant on deck; the battery for this little light is housed
in the hollow handle of the sextant.

Without a doubt a micrometer is a great help, but whether or not
it is worth the greatly increased cost, which is usual for a sextant
possessing it, is a matter for the user to decide.

The more expensive modern sextant also possesses other refinements. The mirrors are large—though in this connection it is interesting to note that the mirrors of the fine Husun Gothic sextant
shown at the bottom of Plate 35 are only very slightly larger than
those of the 1923 model Plath shown at the top of the Plate—and are
hermetically sealed so that damp cannot enter and spoil the silvering,
which was a frequent trouble with earlier sextants. Also the telescopes
have better definition and greater light-gathering power.

Most sextants are provided with two or more telescopes. One of
these will be an erect star telescope of 2× or 4× magnification, the
other an inverting telescope, probably with an additional draw-tube
so that it may have a magnification of either 5× or 10×. In addition
there may be a prism monocular having 6× magnification.

For observations of the sun one should use the highest-power

✓ telescope that it is possible to hold steady, having regard to the motion of the vessel. Generally I have found a 4× telescope to be most suitable. But it should have a limited field of view so that it does not include anything outside the frame of the horizon glass. This is particularly important when the sun is low and the horizon bright with reflected light, for one's eye would then be dazzled by the bright light outside the shaded area of the horizon glass. The erect prism monocular is excellent for sun sights when there is not a lot of motion, and it is essential when using a bubble horizon, see page 204. For star sights high magnification is of little value; light-gathering power and a wide field of view are more important. In common with most small-craft navigators, I find the inverting telescope impossible to use, except when the sea is quite smooth, and even then it is not easy to use because, with the objects inverted, one's reactions have to be reversed.

The reflected image of the sun is too bright to be viewed by the naked eye. A set of coloured glass shades of different densities, known as the index shades (M in Plate 35) is therefore fitted on the sextant frame in such a position that whichever is needed, according to the power of the sun, can be turned up into position between the index glass and the horizon glass. Sometimes, when the sun shines from a thinly clouded sky and appears indistinct and woolly, the use of one of the index shades will make it clear-cut so that a satisfactory observation can be taken. It is important to remember that with most sextants all the index shades have to be turned up into their working position before the instrument can be put away in its box; failure to do this can cause a shade to break or its frame to be bent out of the vertical. Similar shades (N in Plate 35), known as horizon shades, can be turned ✓ up in front of the horizon glass to reduce sun glare on the sea. Usually one or two telescope shades are also provided, but as one does not often want the horizon darkened to the same extent as the sun, their use is limited.

Taking sights

Lift the sextant from its box with your left hand, grasping it not by the index bar or arc, but by the frame, and transfer it to your right hand, holding it by the handle so that the plane of the frame is vertical; all adjustments will be done with your left hand. The telescope you are likely to use most often should be kept shipped in the collar of the rising piece. It should be focused once and for all on some distant object and the position of the draw-tube marked by a sharp file-cut so that it can be focused without loss of time if, as usually happens, it has to be put out of focus to allow the sextant to fit in its box. The file-cut can be felt with the thumbnail.

If you are not familiar with the sextant, you should practise taking sights of the sun before attempting other bodies, and you will find the following method the easiest to start with. Turn all but a medium index and horizon shade out of the way, press the quick-release clamp and swing the index bar back so that the sextant reads approximately zero. Holding the sextant in the vertical plane, tilt it up and look directly at the sun through the telescope, and you will see the direct and reflected images of the sun superimposed, or nearly superimposed, on one another. Alter the index shade as is necessary to prevent glare and to make the reflected image clear and sharply defined. Now separate the direct and reflected images by moving the index bar slowly away from you, at the same time slowly lowering the sextant to follow the reflected image down until the horizon comes into view through the clear portion of the horizon glass. Release your finger pressure on the clamp, and make the final adjustment with the tangent screw, or micrometer head, until the lower rim of the sun (known as the lower limb) just touches the horizon. The reading on the sextant will be the angle between the visible horizon and the lower limb of the sun. This is known as the observed altitude, and for practical use will have to be corrected, as described on page 214, to obtain the true altitude.

Once you have thoroughly accustomed yourself to the use of the sextant, you will probably not use this method except, perhaps, for observing stars. Instead you will turn what you consider to be the most suitable index shade into position, not using an horizon shade unless the sun is low and there is much light reflected from the sea. Looking through the telescope and the clear half of the horizon glass directly at the horizon immediately beneath the sun, you will swing the index bar slowly to and fro until you pick up the reflected image of the sun near the horizon, and then proceed to make contact as before. This is the quicker and more commonly used method.

At the moment when contact is made between the reflected image and the horizon, it is essential that the sextant should be exactly vertical, or the angle measured will be too great. To ensure that the sextant is vertical, it is necessary to swing it a few degrees out of the vertical from side to side, pendulum fashion, with the index glass as centre. The image will then be seen to lift from the horizon, as shown in Fig. 52A, describing the lower part of a circle, and the final adjustment must be so made that the lower limb just kisses the horizon at the bottom of its swing. This is not so difficult as it sounds, and becomes almost automatic with practice.

You may find it easier to let the sun make its own contact with the horizon, i.e. you bring the sun down and adjust the tangent screw until the sun's lower limb is nearly, but not quite, touching the horizon—just below the horizon if the sun's altitude is increasing, and just above it if the altitude is decreasing—and then swing the sextant pendulum fashion until contact is made, and note the time.

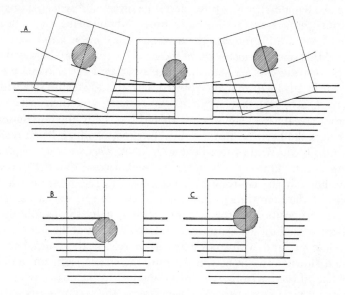

FIG. 52. TAKING SIGHTS

A: Swinging the sextant pendulum fashion to ensure that it is vertical when contact is made. *B*: Observing the upper limb. *C*: When the body is indistinct, it should be bisected by the horizon.

When the sun is at a low altitude it is most likely to be affected by refraction (bending of light rays by the atmosphere): the upper limb should then be observed (Fig. 52B). It would also be necessary to observe the upper limb if the lower limb should be hidden by cloud. The important point to be remembered when correcting such a sight to obtain the true altitude is that the sun's semi-diameter must be deducted instead of being added. If it is essential to take a sight when the sun is indistinct, a more accurate result will be obtained if the sun is brought down so that it is bisected by the horizon (Fig. 52C). In such an event no correction must be made for semi-diameter.

Sights of the moon are taken in the same manner as those of the sun, but of course only the enlightened (visible) limb can be used.

The common method I have described for observing the sun or moon cannot be used for observing a star because of the risk of picking up the wrong star in the index glass. There are, however, three other methods that can be used. One of these is the same as I have already recommended the beginner to use when practising observations of the sun, i.e. to look direct at the star with the index bar at zero, then to move the bar slowly away to separate the direct and reflected images, lowering the sextant at the same time and following the reflected image down until it touches the horizon. But there is a risk with this method that one may lose sight of the reflected image on the way down and pick up some other star. In rough weather a better method is to find out in advance what the approximate altitude of the star will be (see page 236), set that angle on the sextant, and then sweep the horizon beneath the star, when it should be seen in the horizon glass on or near the horizon. But the third method, of which I first read in Captain O. M. Watts's excellent little book *The Sextant Simplified*, is very useful. This consists of taking the horizon up to the star, and as the horizon is a continuous black line, it is easy to see it when it arrives at the star. Proceed as follows. Set the index bar at zero; reverse the position of the sextant, holding it by the handle in the *left* hand so that the arc is uppermost. Look directly at the star through the telescope, and with your right hand release the index bar and move it slowly away from you, keeping the star in sight through the clear half of the horizon glass until the reflected image of the horizon arrives near it. Clamp the index bar, reverse the sextant to its normal position, and make the final adjustment to bring the star to the horizon. As a star is such a tiny point of light there is no question of observing its upper or lower limb.

For sun sights it is usual for the rising piece to be adjusted so that the field of the telescope includes equal parts of clear and silvered horizon glass. But when observing stars, one wants the brightest possible image, so the rising piece should then be screwed in towards the frame of the sextant until the telescope includes as much as possible of the silvered portion of the horizon glass.

There are two accessories that can be fitted to a sextant to make the observation of stars easier. One of these, the Wollaston prism, has the property of doubling the image so that two stars are seen close together one above the other; the index bar is moved until the images are equidistant from the horizon, one above it and one below, and this is of considerable assistance when the horizon is indistinct. The other ✓ accessory, the stellar lenticular, has the property of elongating the image of a star, so that it appears as a fine, horizontal, streak of light. This obviates the need to swing the sextant pendulum fashion to ensure that it is vertical at the time of contact, for one can see immediately whether the streak of light coincides exactly with the horizon, or whether it is at a small angle. Each of these accessories can be fitted to the sextant along with the index shades, as in Plate 35, *bottom*, where immediately to the left of the thin frames of the four index shades can be seen the thicker frames of the prism and lenticular, *Q* and *R*.

The small seagoing yacht does not offer much choice of position from which sights can be taken. The observer needs a position from which he can see the horizon and the body clear of interference from sails and rigging, yet this must be a place where he can wedge himself so that both arms are free and the upper part of his body can swing, and thus keep vertical no matter how wild the motion of the yacht may be. I often take sights in quiet weather from one of the side decks with my feet braced against the bulwarks and my stern against the upturned dinghy; but in rough weather I find it is best to wedge myself in the companionway, for there is then not much risk of being thrown about with resulting damage to the sextant, but it is sometimes necessary for the helmsman to alter course so as to move the sails out of my line of sight, or to ease the motion by bearing away.

The ability to measure accurately the altitude of a celestial body from a small vessel in rough weather can only be got with practice. One has to bring the reflected image down until it kisses the horizon only at the moment when the vessel is perched on top of the highest ✓ sea then running. Between such moments the body will be seen well below the horizon. In such conditions it is a help if the observer can place himself in a high position, but this is not often possible in a yacht. It is sometimes said that one should add to the height of eye half the height of the sea running at the time; but this is incorrect because the horizon is made up of seas of the same height as those affecting the observer, except on the occasion of a single freak sea.

It is usually considered that several sights should be taken and plotted on squared paper against their times, any which are badly out of line being discarded, and the average of the remainder taken. But that is not my practice. If I feel satisfied that the sight I have taken is the best possible in the conditions prevailing, I use it; if I do not feel entirely satisfied I discard it and take another. No doubt each navigator has his own views on this, but it may be worth considering that if he stays on deck for a considerable length of time taking a number of sights, his arm and his eye may tire and the final result then be no better, and maybe worse, than a single sight. It is an excellent plan to practise taking sights from a known position so as to discover one's personal error, and this will give confidence when later on one is out of sight of land. Sights of bodies with an altitude of less than 15° should be avoided because of the abnormal refraction of light rays from a low body. Observations of a body the altitude of which exceeds 80° are difficult to take and may not always be possible from a small vessel.

Until the 1940s the navigator was able to use only the visible horizon when taking sights. Occasions arise when the horizon is indistinct or invisible, though the body may show clearly; also the use of the visible horizon limits the taking of star sights to the short period of twilight at dawn and dusk, when there is still sufficient light to see the horizon, yet not so much light as to prevent the navigational stars from being seen.

In an attempt to get over this difficulty and enable sights to be taken when the horizon is not visible, the bubble horizon was invented. In its marine form this comprises an attachment which can be screwed temporarily to the frame of the sextant immediately in front of the horizon glass, so that the line of sight of the telescope, passing through the clear half of the horizon glass, enters the attachment. By means of mirrors and a lens, rays of light from a bubble within the attachment are projected in a truly horizontal direction. The bubble therefore serves as an artificial true horizon, and to take an observation of a body, one has only to bring the reflected image, seen in the silvered half of the horizon glass, into coincidence with the bubble.

In Plate 35, *bottom*, a bubble horizon attachment (S) is seen shipped in position on the frame, and the prism monocular (T), which is required for use with it, is also shipped. To produce a bubble, tilt the sextant upwards about 45° and screw up the control knob (U), which can be seen on the side of the attachment, until a bubble of the desired size appears; then unscrew the knob and lower the sextant, when the bubble will detach itself from the bottom of the chamber and float free. For sun sights the bubble should be just a little larger than the sun, so that its centre and that of the sun can be brought into coincidence. For star sights a smaller bubble should be used. After use the bubble should be removed by tilting the sextant downwards, screwing up the control knob, and then unscrewing it quickly. A small electric light (V), controlled by a rheostat on the handle of the sextant, can be made to give exactly the right illumination for night use.

In theory the bubble horizon is an ideal arrangement, for the navigator using it is independent of the visible horizon. But in practice it is exceedingly difficult to use in a small vessel because the motion causes the bubble to move about in the chamber, and unless it is kept in the centre of the field of view the resulting altitude will be inaccurate. The makers state that in smooth water the error should not exceed 5′, but no doubt they are thinking in terms of large and steady merchant or naval vessels.

Every sextant is liable to have certain errors, some of which can be eliminated by careful adjustment of the screws provided in the frames of the index and horizon glasses, but the only error that need concern us here is that known as index error (I.E.). When the sextant is set at zero (micrometer or vernier also at zero, of course) the index and horizon glasses should be parallel to one another so that the direct and reflected image of a distant object, such as a star or the horizon, exactly coincide. If on making that test they do not coincide, move the tangent screw or micrometer head until they do, then read the sextant, and that reading will be the I.E. If this reading is on the arc, that is, to the left of zero, all angles measured with the sextant will be that much too great and the I.E. must be deducted from all sextant readings to obtain true readings. If the I.E. is read to the right of zero, that is, off the arc, it must be added to all sextant readings. When reading an angle off the arc, the vernier or the micrometer must be read backwards. Unless the I.E. is considerable, say more than 3′, it is best left uncorrected, for harm may be done by frequent or inexpert tampering with the adjusting screws.

Star recognition

Although this is too large a subject to be dealt with properly here, a few words about it might not be out of place; but the newcomer to navigation need not think that he must master it before he can progress further because, as will be seen later, if Volume I of *Sight Reduction Tables for Air Navigation* is used, it is not essential to be able to identify any of the stars. Nevertheless the ability to recognize some of the brighter ones will be of use at times, and will add much to the pleasure and interest of a voyage.

As children most of us were taught how to recognize one or two of the more obvious constellations, though a fertile imagination was required to picture the objects some of them are supposed to portray; Orion, for example, although he has shoulders, feet, and a belt, has no arms or head; but at least the Plough, which is also known as the

Big Dipper, or as Charles's Wain, does look to my eye something like a plough. With the help of the Plough and Orion it is possible to identify at least half of the stars most useful to the navigator, but as Orion is not visible in the northern summer or southern winter,

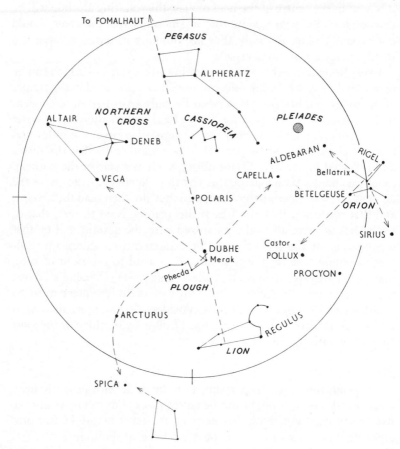

FIG. 53. NORTHERN STARS

because he is then on the same side of the earth as the sun, the new-comer to star recognition would obviously be well advised to learn his way about the night sky during the other half of the year; in mid December, for example, Orion crosses the meridian at midnight.

There are various aids to star recognition: star globes (very expensive), star finders, and star maps such as appear in nautical almanacs. Many years ago when I first started taking an interest in astronomical navigation, I made the primitive little star maps which, now drawn a bit better, are reproduced in Figs. 53 and 54, to help me find my way about the night sky and enable me to put a name to such stars as I wished to use. Fortunately these were few, for although nautical

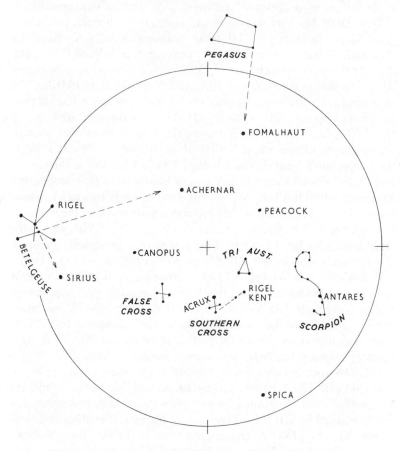

FIG. 54. SOUTHERN STARS

almanacs list about 60, only 22 could be used with the tables I was using, and that number is sufficient for most practical purposes. In Fig. 53, northern celestial hemisphere, I have shown 15 of them, though 3 of these belong in the southern celestial hemisphere as they have south declination (see page 213). All 15 cannot, of course, be seen at any one time because of the rotation of the earth. Their names are in capitals. I have added a few other stars of lesser magnitude and importance to help with the description which follows, and their names are in lower case. I have sketched in with solid lines some of the constellation shapes, and their names are in italic capitals. The three types of lettering used in the Figures are also used in the text of the following paragraphs. Leading or identification lines are dashed.

POLARIS lies very nearly at the celestial north pole, and a line from Merak through DUBHE—the 'pointers' at the fore end of the *PLOUGH*—leads to it, and if that line is continued beyond POLARIS it goes past the constellation of *CASSIOPEIA* (this is like a sprawling W) to one side of the square of *PEGASUS*, in which ALPHERATZ, of second magnitude, is the only useful star. The same line, if projected the opposite side of the *PLOUGH*, cuts through the rump of the *LION*, and REGULUS lies at the fore-paws of that animal. Some people cannot see a lion there at all, but recognize a sickle where the lion's head should be; then REGULUS lies at the handle end. A line drawn from the fore end of the *PLOUGH*, bisecting the angle made by DUBHE, Merak and Phecda, leads to VEGA, which is one of three bright but widely separated stars marking the extremities of a big V. The star at the apex of this V is ALTAIR, and at the third extremity lies DENEB. This last is easy to identify as being the brightest star in the *SWAN* or *NORTHERN CROSS* (some call this constellation the *KITE*). Returning to the *PLOUGH*, if the curve of the handle is lengthened backwards in a generous sweep, it passes through ARCTURUS to SPICA. Near SPICA four stars form a gaff-headed sail, and the peak of the gaff points to SPICA. One final line from the *PLOUGH*, if drawn from Phecda to pass midway between DUBHE and Merak, leads to CAPELLA.

ORION with his glorious bodyguard is the finest spectacle in the night sky. BETELGEUSE and Bellatrix mark his broad shoulders; his narrow waist is spanned by a sloping sword belt, and one of his feet is marked by RIGEL. The slope of his belt, if continued downwards, leads to *SIRIUS*, the brightest star in the sky, though not so bright as the planets Venus and Jupiter. The star at the lower end of the belt in line with Bellatrix points to reddish ALDEBARAN, which lies half-way between *ORION* and the *PLEIADES*, a clearly

PLATE 37

Top: A Waltham watch, like a miniature chronometer, is mounted on gimbals in a case with an inner glass lid. Its size may be compared with the matchbox standing beside it. *Bottom*: A grid compass mounted beneath a hinged cover of unbreakable glass in the bridge deck. Once the verge ring, engraved with degrees, has been set to the desired course, it is only necessary to keep the north-south line parallel to the wires on top of the bowl in order to steer that course. The grid is luminous.

defined patch of stars which is a useful skymark. Incidentally, the
star at the upper end of *ORION*'s belt rises due east and sets due
west of an observer anywhere in the world. A line from Bellatrix
through BETELGEUSE goes close to PROCYON, while a line
from RIGEL through BETELGEUSE points out POLLUX, which
lies with its twin Castor nearly half-way between *ORION* and the
PLOUGH.

The southern celestial hemisphere is not so generously provided
with easily recognizable constellations, and contains only 10 of our
22 required stars. In Fig. 54 these are shown together with one, our
old friend BETELGEUSE, which belongs in the northern hemisphere. The best-known constellation in the southern sky is the
SOUTHERN CROSS, and although this is not so prominent as its
reputation would have one believe, a northerner's first sight of it
always produces a thrill. It lies in the same quarter of the sky as the
PLOUGH, with its longest arm, at the end of which is ACRUX,
pointing towards the south celestial pole, and very close to it is a small,
starless patch of black sky known to seamen as the 'coal sack'. Two
bright stars point towards the *SOUTHERN CROSS* from the side,
the one farthest away being RIGEL KENT, close to which lies a
small triangle, *TRIANGULUS AUSTRALIS*. On the opposite
side of the *SOUTHERN CROSS* to the triangle lies the *FALSE
CROSS*, which as its name implies has often been mistaken for the
real one when seen between clouds. CANOPUS, the second brightest
star, stands alone between SIRIUS, the brightest of them all, and the
celestial pole. A line from BETELGEUSE through the lower end of
ORION's belt goes close to ACHERNAR, far down to the south
and in complete isolation. The two outer stars of the square of *PEGA-
SUS*, which as we have seen in Fig. 53 point to POLARIS, lead in
the opposite direction to FOMALHAUT. PEACOCK, nearly half-
way between FOMALHAUT and RIGEL KENT is only a second-
magnitude star, but there are no others of the same magnitude near
by. An equal distance from RIGEL KENT, but with much smaller
declination, is reddish ANTARES, the brightest star in the un-
mistakable *SCORPION*.

PLATE 38

Top: At anchor in limpid water in the lee of a small, uninhabited island near an untrodden
sandy beach, the yacht lies motionless—Big Major's Spot, Bahamas. *Bottom*: Within a few
hours or days or weeks she will once again be running before the wind in a small, lonely world
of her own, bounded by sea and sky, her vane gear keeping her on course, and the unchecked
seas with flashing crests hurrying up astern.

P

It will be seen that I have failed to discover many worthwhile leading lines for stars in the southern celestial hemisphere, but my attempts to do so, during which I grew more familiar with the night sky, helped to pass many a watch pleasantly.

A bright star which cannot be identified is probably a planet. Four of these can at times be used for navigation, but as their positions in relation to the stars change, they can best be found by consulting a nautical almanac. If, however, movement is detectable, the object is almost certainly a space-craft; one cloudy night I took an observation of one of these instead of CANOPUS, and only realized my mistake when I saw it overtake a star.

Time

To obtain a position line from an observation of a celestial body—except when obtaining the latitude from a meridian altitude of the sun, or from an observation of Polaris—one must know the Greenwich mean time (G.M.T.) of the moment the observation was taken. An error of 1 minute in time could cause an error of up to 15 miles on the earth's surface, 4 seconds an error of 1 mile. For this a chronometer, which is a highly accurate clock compensated for changes in temperature, is used in large ships; but as such an instrument is easily upset by a quick and jerky motion, it is better in a yacht to use a timepiece with a lever movement, such as the Waltham deck watch shown at the top of Plate 37, where its size can be judged from the matchbox standing beside it. Like a chronometer, this watch is slung in gimbals in a wooden box with an inner glass lid. But a good modern watch will probably serve just as well, and although it would be wise to keep this in a safe place where it will not be subject to sudden extreme changes of temperature or rough usage, and to wind it at the same time each day, I have found a self-winding watch worn on the wrist to be remarkably accurate, though resetting it in each time zone (see below) did cause its rate to become irregular for a couple of days. It is, however, usually possible to obtain a time signal with which to check the instrument each day no matter where one may be (see page 94). The gaining or losing rate of the navigator's timepiece—let us call it the chronometer from now on whatever its type may be—does not matter; the important point is that the chronometer should keep a steady rate, and that the rate be known; then in the event of a radio breakdown one will know how much to add or deduct day by day to obtain G.M.T. If a clock or watch stops for the usual reason that it needs cleaning, it can often be made to go again by opening it and rinsing it in a bath of petrol, to which lubricating oil has been added in the proportion of one part of oil to 50 parts of petrol. The only harm this is likely to do is to discolour the face.

It is desirable to have a routine for timing and writing down observations, and the following is the method my wife and I use. My wife holds the stopwatch while I handle the sextant. The instant I have obtained what I believe to be a satisfactory sight, my wife starts the watch. I go below and put the sextant temporarily in a safe place, either in its box or on the lee settee; then, opening the lid of the chronometer box, I watch the second hand until it reaches the end of its minute circuit. 'Stop', I shout, and my wife stops the watch. I at once write down the minute and the hour shown by the chronometer, and beneath it the minutes and/or seconds of time my wife tells me have elapsed since I took the sight. Later, when I come to work the sight out, I have only to deduct that period of time from the chronometer reading to find what chronometer time was at the moment of the sight. By adding or deducting the amount I know from the most recent time signal, or from the accumulated daily rate, the chronometer is slow or fast on G.M.T., I obtain the G.M.T. of the sight. I read the sextant and write down the angle measured. My wife meanwhile has read the patent log, and I write that reading down as well.

If there is no stopwatch on board, or the navigator is single-handed, he should with a little practice be able to get a sufficiently accurate idea of the time that elapses between taking the sight and reading the chronometer by counting the seconds. But if his wristwatch is being used as the chronometer, he can read it himself immediately he has taken the sight, noting first the second, then the minute, and finally the hour. But when later he converts this to G.M.T. he must remember to allow not only for the known error but also for Zone Time, to which it will almost certainly be set.

Zone Time is the system which enables all vessels and people within certain defined limits of longitude to keep the same time. For this purpose the world is divided into 24 zones, each of which covers $15°$ of longitude. In each zone the time differs by one hour from that of the neighbouring zones. The meridian of Greenwich is in the centre of Zone O, which therefore extends from $7\frac{1}{2}°$ E. to $7\frac{1}{2}°$ W. The zones lying to the eastward are numbered in sequence with a negative $(-)$ prefix, those lying to the westward are similarly numbered with a positive $(+)$ prefix. The 12th zone is divided centrally by the 180th meridian, and both prefixes appear in this zone, their position depending on the date line, which is modified in places so as to include all islands of one group on one side or the other of it. As a vessel sails to the westward, noon, i.e. the moment the sun reaches its maximum altitude, will become increasingly later, and in 60° W., for example (West Indies), will occur at approximately 1600 G.M.T. This would appear to be inconvenient only because we are accustomed (or were before the advent of British Standard Time) to the clock indicating 1200 in the middle of the day; but if we stop at a port in that area (Zone +4) it will be awkward to keep in step with the inhabitants, whose watches will be four hours slow on ours. So it is convenient for the voyager to keep Zone Time, and although there is no need for him slavishly to follow the time habits of the other inhabitants of the zone he happens to be in, there is, as Churchill once pointed out, a time called 'stomach time', which at sea agrees fairly well with Zone Time. So, when passing from one zone to another it is usual to alter the ship's clocks accordingly, but the chronometer should, of course, be left untouched to register approximate G.M.T. If the clocks of a vessel making a circumnavigation west-about from the meridian of Greenwich are shifted to Zone Time in each zone, clearly they will be 12 hours slow on G.M.T. by the time she arrives in Zone 12, and 24 hours slow on arriving back at the Greenwich meridian. Therefore, on crossing the date line from east to west the date must be changed ahead so as to skip a day, thus making a six-day week; if crossing from west to east a day must be added, and of course these rules apply to any crossing of the date line from far or near. A small chart, Admiralty No. 5006a, shows the time zones of the world, and indicates in what countries or parts of countries Zone Time is not adhered to. A list of all countries showing the time they keep will be found in *The Nautical Almanac* together with information about daylight-saving time. However, in large countries which extend over several time zones (Australia, Canada, U.S.A., and U.S.S.R.) the zone boundaries are modified to some extent to conform with state or other internal boundaries. In the U.S.A. the times kept in these are spoken of not by zone number but by name: Zone +5 keeps Eastern Standard Time (sometimes known as Atlantic Time); Zone +6 keeps Central Standard Time; Zone +7 keeps Mountain Standard Time; and Zone +8 Pacific Standard Time.

Astronomical navigation

For working any observation of a celestial body so as to obtain a position line, a nautical almanac is essential. There are several from which to choose, but I would recommend the deep-sea yachtsman to use *The Nautical Almanac* (H.M. Stationery Office), which is referred to hereafter as the *N.A.* In 1958 the layouts of this and the *American Nautical Almanac* were changed so that now, under the same title, both are identical in content. Like most almanacs, the *N.A.* is valid for one year, starting on 1 January, but it contains instructions enabling it to be used, though with some inconvenience, for sun and star sights for the following year, to provide for the navigator who finds himself at the end of the year in a part of the world where a new almanac cannot be obtained. The *N.A.* contains instructions for its use, and these will not be repeated here.

Before attempting to explain briefly how a position line is obtained and drawn, I must define some of the terms that have to be used.

Dead reckoning (D.R.). The account kept of a vessel's position, having regard to the course and distance made good since her position was last fixed by observations of terrestrial or celestial objects.

The *Geographical position* (G.P.) of a celestial body is the point on the earth's surface immediately beneath the body, i.e. where a line joining the centre of the body to the centre of the earth would cut the earth's surface.

Declination (Dec.) is the angular distance of a body measured north or south of the equator from the centre of the earth. Dec. therefore corresponds to the body's geographical latitude (Fig. 55).

The *Greenwich hour angle* (G.H.A.) of a body is the angular distance of its meridian measured westward from the meridian of Greenwich (Fig. 56). The Dec. and G.H.A. of the bodies used for navigation can be obtained for any moment of G.M.T. for any day of the year from the nautical almanac. In other words it is possible to ascertain from the almanac a body's G.P. at the moment of observation.

The *Local hour angle* (L.H.A.) of a body is the angular distance of its meridian measured westward from the observer's meridian. It is obtained from the G.H.A. of the body by subtracting the observer's west longitude or adding his east longitude (Fig. 56).

The *Rational Horizon* is a plane through the earth's centre at right angles to a line joining the earth's centre to the observer's zenith.

The *Zenith* of a place is the point in the sky directly overhead.

The *Zenith distance* (Z.D.) of a body from any place is the angular distance between the body and the zenith of the place, or it may be taken as the angle measured at the centre of the earth between the G.P.s of the body and the place (Fig. 57).

The *True altitude* of a body is the angle between its centre and the rational horizon as measured from the centre of the earth (Fig. 57).

An observer obtains this angle as follows. With his sextant he measures
the angle between the body and the visible horizon, and corrects for
index error. Then, turning to the altitude correction tables inside the
front cover of the *N.A.* (inside the back cover if the body is the moon)
he enters the table headed 'Dip' with his height of eye and the table

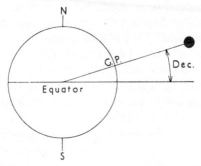

FIG. 55. THE DECLINATION OF A BODY CORRESPONDS TO THE BODY'S
GEOGRAPHICAL LATITUDE

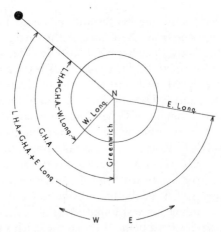

FIG. 56. GREENWICH HOUR ANGLE AND LOCAL HOUR ANGLE

headed 'Sun' or 'Star' with the sextant reading, and thus obtains two
corrections which have to be applied according to their signs to the
sextant reading to obtain the true altitude. These corrections allow
for refraction (bending of light rays in the earth's atmosphere), dip
(the effect of using the visible instead of the rational horizon), and in
the case of the sun for semi-diameter (the lower limb was observed,
but the angle required is to the centre).

The *Azimuth* of a body is its true bearing from a position, measured east or west from the elevated (nearest) pole.

If we know a body's G.P. (Dec. and G.H.A.) and our distance from that position (Z.D.), we must be situated at some point on the circumference of a circle drawn on the surface of the earth with the body's G.P. as centre and the Z.D. as radius. But it is not possible to draw such a large circle on the chart, so the procedure is to calculate what the Z.D. would be from some position close to our real position. It used to be the practice to use the dead-reckoning position for this purpose, but modern tables require something a little different, which

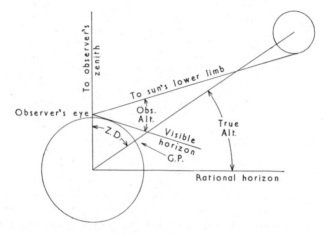

FIG. 57. OBSERVED AND TRUE ALTITUDE

will be mentioned later. We can obtain the body's azimuth from tables, and this we draw through the position which we have chosen and marked on the chart. From calculations we know what the Z.D. would be if we had measured it from that position, and from the observation we have taken we know what the Z.D. is at our real position. The difference between these two Z.D.s is known as the intercept and shows us how far wrong the assumed position is in one direction. Modern practice is to use not the calculated and true Z.D.s, but the calculated and true altitudes. Along the azimuth we measure off the intercept from the assumed position, either towards or away from the body according to which of the altitudes is the greater, and through that point rule a line at right angles to the azimuth. This line is part of the circumference of the big circle mentioned above, having the body's G.P. as centre and the observed Z.D. as radius; but as only a very small part of this circle is being drawn on the chart, it can be drawn as a straight line without error, and we must be situated at some point on it.

From this it will be clear that all that can ever be obtained from any single observation of a body is a position line at right angles to the azimuth of the body. But if we can obtain simultaneous position lines from two bodies, the azimuths of which differ by about 45 degrees or

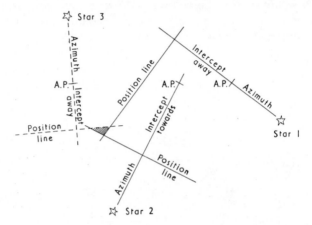

FIG. 58. A FIX BY TWO POSITION LINES (SOLID) WITH A THIRD (DASHED) AS A CHECK. (A.P. = assumed position.)

more, our position will be fixed at their point of intersection as shown in solid lines in Fig. 58. It is for this reason that star sights may result in greater accuracy than sun sights. It is usual to take three star sights as close together in time as possible, the third one, shown dotted in Fig. 58, acting as a check. Ideally the resulting position lines should intersect at one point, but more often they form a triangle, the seaman's 'cocked hat', which is shaded in the figure. This should be small, when the position is presumed to be at the centre of it. If it is large an error in taking or working one or more of the sights has been made.

If only a single body, usually the sun, is being observed, one must wait until its azimuth has changed about 45 degrees, or more, before obtaining a second position line. Where the first position line— brought forward with the parallel rules along the course made good and to the extent of the distance run since the first observation was taken—cuts the second position line, is regarded as the vessel's position. An example of this is shown in Fig. 59. It is, of course, possible for an error of the compass, indifferent steering, an inaccurate patent log, or a current of unknown strength or direction, to affect the accuracy of that position; except in abnormal circumstances, however, such errors during a period of only 2 or 3 hours may reasonably be disregarded in the open sea.

My own practice is to take an observation of the sun in the forenoon, obtain the latitude from a meridian altitude sight at noon, and then carry forward the first position line as described above. This has the

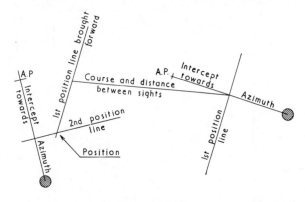

FIG. 59. POSITION OBTAINED BY CARRYING FORWARD
ONE POSITION LINE TO CROSS ANOTHER OBTAINED
FROM A SIGHT TAKEN LATER

advantage in a short-handed yacht that all the navigation work can be completed at noon, which is the traditional time of day for marking the position on the chart, measuring the day's run, and making any needed alteration to the course; if the vessel is well away from the land nothing further need then be done until the following morning. I make use of stars only when the sun has not shone during the day, or when I am getting anxious about a landfall, or some danger that must be avoided, for the period of twilight so often occurs at awkward times, when a meal is being prepared or eaten, or when the watch below is still asleep, and in the tropics twilight is brief. When I do have to use the stars I prefer to take the sights at dawn rather than at dusk, because I can then more easily recognize the constellations while the sky is still dark and choose such stars as are most likely to be of use. But if *S.R.T.*, Vol. I, is used (see page 236) this reason is not valid.

Even though it may not be possible to obtain a fix, the opportunity of getting a single position line should not be neglected when one is near the land, for such a line can often be of great value then. An example of this occurred when on passage from Ushant towards La Coruña on the north-west coast of Spain. The sky was overcast and visibility less than 5 miles as the coast was approached with a fresh north wind. The D.R. position was a little to the west of the direct course for Cabo Prior, on which it was intended to make a landfall.

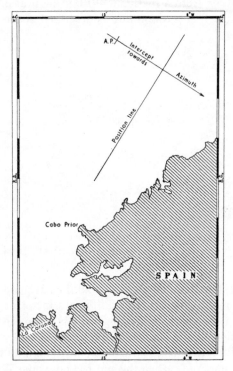

FIG. 60. A POSITION LINE CORRESPONDING
WITH THE COURSE

At about 10.30 I got a shot at the sun through a momentary thinning of the cloud pall, and on drawing the resulting position line on the chart, found that by a happy chance it cleared Cabo Prior by 2 miles (Fig. 60). Course was therefore altered to run confidently down that line, and in the afternoon the cape was sighted a few miles off fine on the port bow. Any single position line is always better than none, for it does fix the position in one direction; and in conditions similar to the above, a position line running parallel to the course is of the greatest value, and when conditions permit should be obtained by observing a body when it is abeam.

Working sights

The simplest and most popular method of obtaining the latitude is by observing the meridian altitude of the sun. This is commonly known as a 'noon lat.' As at the moment of observation the sun will be on the observer's meridian, the position line obtained, being at right angles to the sun's azimuth, will be a parallel of latitude. The observation therefore is independent of a knowledge of G.M.T. As the sun will bear true north or south at noon, a compass bearing of it, corrected for variation and deviation, will give a rough idea as to when the time for observation is approaching. But unless the compass is particularly well situated and is provided with an azimuth mirror, such a bearing is not easy to take, especially when the altitude is great, and time and trouble will be saved if we make use of G.M.T. and work out in advance the time of the sun's meridian passage.

As the sun moves 15 degrees to the westward in 1 hour (one degree in 4 minutes), we can convert arc into time by dividing by 15, or more conveniently by consulting the 'Conversion of Arc to Time' table in the *N.A.* If we convert the ship's longitude into time, and add it to 1200 if we are in west longitude, or deduct it if in east longitude, we would obtain the time of the sun's passage across the ship's meridian, provided the sun kept precisely to G.M.T. in its movements. But only occasionally does the sun cross the meridian of Greenwich at 1200 G.M.T., and it may be as much as 20 minutes fast or slow. We must therefore ascertain the time it does cross the Greenwich meridian on the day we wish to take our sight. This information, in hours and minutes, is given in the *N.A.*, where under the heading 'Mer. Pass.' it will be found at the foot of each right-hand page. To obtain the G.M.T. of the sun's passage across the ship's meridian, convert the D.R. noon longitude into time, and add it to the sun's meridian passage time if the longitude is west, deduct it if east.

Example. On 31 October 1967, in D.R. longitude 15° 45′ W.; required the G.M.T. of the sun's passage across the ship's meridian.

From conversion table in the *N.A.*	15° 45′ = 1h. 3m.
Sun's mer. pass. at Greenwich on 31 Oct. 1967 from daily page in the *N.A.*	+11h. 44m.
	12h. 47m.

To take the noon sight, go on deck a little before that time and start observing the sun while it is still rising; adjust the tangent screw or micrometer head as necessary to keep the reflected sun's lower limb just kissing the horizon, until the sun rises no higher and appears to maintain a steady altitude for a minute or so before commencing its downward path. Read the sextant and apply the index error to obtain the observed altitude. Convert this to true altitude by applying the two corrections from the *N.A.*, the 'Sun' table being entered with the observed altitude, and the 'Dip' table with the height of eye. Subtract the true altitude from 90° to obtain the Z.D. Obtain from the *N.A.* the Dec. of the sun at the date and G.M.T. of the sight, and write it beneath the Z.D. If the latitude and the Dec. are of the same name, that is, both north or both south, add the Z.D. and the Dec. together to obtain the latitude (Fig. 61); if they are of contrary names, that is one is north and the other south, their difference will give the latitude (Fig. 62).

Example. On 5 July 1967, in the mouth of the English Channel in D.R. lat. 49° 32′ N., long. 5° 49′ W., the meridian altitude of the sun was observed. The sextant read 63° 02′; I.E. was − 2′, and H.E. 6 feet. Required the latitude.

Sextant	63° 02′
I.E.	− 2′
Observed altitude	63° 00′
Correction for sun	+15′·5
,, ,, dip (H.E. 6 feet)	− 2′·5
True altitude	63° 13′
	90
Z.D.	26° 47′
Dec. at time of sight from *N.A.*	22° 49′ N.
Latitude	49° 36′ N.

As Dec. and latitude were the same name (both north in this example) Dec. and Z.D. were added to obtain the latitude.

Example. On 9 September 1954, when on passage from Suva, Fiji, towards Russell, New Zealand, and in D.R. lat. 28° 16′ S., long. 174° E., the meridian altitude of the sun was observed. The sextant read 56° 05′; I.E. − 2′, H.E. 6 feet. Required the latitude.

Obs. Alt.	56° 05′
Total correction	+11′
True Alt.	56° 16′
	90
Z.D.	33° 44′
Dec.	5° 31′ N.
Latitude	28° 13′ S.

Here Dec. and latitude were of different names, so the difference between Z.D. and Dec. gave the latitude.

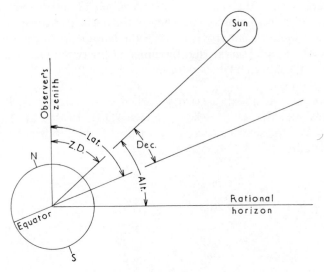

FIG. 61

When latitude and declination are of the same name, the sum of the zenith
distance and the declination gives the latitude.

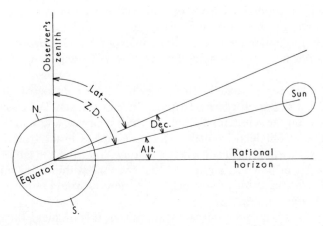

FIG. 62

When latitude and declination are of contrary names, the difference between
zenith distance and declination gives the latitude.

There is one other variation of this problem, and it only affects navigators in low latitudes. If, when in the northern hemisphere, the Dec. is also north but is greater than the latitude, then the sun is to the north of the observer, and the difference between Dec. and Z.D. will give the latitude. Of course a similar thing applies in the southern hemisphere.

Example. On 9 February 1953, when on passage from Panama towards the Marquesas Islands, and in D.R. lat. 3° 55′ S., long 98° 05′ W.; the sextant read 79° 04′; I.E. −2′, H.E. 6 feet. Required the latitude.

Obs. Alt.	79° 04′
Total correction	+11′
True Alt.	79° 15′
	90
Z.D.	10° 45′
Dec.	14° 35′ S.
Latitude	3° 50′ S.

When the sun is nearly overhead at noon, i.e. when its Dec. is nearly the same as the observer's latitude, it cannot be observed from a small vessel at sea. As on such an occasion the sun's azimuth will be nearly east all the forenoon and nearly west all the afternoon, position lines obtained from the sun will not cross at a sufficiently wide angle to give a fix. Stars will then have to be used instead, or at least for obtaining one position line, for which purpose, when in the northern hemisphere, Polaris (the Pole star) is particularly useful.

If Polaris were situated exactly over the north pole its true altitude would equal the observer's latitude. In fact the star lies about one degree from the celestial pole, round which it circles, and some simple corrections, obtained from the *N.A.* and applied to the true altitude of this star, will give the observer's latitude. To enter the tables some knowledge of G.M.T. is required, but the time within a few minutes is sufficient; an error of as much as 15 minutes would only cause an error of 2 or 3 miles.

As Polaris is only a moderately bright star, it is frequently invisible to the naked eye in the evening until the night has become too dark to permit the horizon to be seen clearly. The procedure, therefore, is to clamp the index bar of the sextant at the D.R. latitude, and while the night is still light enough for the horizon to be seen, sweep the horizon with the sextant in the direction of true north, when the star will usually be seen through the telescope near the horizon; the final adjustment to bring it to the horizon can then be made. Note the time of the sight, read the sextant, apply the index error and the corrections for dip and refraction, using the correction table for stars. Find in the *N.A.* what the G.H.A. of Aries was for the date, hour, and nearest minute of the G.M.T. of the sight, and write it down. (Aries is a point in the sky to which all the stars can be related.) Beneath the G.H.A. of Aries write the D.R. longitude, and subtract it if it is west, add it if it is east, to obtain the local hour angle (L.H.A.) of Aries. Turn to the Polaris tables in the *N.A.* These have to be entered in three places, as explained at the foot of the tables, with the L.H.A. of Aries, the D.R. latitude, and the month. Add the three corrections so obtained to the true altitude, *deduct one degree*, and the result is the latitude.

   Example. On 15 January 1962, when on passage in the Red Sea from Port Sudan towards Suez, and in D.R. lat. 23° 29′ N., long. 37° 13′ E., an observation of Polaris was taken at 15h. 48m. G.M.T.; the observed altitude (correction for index error having been made) was 24° 33′; H.E. 5 feet. Required the latitude.

G.H.A. Aries at 15h.	339° 34′·8
Increment for 48m.	12 00
	351° 35′
Assumed longitude	+37 25′ E.
	389°
	−360
L.H.A. Aries	29°
Observed altitude	24° 33′
Correction for dip and refraction	−4′
True alt.	24° 29′
Correction for L.H.A. Aries	+4′·1
„ „ latitude	+0′·6
„ „ month	+0′·8
	24° 34′
	−1°
Latitude	23° 34′ N.

   For the working of all sights other than meridian altitude and Polaris sights, some books of tables will be needed in addition to the nautical almanac. Tables which reduce the work to the minimum and eliminate the use of logarithms in solving the spherical triangles, which occur in most navigational problems, have been available to the yachtsman for some years. Unlike the nautical almanac, these do not alter from year to year, and once the navigator has bought the volumes he needs they will, with the exception noted on page 236, last him for the rest of his life.

Astronomical Navigation Tables (*A.N.T.*), which have been used successfully by so many ocean-going yachtsmen, including Goldsmith, Worth, Pye, Carr, and myself, are, unfortunately, no longer in print, their place having been taken by a publication known as *Sight Reduction Tables for Air Navigation*, referred to hereafter as *S.R.T.* Volumes II and III of this provide the same information as *A.N.T.*, and in the same simple fashion, but are not so boldly printed, and the large page size (12½ inches by 10 inches) is inconvenient to use and to stow. Volume II is for any body with a declination of up to 29°, and for latitudes 0° to 39°, north or south; Volume III, with the same limit of declination, is for latitudes 40° to 89°, north or south. These are the tables to which reference is made in this section. Although they were devised for use in air navigation they are equally suitable for surface navigation, and it should not be thought that they can only be used with the *Air Almanac*; on the contrary, the yacht navigator would do well to continue to use the *N.A.*, one volume of which covers the whole year, whereas three volumes of the *Air Almanac* are needed for the same period. *S.R.T.* Volume I is for selected stars only.

Tables of Computed Altitude and Azimuth (H.D. 486) do the same job as *S.R.T.*, but with a greater accuracy which is not required by the yacht navigator. They are not quite so easy to use, and they are more bulky and more expensive.

As was explained in the section on astronomical navigation, if at any moment we know the altitude of a body from some position close to us, and we know the body's altitude from our own position, the difference between the two altitudes is our distance in one direction from that place, the resulting position line being drawn at right angles to the azimuth of the body. *S.R.T.* make it unnecessary for us to

---

calculate the altitude at that place, for they tell us straight away what the altitude and azimuth of any body with a declination of up to 29° is for every whole degree of latitude from 89° N. to 89° S., and for every whole degree of L.H.A.

From this it will be apparent that we cannot make use of the D.R. position, except in the very unlikely event of the latitude being in whole degrees and the longitude being such that when it is applied to the G.H.A. of the body the resulting L.H.A. is also in whole degrees. It therefore becomes necessary to assume a position such that the latitude *is* in whole degrees, and the longitude *is* a figure such that when applied to the G.H.A. (added if east, subtracted if west) it gives a L.H.A. in whole degrees. Naturally we will assume a position as close as possible to the D.R. position, otherwise the intercept might be inconveniently long.

Suppose, for example, the D.R. latitude is 48° 45′ N., the D.R. longitude 22° 36′ W., and the G.H.A. of the body 323° 50′. The nearest whole degree of latitude is 49° N., so that must be used as the assumed latitude. As west longitude has to be deducted from G.H.A. to obtain the L.H.A., the assumed longitude will have to be 22° 50′ W., so that the minutes of arc cancel out and leave a L.H.A. in whole degrees, as below:

G.H.A.	323° 50′
Assumed West longitude	− 22° 50′
L.H.A.	301° 00′

To take another example: if the D.R. position is latitude 54° 26′ N., longitude 3° 21′ E., and the G.H.A. of the body is 294° 46′; the assumed latitude will have to be 54° N., and the assumed longitude 3° 14′ E., as below:

G.H.A.	294° 46′
Assumed East longitude	+3° 14′
L.H.A.	298° 00′

PLATE 40

Top: The palms growing on an atoll may lift over the horizon only when 10 to 15 miles away, and there will probably be no more than the usual trade wind clouds sailing above it—Takaroa, one of the atolls of the Tuamotu, or Dangerous Archipelago, South Pacific. *Middle*: Cloud, streaming away like smoke, could be seen to leeward of Rodriguez (1,300 feet high) when the island was well out of sight. *Bottom*: In clear weather a high island should be visible many miles away, and land birds may tell of its presence before it is seen. After 16 days out of sight of land, St. Helena appears ahead.

Q

DECLINATION (15°–29°) SAME NAME AS LATITUDE

LHA	15° Hc d Z	16° Hc d Z	17° Hc d Z	18° Hc d Z	19° Hc d Z	20° Hc d Z	21° Hc d Z	22° Hc d Z	23° Hc d Z	24° Hc d Z	25° Hc d Z	26° Hc d Z	27° Hc d Z	28° Hc d Z	29° Hc d Z	LHA
0	61 00 +60 180	62 00 +60 180	63 00 +60 180	64 00 +60 180	65 00 +60 180	66 00 +60 180	67 00 +60 180	68 00 +60 180	69 00 +60 180	70 00 +60 180	71 00 +60 180	72 00 +60 180	73 00 +60 180	74 00 +60 180	75 00 +60 180	360
1	60 59 60 178	61 59 60 178	62 59 60 178	63 59 60 178	64 59 60 178	65 59 60 178	66 59 60 178	67 59 60 177	68 59 60 177	69 59 60 177	70 59 60 177	71 59 60 177	72 59 60 177	73 59 60 177	74 59 60 177	359
2	60 57 60 176	61 57 60 176	62 57 60 176	63 57 60 176	64 57 60 176	65 57 60 175	66 56 58 175	67 56 59 175	68 56 60 175	69 56 60 175	70 56 60 175	71 56 59 174	72 56 59 174	73 56 60 174	74 56 59 174	358
3	60 53 60 174	61 53 60 174	62 53 60 174	63 53 59 174	64 53 60 173	65 52 60 173	66 52 60 173	67 52 52 173	68 51 60 172	69 51 60 172	70 51 60 172	71 50 60 171	72 50 60 171	73 49 60 170	74 49 60 170	357
4	60 48 60 172	61 48 59 172	62 47 59 172	63 47 60 171	64 47 59 171	65 46 60 171	66 46 59 171	67 45 60 170	68 45 60 170	69 44 59 169	70 43 60 169	71 43 59 169	72 42 59 168	73 41 59 167	74 40 59 167	356
5	60 41 +60 170	61 41 +59 170	62 40 +60 169	63 40 +59 169	64 39 +59 169	65 38 +60 168	66 37 +59 168	67 37 +59 168	68 36 +59 167	69 35 +59 167	70 34 +58 166	71 33 +59 166	72 32 +59 165	73 30 +59 164	74 29 +58 163	355
6	60 33 60 168	61 33 59 168	62 32 59 168	63 31 59 167	64 30 59 167	65 29 59 166	66 28 59 166	67 27 59 166	68 26 59 165	69 25 59 164	70 23 58 164	71 22 58 163	72 20 58 162	73 18 57 161	74 15 58 160	354
7	60 24 59 166	61 23 59 166	62 22 59 165	63 21 59 165	64 19 59 165	65 18 59 164	66 17 59 164	67 15 58 163	68 14 58 162	69 12 58 162	70 10 58 161	71 08 57 160	72 06 57 159	73 03 58 158	74 00 56 157	353
8	60 13 58 164	61 11 59 164	62 10 59 163	63 09 58 163	64 07 58 162	65 05 59 162	66 03 58 161	67 01 58 161	67 59 58 160	68 57 58 159	69 55 57 159	70 52 57 158	71 49 57 157	72 46 56 156	73 42 56 155	352
9	60 00 58 162	60 59 58 162	61 57 58 162	62 55 58 161	63 53 58 160	64 51 58 160	65 49 58 159	66 47 57 158	67 44 57 158	68 42 57 157	69 39 57 156	70 35 56 155	71 33 55 154	72 27 55 153	73 22 55 151	351
10	59 47 +58 161	60 45 +58 160	61 43 +57 160	62 40 +58 159	63 38 +58 158	64 36 +57 158	65 33 +57 157	66 30 +57 156	67 27 +56 155	68 23 +57 154	69 20 +56 154	70 16 +55 153	71 11 +55 151	72 06 +53 150	73 01 +54 149	350
11	59 32 57 159	60 29 58 158	61 27 57 158	62 24 57 157	63 21 57 156	64 18 57 156	65 15 57 155	66 12 56 154	67 08 56 153	68 04 56 152	69 00 55 151	69 56 55 150	70 50 54 149	71 44 54 148	72 38 52 146	349
12	59 15 56 157	60 13 57 156	61 10 56 156	62 07 56 155	63 04 56 154	64 00 56 154	64 56 56 153	65 52 56 152	66 48 55 151	67 43 55 150	68 38 55 149	69 33 54 148	70 27 53 146	71 20 53 145	72 13 51 144	348
13	58 58 57 155	59 55 56 155	60 51 57 154	61 48 56 153	62 44 56 152	63 40 56 152	64 36 55 151	65 31 55 150	66 26 55 149	67 21 54 148	68 15 54 147	69 09 53 145	70 02 52 144	70 55 52 142	71 47 51 141	347
14	58 39 56 153	59 35 57 153	60 32 56 152	61 28 55 151	62 23 56 150	63 19 55 150	64 14 55 149	65 09 54 148	66 03 54 147	66 57 54 146	67 51 53 145	68 44 52 143	69 36 52 142	70 28 50 140	71 18 50 139	346
15	58 19 +56 152	59 15 +55 151	60 11 +55 150	61 06 +55 149	62 01 +55 149	62 56 +55 148	63 51 +54 147	64 45 +54 146	65 39 +53 145	66 32 +53 144	67 25 +52 142	68 17 +52 141	69 09 +50 140	69 59 +50 138	70 49 +48 136	345
16	57 58 55 150	58 53 55 149	59 49 55 148	60 44 54 148	61 38 55 147	62 33 54 146	63 27 54 145	64 20 54 144	65 14 53 143	66 06 52 142	66 58 51 140	67 50 50 139	68 40 50 138	69 30 49 136	70 19 48 134	344
17	57 36 55 148	58 31 54 147	59 25 55 147	60 20 54 146	61 14 54 145	62 08 53 144	63 01 53 143	63 54 53 142	64 47 52 141	65 39 51 139	66 31 51 138	67 22 50 137	68 12 49 135	69 01 48 134	69 50 47 132	343
18	57 12 55 147	58 07 54 145	59 01 54 145	59 55 54 144	60 49 53 143	61 42 53 142	62 35 52 141	63 27 52 140	64 19 51 139	65 10 51 138	66 01 50 137	66 50 49 135	67 40 48 134	68 28 47 132	69 15 46 130	342
19	56 48 54 145	57 42 54 144	58 36 53 143	59 29 53 142	60 22 53 141	61 15 52 141	62 07 52 140	62 59 51 139	63 50 51 138	64 41 50 137	65 30 50 135	66 20 48 134	67 08 47 132	67 55 47 130	68 42 45 128	341
20	56 23 +53 143	57 16 +54 143	58 10 +53 142	59 03 +52 141	59 55 +52 140	60 47 +52 139	61 39 +51 138	62 30 +50 137	63 20 +50 135	64 10 +49 134	64 59 +49 133	65 48 +47 131	66 35 +47 130	67 22 +45 128	68 07 +45 127	340
21	55 57 53 141	56 50 52 141	57 42 52 140	58 35 52 139	59 27 51 138	60 18 51 137	61 09 51 136	62 00 49 135	62 49 50 134	63 39 48 133	64 27 48 131	65 15 47 130	66 02 46 128	66 48 44 127	67 33 44 125	339
22	55 29 53 139	56 22 52 139	57 14 52 138	58 06 51 137	58 57 51 136	59 48 50 135	60 39 50 134	61 29 49 133	62 18 48 131	63 06 48 130	63 54 46 129	64 41 46 128	65 28 44 126	66 14 44 125	66 58 42 123	338
23	55 00 53 138	55 53 51 137	56 45 52 136	57 37 50 135	58 27 50 134	59 18 50 133	60 08 48 132	60 56 48 131	61 44 48 130	62 32 46 128	63 18 46 127	64 04 45 126	64 49 44 124	65 33 42 123	66 15 42 121	337
24	54 33 51 137	55 24 51 135	56 15 51 134	57 06 50 133	57 57 49 133	58 45 49 131	59 34 48 130	60 23 48 129	61 10 47 128	61 57 46 127	62 43 44 125	63 28 44 124	64 12 43 122	64 55 42 121	65 37 40 119	336
25	54 03 +51 135	54 54 +51 134	55 45 +50 133	56 35 +50 132	57 25 +49 131	58 14 +49 131	59 03 +48 130	59 51 +47 128	60 38 +47 127	61 25 +46 126	62 11 +45 125	62 56 +44 123	63 40 +41 122	64 23 +41 121	65 04 +40 119	335
26	53 32 51 134	54 23 50 133	55 13 50 132	56 03 50 130	56 53 48 129	57 41 48 129	58 29 48 128	59 17 47 127	60 04 46 126	60 50 45 125	61 35 44 123	62 20 43 122	63 04 43 121	63 46 41 119	64 28 39 118	334
27	53 01 51 133	53 52 50 132	54 42 49 131	55 31 49 130	56 19 49 129	57 08 47 128	57 55 47 127	58 42 46 126	59 28 46 125	60 14 45 123	60 59 44 122	61 43 43 121	62 26 41 119	63 08 41 118	63 49 40 116	333
28	52 30 50 132	53 19 50 131	54 09 48 130	54 57 49 129	55 46 48 128	56 33 48 127	57 21 46 125	58 07 46 125	58 53 45 123	59 38 44 122	60 22 43 121	61 05 43 119	61 48 41 118	62 29 39 116	63 09 39 115	332
29	51 57 49 131	52 46 50 130	53 35 49 128	54 24 47 127	55 11 48 127	55 59 46 126	56 45 46 125	57 31 45 124	58 16 45 123	59 01 43 121	59 44 43 120	60 27 41 118	61 08 40 117	61 48 40 115	62 28 39 114	331
30	51 24 +49 129	52 13 +48 128	53 01 +48 127	53 49 +47 126	54 36 +47 125	55 23 +46 124	56 09 +46 124	56 55 +45 122	57 40 +44 121	58 24 +43 120	59 07 +42 118	59 49 +42 117	60 31 +41 115	61 11 +39 114	61 50 +39 112	330
31	50 50 48 128	51 38 48 127	52 27 47 126	53 14 47 125	54 01 46 124	54 47 46 123	55 33 45 122	56 18 44 121	57 02 44 119	57 46 43 118	58 29 42 117	59 11 40 115	59 51 39 114	60 31 39 113	61 10 37 111	329
32	50 16 48 127	51 04 48 126	51 52 46 125	52 38 47 124	53 25 46 123	54 11 45 122	54 56 44 121	55 40 43 120	56 23 44 118	57 07 42 117	57 49 41 116	58 30 40 114	59 10 39 113	59 50 38 111	60 28 37 109	328
33	49 41 48 126	50 29 46 125	51 15 46 124	52 02 46 123	52 48 45 122	53 33 45 121	54 19 44 120	55 03 43 118	55 46 43 117	56 29 42 116	57 11 41 114	57 52 39 113	58 31 39 111	59 10 37 110	59 47 37 108	327
34	49 06 47 124	49 53 47 123	50 41 45 123	51 26 45 122	52 12 45 120	52 56 45 119	53 41 43 118	54 24 43 117	55 07 42 116	55 49 41 115	56 31 40 113	57 11 40 111	57 51 38 110	58 30 37 108	59 08 37 107	326
35	48 30 +47 123	49 17 +46 122	50 03 +46 121	50 49 +45 120	51 34 +44 119	52 18 +44 118	53 02 +43 117	53 46 +42 116	54 28 +42 115	55 10 +40 113	55 51 +39 112	56 30 +39 110	57 11 +38 109	57 48 +37 107	58 27 +36 105	325
36	47 54 46 122	48 40 46 121	49 26 45 120	50 11 45 119	50 56 44 118	51 40 43 117	52 24 43 116	53 07 42 115	53 50 41 113	54 31 41 112	55 12 39 111	55 52 39 109	56 31 37 108	57 09 37 106	57 46 35 105	324
37	47 17 46 121	48 03 45 120	48 48 45 119	49 33 44 118	50 17 43 117	51 00 44 116	51 44 42 114	52 26 42 113	53 08 41 112	53 50 40 111	54 31 39 109	55 10 38 108	55 49 38 106	56 27 37 105	57 04 35 104	323
38	46 40 45 119	47 25 45 118	48 11 43 118	48 55 44 117	49 39 44 116	50 23 42 115	51 05 42 113	51 48 41 112	52 29 40 111	53 09 40 110	53 50 39 108	54 28 38 107	55 07 36 105	55 44 36 104	56 22 35 102	322
39	46 02 45 118	46 47 44 117	47 32 44 116	48 16 44 116	49 00 42 115	49 43 42 114	50 26 42 113	51 08 41 111	51 49 40 110	52 29 39 108	53 08 39 107	53 48 38 106	54 25 36 104	55 03 34 103	55 40 35 102	321
40	45 24 +45 118	46 09 +45 117	46 54 +43 116	47 37 +44 114	48 21 +43 114	49 04 +42 113	49 46 +41 112	50 27 +41 111	51 08 +40 109	51 49 +38 108	52 28 +39 107	53 07 +36 106	53 45 +36 104	54 21 +37 103	54 58 +35 102	320
41	44 46 44 117	45 30 43 116	46 15 43 115	46 58 43 114	47 41 42 113	48 24 42 112	49 06 40 111	49 47 41 110	50 28 39 109	51 07 39 108	51 47 38 106	52 25 36 105	53 04 36 104	53 41 35 102	54 15 35 101	319
42	44 07 44 116	44 51 44 115	45 35 43 114	46 18 42 113	47 01 42 112	47 43 41 111	48 24 41 110	49 06 40 109	49 45 40 108	50 25 38 106	51 03 38 105	51 42 36 103	52 18 36 102	52 56 35 101	53 31 34 99	318
43	43 28 44 115	44 12 43 114	44 55 42 113	45 39 42 112	46 21 41 111	47 03 41 110	47 44 41 109	48 25 40 108	49 05 39 107	49 44 38 106	50 22 37 104	51 00 37 103	51 37 35 102	52 13 35 100	52 48 34 98	317
44	42 49 44 115	43 33 43 113	44 16 43 112	44 59 41 111	45 41 41 110	46 22 41 109	47 03 41 108	47 44 40 107	48 23 39 106	49 02 38 105	49 41 37 104	50 18 36 102	50 54 35 101	51 30 34 99	52 04 34 98	316
45	42 09 +44 114	42 53 +43 112	43 36 +42 111	44 18 +42 110	45 00 +41 109	45 41 +41 108	46 22 +40 107	47 02 +40 106	47 42 +39 105	48 21 +38 103	49 00 +37 102	49 37 +37 101	50 14 +36 100	50 50 +35 99	51 25 +34 97	315
46	41 29 43 112	42 12 43 111	42 55 42 110	43 37 42 110	44 19 41 109	45 00 41 108	45 41 40 106	46 21 39 105	47 00 39 104	47 39 38 103	48 17 37 102	48 54 37 100	49 31 36 99	50 07 35 98	50 42 34 97	314
47	40 49 43 112	41 32 43 110	42 15 41 109	42 56 42 108	43 38 41 108	44 19 40 107	44 59 40 106	45 39 39 104	46 18 39 103	46 57 38 103	47 34 37 101	48 11 36 100	48 48 35 98	49 23 34 97	49 57 34 96	313
48	40 09 42 111	40 51 43 110	41 34 41 109	42 15 41 108	42 57 40 107	43 37 40 106	44 18 40 105	44 58 39 104	45 37 38 102	46 15 38 101	46 53 37 100	47 30 36 99	48 07 35 98	48 42 35 96	49 17 34 95	312
49	39 28 43 109	40 11 42 108	40 53 41 107	41 34 41 106	42 15 40 105	42 56 40 104	43 36 40 104	44 16 38 103	44 55 38 102	45 33 37 100	46 10 37 99	46 47 35 98	47 23 35 97	47 58 35 96	48 33 34 94	311
50	38 47 +42 108	39 29 +42 107	40 11 +42 107	41 33 +41 106	41 33 +41 105	42 14 +40 104	42 54 +40 103	43 33 +38 102	44 11 +39 100	44 50 +37 99	45 27 +37 98	46 04 +36 97	46 40 +35 96	47 15 +35 95	47 50 +34 94	310
51	38 06 42 108	38 48 42 107	39 30 41 106	40 12 40 105	40 52 41 104	41 34 40 103	42 14 39 102	42 53 39 101	43 31 38 100	44 09 38 99	44 46 37 98	45 23 35 96	45 58 35 95	46 33 34 94	47 08 34 93	309
52	37 24 42 107	38 07 41 106	38 48 41 105	39 29 41 104	40 10 40 103	40 50 40 102	41 30 39 101	42 09 39 100	42 48 38 99	43 26 37 98	44 03 37 97	44 40 36 95	45 16 35 94	45 51 34 93	46 26 34 92	308
53	36 43 42 106	37 25 41 105	38 06 41 104	38 47 40 103	39 27 40 102	40 07 40 101	40 47 39 100	41 26 38 99	42 04 38 98	42 42 37 97	43 19 36 96	43 55 35 94	44 31 35 93	45 05 34 92	45 40 33 91	307
54	36 01 41 105	36 43 41 104	37 24 41 103	38 05 40 102	38 45 40 101	39 25 39 100	40 04 39 99	40 43 38 98	41 21 38 97	41 59 37 96	42 35 36 95	43 11 36 94	43 46 35 92	44 22 34 91	44 56 33 90	306
55	35 20 +41 104	36 01 +41 103	36 42 +40 102	37 22 +40 101	38 02 +39 100	38 41 +39 99	39 20 +39 98	39 59 +38 97	40 37 +38 96	41 15 +37 95	41 52 +36 94	42 29 +35 92	43 04 +35 91	43 40 +34 90	44 14 +34 89	305
56	34 38 41 103	35 19 41 102	36 00 40 101	36 40 40 100	37 20 39 99	37 59 39 98	38 39 38 97	39 17 38 96	39 55 37 95	40 33 37 94	41 10 36 93	41 46 35 92	42 22 34 91	42 57 34 89	43 31 33 88	304
57	33 56 41 102	34 37 40 101	35 17 40 100	35 57 40 99	36 37 39 98	37 16 39 97	37 55 38 96	38 33 38 95	39 11 37 94	39 48 37 93	40 25 36 92	41 01 35 91	41 37 35 90	42 12 34 89	42 46 33 87	303
58	33 14 41 101	33 55 40 100	34 35 40 99	35 15 40 98	35 54 39 97	36 33 39 96	37 12 38 95	37 50 37 94	38 27 37 93	39 04 37 92	39 41 36 91	40 17 35 90	40 52 34 89	41 27 33 88	42 00 33 86	302
59	32 31 41 101	33 12 40 100	33 52 40 99	34 32 39 98	35 11 39 97	35 50 39 96	36 29 38 95	37 07 38 94	37 45 37 93	38 22 36 92	38 59 36 90	39 35 35 89	40 11 34 88	40 46 33 87	41 19 33 85	301
60	31 49 +41 100	32 30 +40 99	33 10 +40 98	33 50 +39 97	34 29 +39 96	35 08 +39 95	35 47 +38 94	36 25 +38 93	37 02 +37 92	37 40 +37 91	38 17 +36 90	38 53 +35 89	39 29 +34 87	40 04 +33 86	40 37 +33 85	300
61	31 07 40 99	31 47 40 98	32 28 40 97	33 07 39 96	33 47 39 95	34 25 38 94	35 04 38 93	35 42 38 92	36 20 37 91	36 57 36 90	37 33 36 89	38 09 35 88	38 45 34 86	39 20 33 85	39 53 33 84	299
62	30 24 40 98	31 04 40 97	31 44 40 96	32 24 39 95	33 03 39 94	33 42 38 93	34 20 38 92	34 58 38 91	35 37 37 90	36 13 37 89	36 50 36 88	37 26 35 87	38 01 34 85	38 36 33 84	39 10 33 83	298
63	29 41 40 98	30 22 39 97	31 02 39 96	31 41 39 95	32 20 38 93	32 58 38 92	33 37 38 91	34 15 38 90	34 53 37 89	35 30 36 88	36 06 36 87	36 42 35 86	37 17 34 85	37 52 33 84	38 25 32 82	297
64	28 58 41 97	29 39 39 96	30 18 40 95	30 58 38 94	31 36 39 93	32 15 38 92	32 53 37 91	33 31 37 90	34 08 37 89	34 45 36 88	35 21 36 87	35 57 35 86	36 33 34 84	37 07 33 83	37 41 32 81	296
65	28 16 +40 96	28 56 +40 95	29 36 +39 94	30 15 +39 93	30 54 +39 92	31 33 +39 92	32 12 +38 91	32 50 +37 90	33 27 +37 89	34 04 +36 88	34 40 +36 86	35 16 +35 85	35 52 +34 84	36 26 +34 83	37 00 +33 82	295
66	27 33 40 95	28 13 40 94	28 53 39 93	29 32 39 92	30 11 38 91	30 49 38 90	31 27 38 89	32 04 38 88	32 42 36 87	33 18 37 86	33 54 36 85	34 30 35 84	35 05 34 83	35 40 34 82	36 14 33 81	294
67	26 50 40 94	27 30 39 93	28 09 39 92	28 48 39 91	29 27 38 90	30 05 38 89	30 43 37 88	31 20 38 87	31 58 37 86	32 35 36 85	33 11 36 84	33 47 35 83	34 22 34 82	34 57 34 81	35 31 32 80	293
68	26 07 40 93	26 47 39 93	27 26 39 92	28 05 38 91	28 43 39 90	29 22 37 89	29 59 38 88	30 37 37 87	31 14 36 86	31 51 36 85	32 27 35 84	33 02 35 83	33 38 34 82	34 12 33 81	34 45 33 80	292
69	25 24 40 93	26 04 39 92	26 43 39 91	27 22 38 90	28 00 38 89	28 38 38 88	29 16 37 87	29 53 37 86	30 30 36 85	31 06 36 84	31 42 35 83	32 17 35 82	32 52 34 81	33 27 33 80	34 00 32 79	291

DECLINATION (15°–29°) SAME NAME AS LATITUDE

TABLE 5—CORRECTION TO TABULATED ALTITUDE FOR MINUTES OF DECLINATION

These tables, reduced, are taken from *Sight Reduction Tables for Air Navigation* with the permission of the Controller of H.M. Stationery Office.

In certain circumstances it is possible for the L.H.A. to exceed 360°; 360 must then be deducted from it as in the working of the Polaris sight on page 223. If the assumed west longitude is greater than the G.H.A., the L.H.A. will be a minus quantity; 360 must then be added to it.

Having arranged an assumed position such that *S.R.T.* can be entered with it, the appropriate volume is opened at the pages headed with the assumed latitude. It will be noticed that some pages are headed 'Declination *same* name as latitude', and others 'Declination *contrary* name to latitude'. It is essential to use the correctly labelled page. From the reproduction of a page of *S.R.T.* on page 226, it will be seen that it is divided into vertical columns, one for each whole degree of declination, the declination figures being boldly printed at top and bottom. Degrees of L.H.A. are printed down the sides. Entering the table with the whole degree of declination numerically less than that of the body at the time of the sight, and with the L.H.A., three sets of figures will be found in a row; from left to right these are Hc, *d*, and Z. Hc stands for height calculated, i.e. the altitude of the body calculated for the assumed position. The *d* figure (the difference in Hc for one degree difference in declination), prefixed by a plus or minus sign, enables a correction to be made to the Hc figure to allow for the minutes of declination, as explained below. The Z figure is the azimuth. This is given in degrees from the elevated (nearest) pole, and must be named east or west according to whether the body has not yet crossed the observer's meridian, or has crossed it. If preferred, it may be converted to a true bearing, i.e. expressed in notation 'A' (0° to 360° clockwise from true north) by following the rules printed on each page of *S.R.T.*

The *d* figure is used as follows for correcting Hc for the minutes of declination. Suppose, for example, the body's declination at the time of the sight was 29° 33′, and the *d* figure was +45. Turn to the correction table at the end of *S.R.T.* (this is reproduced on page 227), and entering it across the top with the *d* figure, 45, and down the side with the minutes of declination, 33, the figure 25 will be found. This is the correction in minutes of arc, and as the *d* figure was preceded by a + sign, 25′ must be added to the Hc figure to correct it for those 33′ of declination.

To demonstrate the simplicity of *S.R.T.*, here is an example of a sun sight showing all the work that is needed, and it is followed by a brief explanation of each step to show how or where the figures were obtained.

Example. On Saturday, 28 June 1958, in D.R. lat. 43° 48′ N., long. 20° 45′ W. the sun was observed in the forenoon at 10h. 24m. 47s. G.M.T. The observed altitude was 47° 38′, and H.E. 5 feet. Required the assumed position, the intercept, and the azimuth.

Obs. Alt.	47° 38′	Sun's G.H.A. at 10h.	329° 14′·4
Sun Corr.	+15′·1	Increment for 24m. 47s.	+6° 11′·8
Dip	−2′·3		
		G.H.A.	335° 26′
True Alt.	47° 51′	Assumed longitude	−20° 26′ W.
		L.H.A.	315°

Assumed latitude	44° N.	Dec. 23° 18′ N.	Same name.
Hc	47° 42′	d+39	Z 105 = N. 105° E.
d corr.	+12′		

Calc. Alt.	47° 54′
True Alt.	47° 51′

Intercept	3′ away

Obs. Alt. The observed altitude is the angle measured between the visible horizon and the sun's lower limb, index error of the sextant having been allowed for when reading the instrument.

Sun Corr. The correction for semi-diameter and refraction, taken from inside the front cover of the *N.A.*, the table being entered with the observed altitude.

Dip. The correction necessary because the visible, not the rational, horizon was used. The dip table is inside the front cover of the *N.A.*, and was entered with the height of the observer's eye above the sea, which was 5 feet.

True Alt. The altitude between the rational horizon and the sun's centre. Note that although decimals of a minute of arc are included in the corrections for refraction, semi-diameter, and dip, the true altitude is written to the nearest whole minute only, because Hc is given for the nearest whole minute only in *S.R.T.* The same applies when taking the G.H.A. and Dec. from the *N.A.*

Sun's G.H.A. at 10h. Obtained from the appropriate daily page (28 June) in the *N.A.*

Increment for 24m. 47s. Obtained from the increments table headed 24m., found on the coloured pages near the end of the *N.A.*, and entered with 47 seconds.

G.H.A. The angular distance of the meridian of the sun measured westward from the meridian of Greenwich at the G.M.T. of the sight, i.e. at 10h. 24m. 47s. on 28 June 1958.

Assumed longitude. This is as near as possible to the D.R. longitude, but is such that when subtracted from the G.H.A. the seconds cancel out.

L.H.A. The local hour angle, i.e. the angular distance of the sun's meridian measured westward from the assumed longitude; it is obtained by subtracting west longitude from, or adding east longitude to, the G.H.A. of the body.

Assumed latitude. The whole degree of latitude nearest to the D.R. latitude.

Dec. The declination (angular distance of the sun north or south of the equator) at the G.M.T. of the sight is obtained to the nearest minute of arc from the daily page of the *N.A.*

Same name. A reminder that as latitude and declination are both north, a page in *S.R.T.* headed '*Same* name' must be used.

Hc. Height calculated is obtained from *S.R.T.* by entering the table headed 'Lat. 44°, declination *same* name as latitude' with Dec. 23° and L.H.A. 315°. This table is reproduced on page 226.

d. This is found in the same column and next to the Hc figure. It will be used to correct Hc for the 18′ of declination. Note that it is given a + sign in the table.

Z, the azimuth, is given in the same column and same line as the Hc and *d* figures. This sight was taken in the northern hemisphere and in the forenoon, so the sun's bearing must be reckoned from north in an easterly direction, N. 105° E.

d Corr. The correction for minutes of declination is obtained from Table 5 in *S.R.T.*, which is reproduced on page 227. The table is entered across the top with *d* 39, and down the side with the minutes of declination, 18, and gives a correction of 12′. As *d* had a + sign, the correction must be added to the Hc figure.

Calc. alt. The altitude of the sun calculated for the assumed position at the time of the sight, and corrected for the minutes of declination.

Intercept. The difference between the calculated altitude at the assumed position and the true altitude at the observer's position, i.e. the distance the position line must be moved away from the assumed position along the azimuth either towards or away from the sun according to the following rule: If the calc. alt. is the greater, the intercept must be measured away from the body; if the observed altitude is the greater the intercept must be measured towards the body. An easy way to remember this rule is by means of the word GOAT—greater observed altitude towards.

Here is another example of a sun sight, taken during the afternoon of 6 September 1954 when nearing the island of Rodriguez in the Indian Ocean. It differs from the previous example only in the following respects. As the assumed longitude was east, it had to be *added* to the G.H.A. to obtain the L.H.A., and as the L.H.A. exceeded 360°, 360 had to be taken from it; as latitude was south and declination was north, a page in *S.R.T.* headed 'Declination *contrary* name to latitude' had to be used.

At 9h. 44m. 31s. G.M.T. on 6 September 1954, in D.R. lat. 19° 23′ S., long. 64° 40′ E., the observed altitude of the sun was 49° 37′; H.E. 5 feet. Required the intercept and azimuth.

Obs. Alt.	49° 37′	Sun's G.H.A. at 9h.	315° 22′·3
Sun Corr.	+15′·2	Increment for 44m. 31s.	11° 07′·8
Dip	−2′·3		
		G.H.A.	326° 30′
True Alt.	49° 50′	Assumed longitude	+64° 30′ E.
			391°
			−360°
		L.H.A.	31°

Assumed latitude 19° S.   Dec. 6° 34′ N.   Contrary names.
Hc   50° 32′   *d*−39   Z 126 = S. 126° W. or 306° notation 'A'.
d corr.   −22′

Calc. Alt.	50° 10′
True Alt.	49° 50′

Intercept   20′ away

To show all the arithmetic that is necessary in 24 hours, when two sun sights—one in the forenoon and one at noon—are taken, a page from my navigating work-book is reproduced on page 232. At the time *Wanderer III* was in the southern hemisphere on passage from Cape Town towards St. Helena and was in the Greenwich time zone. So that the reader may understand what each figure represents, I have added abbreviations, which do not normally appear in the work-book, and have included a key at the foot of the page to such abbreviations as have not previously been used in this chapter.

All the plotting that was done during those 24 hours is shown on the position line chart, Fig. 63 (page 234), and here again abbreviations, which are not used in practice, have been added. As the yacht was in latitude 25° 37′ S., longitude 6° 26′ E. at noon on 25 March, the central horizontal line on the position line chart was named 25° S., and a meridian towards the east side of the chart was named 6° E.

Using the longitude scale on the chart, and the latitude scale labelled 25°-26° on one of the folding flaps (in the figure this scale is indicated by arrows and the letter *S*.), the observed noon position (*N.P.*) was plotted, and the date, 25 March, written beside it. (A note of the patent log reading, 763′, was made in the work-book.) From that position the course steered, 310° true, was laid down with the parallel rules, using the circle marked in degrees.

Saturday 26 March 1955

a.m. Sun

Ch.	07h. 48m. 58s		L	845
S.W.	—	41		- 763
	7	48	17	Run	82
E. Fast	—	3	15		
	7	45	02		

D.R. 24° 40′ S., 5° 21′ E.

Obs alt	25° 58′
Sun corr.	+ 12
True alt	26 10

G.H.A. at 07h.	283° 30′·1
Increment for 45s. 02m	11 15·5
G.H.A	294 46
Assumed Long.	5 14 E.
L.H.A.	300

Assumed Lat. 25° S.   Dec 1° 56′ N. Contrary names

Hc   26° 30′   d — 29.   Z 105° = N. 75° E.

d. corr	— 27
Calc. Alt.	26 03
True Alt.	26 10
Intercept	7′ towards

Noon Lat.		L862
Obs. alt.	63° 24′
Sun corr.	+ 11
True alt.	63 35
	90
Z.D.	26 15
Dec.	- 2 00 N.
Lat.	24° 25′ S

Key. Ch. = chronometer reading.   S.W. = stopwatch reading.
      E = chronometer error.   L = patent log reading.

A page from the author's navigation work-book, showing the small amount of arithmetic that is required for obtaining an a.m. position line and a noon latitude. The abbreviations have been added for the reader's benefit and are not used in practice.

At the time of the a.m. sight on 26 March, the patent log read 845′; the reading at noon the previous day was deducted, leaving 82′, the distance run. As no current was suspected, that distance was measured off with the dividers along the course line, and the resulting D.R. position at the time of the a.m. sight was marked on the chart (*D.R.*) and noted in the work-book. The sight was worked (see opposite page), the assumed position for that sight was marked on the chart (*A.P.*), the azimuth (*Az.*) drawn through it, and the intercept (*Int.*), 7′ towards, measured along it towards the sun. Through that point the position line (*P.L.*) was drawn at right angles to the azimuth.

At noon the meridian altitude of the sun was observed and the sight worked out (see reproduction of work-book page opposite), and the patent log reading, 862′, was noted. The observed latitude, 24° 25′ S., was ruled on the chart (*Lat.*). The distance run from the a.m. sight, 17′, was measured along the course steered from the point where the morning position line cut it, and marked (*D.R. 2*). The a.m. position line was then carried forward with the parallel rules and drawn through that point (*P.L. 2*). Where that line crossed the noon latitude line was the noon position (*N.P.*). To find the day's run, the direct distance between the noon position on the 25th and the noon position on the 26th was measured on the position line chart, using, of course, the appropriate scale of latitude. The run was found to be 106′; but by patent log the run was only 99′; the presence of a favourable current of 7′ per day was therefore indicated. Finally, the noon position was transferred to the South Atlantic chart so that the course could be checked and altered if necessary. I have found the older type of patent log, with a rotator on a line astern, to be remarkably accurate, except in waters where there is weed floating on the surface, as in and near the Sargasso Sea and the Gulf Stream, for the weed tends to slide down the line and foul the rotator. The impeller of the modern electronic log is less likely to get fouled, but the instrument will get through a surprisingly large number of batteries if it is kept in constant use throughout a long passage.

The working of a planet sight is identical with that of a sun sight, except that the correction for refraction must be taken from the correction table in the *N.A.* headed 'Stars and Planets'.

A star sight is similar except that as the G.H.A. of the individual stars is not given in the *N.A.* it has to be calculated. The *N.A.* gives the G.H.A. of a point in the heavens known as Aries, to which the stars are related; it also gives the sidereal hour angle (S.H.A.) of all the navigational stars. S.H.A. is the angular distance of the meridian

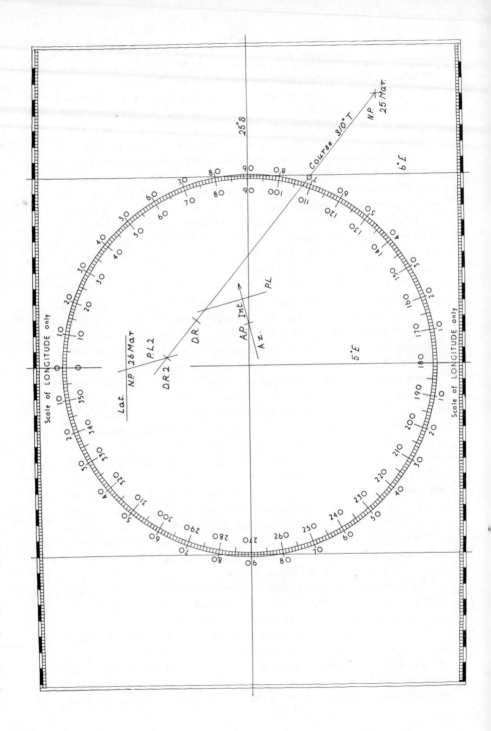

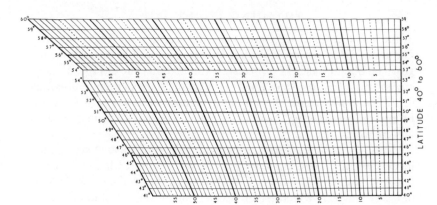

Scale of MILES or MINUTES of LATITUDE

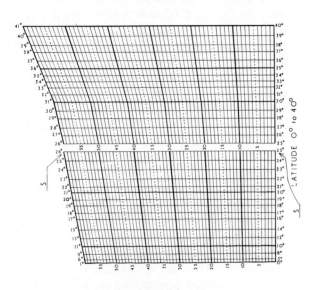

FIG. 63. BAKER'S POSITION
LINE CHART

The upper part shows the chart with a day's work drawn on it; the lower part shows the latitude scales, which normally are printed on folding flaps at the west and east edges of the chart. In use the longitude scale remains constant, while the scale of latitude and distance for the latitude in which the ship lies is always used, thus making the chart into a blank Mercator chart for any latitude from 0° to 60°. (Reproduced by courtesy of Imray, Laurie, Norie & Wilson.)

of the star measured westward from the meridian of Aries, and this, like a star's declination, does not change appreciably throughout the year. To obtain the G.H.A. of a star, add together the G.H.A. of Aries and the S.H.A. of the star. The L.H.A. of the star can then be obtained in the same way as for the sun, and Volume II or III of *S.R.T.* can be entered in the usual manner, provided the star's declination does not exceed 29°.

However, if the navigator intends to make much use of star sights, he may decide to obtain a copy of *S.R.T.* Volume I, which is devoted entirely to stars and has certain advantages, but unlike *S.R.T.* Volumes II and III, which never become out of date, Volume I needs to be recomputed after approximately 5 years, and then a new edition is published. Its method of use is different, for instead of being entered with the assumed latitude and the declination and L.H.A. of the observed star, it is entered with the assumed latitude and L.H.A. of Aries; when it at once indicates which 6 of the 34 stars tabulated may be used for navigation from the ship's position at the chosen time. For each entry of assumed latitude and L.H.A. of Aries, the Hc (altitude) and Z (azimuth) of the 6 selected stars will be found in adjoining columns on the same page. Before taking any sights one should first consult the tables to find what stars can be used. This is done by making in advance an approximate determination of the L.H.A. of Aries at the intended G.M.T. of the observations, using the D.R. and the *N.A.* With the information so obtained, the tables are entered, and the navigator can make his selection of 3 of the 6 available listed stars, noting for each the Hc and Z. From this it will be understood that Volume I serves also as a star finder for a limited number of stars. In making his selection from the 6 stars available, the navigator will be guided by considerations of magnitude (apparent brightness) and azimuth, choosing as far as possible stars easily seen and evenly spaced around him so that the resulting 'cocked hat' may be approximately in the form of an equilateral triangle. At twilight he will set his sextant to the Hc of the first star, sweep the horizon in the direction indicated by the azimuth, pick up the star in the telescope, make contact, and note the time in the usual way, repeating the procedure with the other two stars. The L.H.A. of Aries at the G.M.T. of each sight is then obtained from the *N.A.*, and with these figures and the assumed latitude the volume of tables is again entered to obtain the Hc and Z of each star, these figures of course being needed for working and plotting the three observations. The need to enter the tables in advance, and again after the sights have been taken, may seem to be a complication, but in practice much time is saved in identifying and observing and, particularly in the tropics, where twilight is brief, the results will be more certain.

When the moon is visible during daylight and the azimuths of moon and sun differ by about 45° or more, simultaneous sights of moon and sun will give a fix. The working of a moon sight differs from that of a sun sight only in the manner of obtaining the true from the observed altitude. Having corrected for dip in the usual way, the altitude correction table for the moon (inside the back cover of the *N.A.*) is entered in two places. First the upper part of the table is entered with the altitude corrected for dip, then the lower part is entered in the same column with the moon's horizontal parallax (obtained from the daily page of the *N.A.*). The two corrections are added to the altitude which has been corrected for dip, to obtain the true altitude. If the upper limb is observed, 30′ must be deducted. There is no need to memorize this procedure as an explanation is printed with the correction tables in the *N.A.*

Making a landfall

When the time approaches to make a landfall, the only sights that really matter are the last pair the navigator has taken, and if he was satisfied with them and made no error in their working, and was certain of the G.M.T., the landfall should be a good one, that is, the right piece of land should appear ahead at the estimated time. This is the common experience of most yachtsmen. Nevertheless there will be a feeling of excitement aboard as the time draws near for the land to lift above the horizon, or materialize through the haze, for the thrill of making the land, especially after one has been for many days out of sight of it, never grows stale. But the navigator, no matter how experienced he may be, will feel more tense than his shipmates, for the responsibility is his, and it is probable that he will start looking too early, and perhaps a little anxiously, for land, especially if the landfall is to be on a low and featureless coast, or a comparatively small island which he is fearful of missing, when it might be necessary to sail on a great distance to find a larger target. Such a happening has occurred on many occasions in the past, when, before the advent of radio, the navigator had no means of telling that his chronometer had changed its rate, and that his longitude was therefore badly in error. The risk of that happening today is small; but if for some reason G.M.T. is not known, the navigator can resort to latitude sailing, that is, get on to the latitude of the island some distance east or west of it, and sail along that latitude, checking it with meridian altitude sights (which are independent of G.M.T.) until he reaches his objective, just as his forefathers did before the advent of the chronometer.

A landfall on a high island should present no problem if the weather is clear, for the land will be visible a long way off. Even before it is visible there will probably be signs of it, such as the appearance of land birds, or an accumulation of cloud standing above it (Plates 39, *bottom*, and 40, *bottom*), or perhaps a cloud trail driven by the wind streaming away like smoke for many miles to leeward. I have seen this on several occasions, but it was most marked to leeward of the 1,300-foot-high island of Rodriguez, where it could clearly be seen while the island was well out of sight (Plate 40, *middle*). But one must not always expect the land to be cloud-capped, or to be able to see high land at a great distance, even though visibility may appear to be excellent, a lesson I learnt when making the island of San Miguel, one of the Azores. The day was bright and sunny with a vivid blue sea and sky and a sharp horizon, and I guessed the visibility to be about 40 miles, or more. The island rises to 3,600 feet, and at 0600 it was by account less than 40 miles away, so I naturally expected to sight it soon. But the morning wore on with an empty horizon, and although terns visited us, and my wife, who is a non-smoker, could smell pines and sweet-scented flowers, we had no sight of land or of any cloud in the sky to suggest the proximity of land; and not until we had sailed to within 15 miles of it did the island appear faintly through its unsuspected shroud of mist, which was the same colour as the sky. Then slowly, as we approached, it took shape (Plate 39, *top*); the volcanic peaks, the dark green areas of trees, a patchwork of tiny fields, and clusters of whitewashed dwellings hanging on the slopes, gathered colour and substance.

If a landfall has to be made on a small, low island, there may be no intimation of the island's nearness, for it may be lacking in bird life, and it will probably have no cloud cap, though if it lies beside a lagoon there is just a chance that light reflected from the shallows may tint the atmosphere above with a green hue—I have seen this phenomenon several times among the Tuamotu, and once on approaching South Caicos in the Bahamas. Because of the island's low elevation any palms growing on it will be visible from the deck of a yacht only 10 to 15 miles off (Plate 40, *top*). Every possible opportunity should be taken of fixing the position, but if one is uncertain of the position in one direction, then the island should be approached on a line of which one is more certain. A lookout should be kept from aloft, because of the increased distance seen with the extra height of eye; it is probable that *Driac II* would have missed Keeling Cocos had her crew not taken that precaution, and the trouble about missing Cocos is that the next land to leeward is 2,000 miles farther on.

The good seaman will study all the available information before deciding whether or not it is prudent to attempt to close with the land in the weather conditions prevailing at the time, or during the hours of darkness, and if he is accustomed to cruising in well-marked European or American waters, will remember that some of the coasts he may wish to visit across the oceans are devoid of all lights, buoys, fog signals, and radio beacons, and that the loom of a city's lights in the sky may be rare. As *Reed's Nautical Almanac*, which most yachts in U.K. waters carry as a matter of course, does not give lights or tides south or west of Brest, the foreign-going yacht will need to have the appropriate volume or volumes of *The Admiralty List of Lights* and *The Admiralty Tide Tables*. The former is in twelve volumes, the latter in three, and details of the areas covered by each volume will be found in *The Catalogue of Admiralty Charts*.

Pilotage among coral

The most practical methods of fixing a small vessel's position when within sight of terrestrial objects, together with the more usual solutions of the various pilotage problems which arise in coastal waters, were discussed at some length in *Cruising Under Sail*; but a few words on pilotage among coral reefs should be included here.

Some corals, singly or in small colonies, are found in all oceans even in cold water, but the reef-building corals can live only in clear water the temperature of which seldom drops below 70 °F. They are therefore mostly restricted to a belt a little wider than the tropical zone, but due to warm currents they may thrive elsewhere, as, for example, around Bermuda, because of the Gulf Stream. Most coral reefs will be found on the windward sides of continents and islands, i.e. on their eastern sides, having regard to the direction of the trade winds; but they may also be found on the lee side of an island if a current carries there the plankton on which they live.

The first essential in coral waters is vigilance. No opportunity of fixing or confirming the ship's position, or of obtaining a single position line, should be neglected, and all possible forethought and cunning must be brought into use, changing course, or heaving-to, so as to avoid getting into a dangerous position in darkness or bad visibility. Fog in coral waters is almost unheard of, but rain can reduce visibility to a few yards and last for a long time, though generally one has warning of its approach. The coral-encumbered waters in some parts of the world have not been surveyed in detail, and even where a thorough and detailed survey has been made, the charts in time become inaccurate because of the growth of coral; incidentally, coral never grows above the surface of the sea, though blocks of dead coral may be thrown by the breakers up on to a reef. Successful pilotage in such waters is therefore often dependent on the eye, and provided that conditions are right, this does not present any great difficulty, and confidence is gained with practice.

A barrier reef may front a coastline, as along the east coast of Australia, or encircle an island or group of islands, as in the Society Islands, leaving a channel or a lagoon between it and the land. An atoll comprises one or more coral islands of little elevation situated on a strip or ring of coral reef surrounding a central lagoon. Such reefs as the above frequently rise steeply without warning from a depth of 1,000 fathoms or more, and on them the sea breaks with a fury that has to be seen to be believed. Usually the swell produced by gales in the southern part of the Pacific and Indian Ocean causes heavier breakers on the leeward side of a reef than the trade wind causes on the weather side; when approaching a reef down wind by night, which is an exceedingly dangerous thing to do and has been the cause of many wrecks, the breakers cannot easily be distinguished from the harmless crests caused by a fresh wind until they are too close for avoiding action to be taken, nor can their roar be heard up wind above the noises of the ship. The pass through a barrier reef is generally opposite a valley in the land, but the pass of an atoll (not all atolls have passes) is generally on the lee side of the lagoon, and as that may be the only exit for the surplus water thrown by the breakers over the reef into the lagoon, there is frequently a strong out-running current.

The lagoon, or the channel, protected by a coral reef, may be free of dangers, it may have isolated dangers, or it may contain a maze of reefs and coral heads. Reefs are always more plainly seen from aloft than from on deck, and when the sun is higher rather than when it is low; also, the sun should be abaft the observer's beam, and the water rippled by a breeze; if there is no wind, nothing can be seen but the reflection of the sky. The wearing of polaroid spectacles is strongly recommended because they minimize light reflected from the surface of the water. Given the right conditions, the scene from aloft is very beautiful, and the depth of water can be judged by the colour. Approaching a channel, lagoon, or bank from the deep blue—almost mauve—of the ocean off soundings, the colour changes abruptly to a light blue where the depth over a clean bottom is 10-15 fathoms; but if the bottom there is rocky, the colour will be dark green. As the depth decreases so the colour changes, and in 4-5 fathoms over a sand bottom will be light green, in 1½-2 fathoms pale green, and in less than 1 fathom the water is almost colourless, indeed, in the Bahamas this is known as 'white water'. On a sand bottom coral heads, often called 'nigger heads', show up clearly as black, brown, or yellow patches; but if the bottom is grassy they will be hard to see. A pale mauve or russet colour means there is very little water. My wife and I did not find pilotage by eye in the Pacific particularly difficult, and in most of the lagoons we visited it was possible to sail round any apparent dangers; but where such dangers have to be crossed, as for example on some parts of the Bahama banks, skill in judging the depth by colour to within a few inches may be needed, and it is said that to acquire this skill takes 18 months or more.

R

10

WELFARE OF THE CREW

Provisioning—Sleep—Health—Pest control—Pets—Shooting and fishing

I T is a commonplace that a seaworthy, well-found, cruising yacht can withstand more bad weather than her crew; and when a yacht does get into trouble, it is usually due to the physical or mental failure of her people; indeed, many a yacht abandoned at sea has later been towed into port undamaged. The well-being of everyone on board is therefore of the first importance, so that all necessary actions may be taken and wise decisions made, and this will depend largely on proper food and sound sleep, both of which can be ensured by good organization, determination, and attention to detail. But mental fatigue is not so easily prevented, although food and sleep are essential attributes. In bad weather, for example, once everything possible has been done for the safety of the ship and there is no further immediate action to be taken, it is possible for one's mental uneasiness —probably caused by the noise more than anything else—to develop into serious worrying, which can be very fatiguing. Some people find that on such occasions reading light fiction acts as a mental sedative, others that listening to a radio broadcast has an equally tranquillizing effect. Again, it is possible for the voyage itself to cause worry—what will happen, one may wonder, if bad weather is encountered, if the navigation is faulty, or the anchorage, when it is reached, is untenable? This kind of worry is more serious because it is almost continuous, and there have been instances of people making themselves ill by it. It is difficult to offer advice; certainly you should read in advance the accounts of other small vessels' voyages along your route, and con- sider that as all went well with them, so it should with you; endeavour to relax with outside interests when in port, getting right away from the ship for a few days if possible, and, above all, bear in mind the wise advice which was given to me by Arthur Ransome when I was worrying about time and route planning and the difficulties of naviga- tion among coral reefs: 'Too many of you ocean-voyaging people wear yourselves out,' he wrote, 'enjoyment, not endurance, should be your watchword.'

Provisioning

It is not possible to give any precise information about provisioning for an ocean passage, as so much will depend on the tastes and appetites of the crew—some, such as Frank Wightman in *Wylo*, may be content with a diet of raisins, nuts, honey, and dehydrated cabbage, while others require large quantities of meat and fish—the amount of stowage space, the types and qualities of the provisions available at the ports of call, whether or not the yacht has any refrigerated space, and the financial aspect. Some years ago my wife kept an account of all that the two of us consumed in one week, and when compiling the provision list for our first long trip, she multiplied that quantity of food by the number of weeks we reckoned we would be at sea, and then allowed something extra for emergencies. Provisioning for our subsequent voyages was done in a similar manner, but there were occasions when the quantity we wished to take was limited by the space available. If the reader wishes to follow this plan, but feels unable to hazard a guess at the length of time he is likely to be at sea, he should consult Appendix I, where a variety of ocean passages is listed showing the times taken on them by yachts of various sizes at different times of the year.

It is important that the diet should be palatable and varied, and include as much fresh food as possible, for although vitamin tablets are available, they do not appear to be the equal of fresh food in supplying all the requirements essential for good health; nevertheless it is advisable to take these, either in separate or combined form, particularly after the fresh food has been exhausted. As Conor O'Brien once remarked, the world is not a desert; all kinds of provisions can be had in any fair-sized seaport in any country, but the diet may have to be altered when relying on provisions obtained at many of the smaller islands, which make such convenient and pleasant stopping-places for the ocean-going yacht; on the islands of the central South Pacific, for example, eggs, fresh meat, potatoes, and green vegetables are hard to come by, except at Papeete, where they are very expensive; rice, fish, corned beef, coconuts, paw-paw, bananas, breadfruit, yams, and taro (a starchy root known in Fiji as *ndalo*) being the staple foods.

Unless the yacht has refrigerated space, the fresh provisions which will keep in good condition aboard in the tropics are limited. Potatoes and onions of good quality will keep for up to three months, provided they are kept dry, and the sprouts and those which are not keeping properly are removed occasionally. Lemons wrapped in tinfoil and stowed in a tin will keep almost indefinitely, while oranges, grapefruit, and limes will keep for a month or more, depending on their quality. Bananas are best bought on the stem, and this should be lashed in such a manner that the fruit cannot become bruised. Depending on their state of ripeness when shipped, and on the temperature, they may start to ripen after one or two weeks, and will provide a welcome supply of fresh fruit; but once it starts, the whole stem will ripen rapidly and give a surfeit of bananas for a short time. It is recommended that all uncooked fruit and vegetables bought abroad should be washed in water in which a few crystals of permanganate of potash have been dissolved, and that lettuce should not be eaten in countries where dysentery is common.

Fresh eggs should be sealed with grease within twenty-four hours of being laid—petroleum jelly serves well for this purpose; lard is too soft and will run off in hot weather—and they will then keep in good condition for at least three months, and can be stowed in cardboard egg-boxes holding 6 or 12 each, the boxes being turned over once a week. Eggs should not be packed in sawdust as this may impart a taste to them.

Meat may be kept in brine in the traditional manner—large, wide-necked, glass sweet-bottles with ground-glass stoppers make suitable containers. Howell and McNulty, during their Atlantic crossing in *Wanderer II*, preserved meat in the following manner. Beef without bone or fat was cut into thin steaks and thoroughly cooked, either by frying or in a pressure cooker; it was then placed in a deep biscuit tin and covered with melted lard. Their meat so treated kept for six weeks. When a meal was to be prepared, sufficient meat was removed from the lard and placed in a pressure cooker together with vegetables, and cooking time for the latter only was allowed. The keeping property of bacon depends on the curing. A 9-lb. side of bacon bought at Balboa and hung in a muslin bag in the forepeak lasted for six weeks and was good to the end. Sometimes rashers were cut from it and fried; at other times a pound or so was cut off and boiled. If there is any doubt about the curing, bacon should be kept buried in salt.

The British type of white or brown baker's bread will keep for up to ten days if stowed in a dry, airy place, the mildew being cut off as it appears; but most other types of bread obtained abroad have poor

PLATE 41

A: If there is a cat aboard, one of the hatchway screens should be provided with a two-way flap.
B and C: A simple form of portlight screen with the netting sewn to a ring of springy wire.
D and E: To black out a portlight, a piece of card secured to a small suction cup serves well.

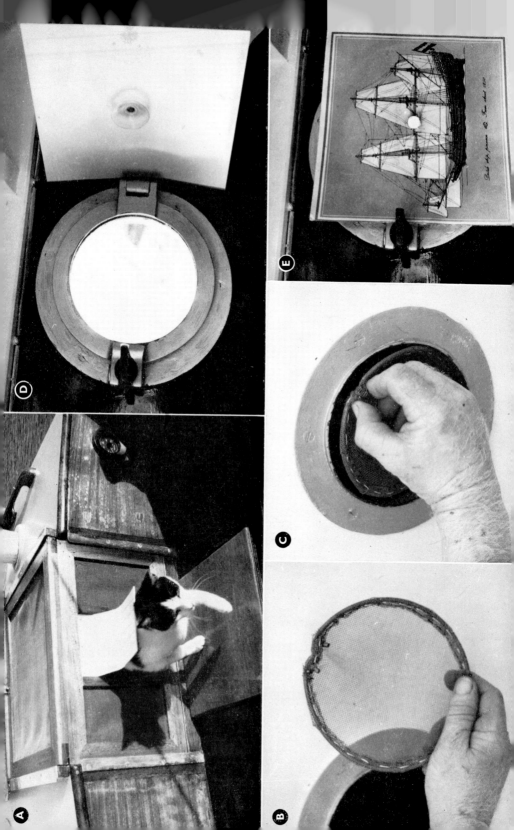

A

B

C

D

E

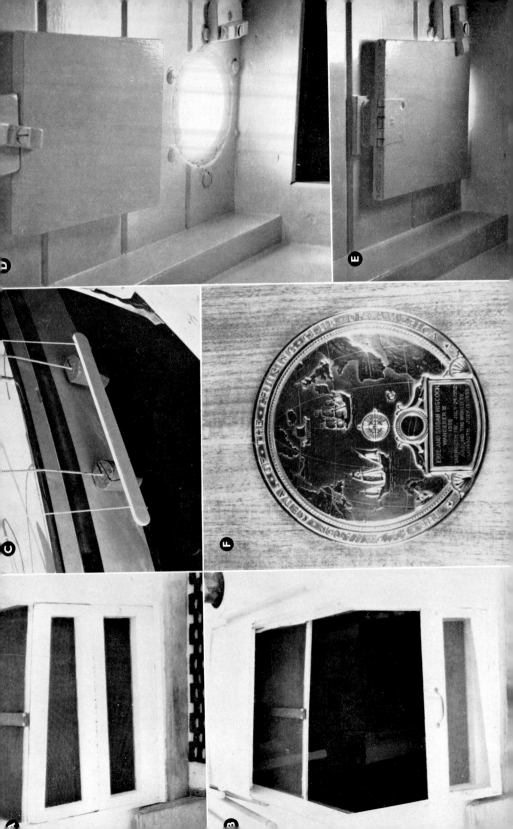

keeping qualities and quickly turn sour. With all manner of excellent biscuits and crispbread to choose from, there is no need to take a large supply of the hard and not very palatable ship's biscuit, unless someone on board has a particular fancy for it. If biscuits cannot be obtained in air-sealed tins, they should be from fresh stock and must at once be stored in tins sealed with adhesive tape. Flour, rice, sugar, and other dry stores bought in small places in the tropics are often already infected, when they will soon become weevily or mildewed. An occasional loaf of fresh bread is a great luxury. This may be soda bread or yeast bread, and although the latter is more trouble to make, it is more like the real thing. Dried yeast can be had in cans, and unopened will keep for a long time.

Powdered milk and evaporated milk are both excellent, the choice being a matter of taste. A supply of bottled sterilized milk, which will keep indefinitely, is a good investment, as in tea or coffee it is indistinguishable from fresh milk, and a bottle opened occasionally is a treat for all hands.

Salted-down butter should last well in the tropics, but I have no experience of it except in temperate climates, where it remains fresh for many weeks. Canned butter is satisfactory, provided the stock is fresh, but although some cans are date-stamped, more often they are not, and then there is no means of ensuring that it is fresh, and only in three places have I obtained it untainted—Cowes, Cape Town, and Horta. It is generally so unsatisfactory that margarine, specially prepared and canned for the tropics (Oleomargarine), is often to be preferred, and has the merit of remaining firm at a higher temperature.

There is such a variety of excellent canned foods on the market, both at home and abroad, that nothing need be said of them here, except that if the cans have to be stowed in a damp place, their labels should be removed, and the stamped identification marks, which most cans possess, noted, or a code letter painted on. The only losses I have ever experienced were cans of ham, some of which became inflated with gas. Of all the available canned meats, corned beef usually proves to be the most popular in the end because of the variety of ways in which it can be served.

PLATE 42

A: Aboard *Altair* the companionway screens are hinged, *B*, for ease of access. *C*: When lying alongside piles or other projections, a fender-board for each pair of fenders is required. *D*: To black out a decklight, a plywood flap (a 'tinker trap') may be arranged so that, *E*, it can be hinged over and held with a turn-button. *F*: The Blue Water Medal of the Cruising Club of America.

Aboard *Wanderer III* every locker was numbered and every shelf lettered with waterproof ink; a store-book was kept with similarly numbered and lettered pages, in which was shown the quantity and stowage place of all provisions at the start of a passage. Other pages were ruled with columns, each headed with a type of provision, the days being written at the side; there every item was entered as it was consumed, so that we could see at any time the exact amount of any commodity remaining.

The total amount of stores consumed by my wife and me during a 5,847-mile voyage from Cape Town to Horta, Azores, may be of interest, and is listed below. We were unfortunate in finding the doldrum belt unusually wide, and in experiencing a large proportion of light headwinds in the horse latitudes, and the entire trip took 84 days. Of these, 10 days were spent at anchor at St. Helena and Ascension: during 7 of them ship's stores were consumed; for the remaining three we were guests of the Resident Magistrate at Ascension.

Stores consumed by two people in 81 days

Potatoes, 90 lb. (70 lb. from Cape Town lasted throughout the voyage; but of 20 lb. from St. Helena there was 50 per cent. wastage.)

Onions, 20 lb. (Lasted entire voyage.)

Apples, 64. (Those from Cape Town kept excellently until finished after 6 weeks. Others from St. Helena soon deteriorated as they had been refrigerated.)

Lemons, 24. (Wrapped in foil and stowed in tin kept perfectly and were not used for first 6 weeks.)

Eggs, 14 dozen. (12 dozen were greased when less than 24 hours old. There were no losses, and eggs were still good for boiling after 3 months. Two dozen were obtained at St. Helena and Ascension, where they were very scarce.)

Butter, 14 lb. (From Cape Town 4 × 1-lb. cans; also 3 lb. fresh, which lasted 3 weeks, as sea temperature during that time did not exceed 70 °F. From St. Helena 4 lb. fresh, and from Ascension 3 lb. fresh.)

Lard, 2 lb. (Used mostly for frying potatoes.)

Cheese, 5 lb. (4 × 1-lb. cans, and 1 lb. in tinfoil.)

Flour, 20 lb. (Used for yeast bread, cakes, scones, and sauces.)

Fruit cake, 8 lb.

Rice, 1 lb. (Used for occasional pudding.)

Quaker oats, 6 lb. (In air-sealed cans.)

Macvita, 4 lb. (2 × 2-lb. air-sealed tins, bought in England and on board for 2½ years. In perfect condition.)

Biscuits, plain, 7 lb.
Ginger biscuits, 4 lb.
Breakfast cereals, 4 packets. (Crisped in oven before use.)
Sugar, 12 lb.
Nescafé, 9 × 2-ounce cans.
Milk chocolate, 18 lb. (In 2-ounce slabs; slight loss from mildew.)
Barley sugar, 8 lb.
Pickled onions, 2 lb.
Meat, 58 cans. (Pork sausages in 1-lb., corned beef in 12-ounce, and steak
    and kidney pudding in 1-lb. cans.)
Crayfish, 12 × 6-ounce cans.
Sardines, 12 × 4½-ounce cans.
Soup, 40 × 10-ounce cans. (Mostly tomato and cream of chicken.)
Evaporated milk, 28 × 1-lb. cans.
Dried milk, 2 × 1-lb. cans. (Klim.)
Sterilized milk, 11 × 1-pint bottles.
Vegetables, 15 × 10½-ounce cans. (Peas and carrots.)
Baked beans, 4 × 6-ounce cans.
Sweet corn, 7 × 6-ounce cans.
Fruit, 16 × 15-ounce cans.
Cream, 16 × 4-ounce cans.
Tomato juice, 45 × 15-ounce cans. (Regarded as a source of vitamins, and
    started after first fortnight.)
Nuts, 4 × 4-ounce cans.
Orange juice, 4 × 1-pint bottles.
Cape wine, 16 litres.
Multivite, 280 tablets.
Fortified yeast, 140 tablets.

In addition small quantities of the following were used: Baking powder, Dri-barm yeast, salt, pepper, mustard, curry powder, mixed herbs, oil, vinegar, and cocoa. For cooking and lighting 7 gallons of paraffin were consumed, and half a gallon of methylated spirit for preheating the galley stove.

Sleep

Sound sleep is so important that one must be prepared to make some sacrifice of speed if necessary. If the watch below is frequently roused out to make or take in sail, or, because of hard driving, the motion and the noise are excessive, there will be a general lack of sound sleep; this at first may show itself in frayed tempers and poor appetites, and will soon lead to loss of efficiency. I do not advocate shortening sail for the night as a matter of course, for such an action would make night sailing dull and would spoil each day's run and the pleasure to be had from swift and efficient passage-making; but judgement should be used to ensure, so far as is possible, that sail-handling is done only at the change of watches. Even then in a short-handed yacht there may come a time when her people show signs of suffering from lack of a long, undisturbed sleep, and if on such an occasion the yacht cannot be got to steer herself, it will be wise to heave-to and let all hands have a night in their bunks. The resulting general cheerfulness and increased efficiency which will inevitably follow will certainly be worth the miles sacrificed. With a complement of three or more, once watches have been arranged it will be best to keep to them, prompt relief being a matter of self-respect. But with only two people aboard, it is sometimes a good policy for the man on deck to let the other sleep on for an hour or so in the morning, if he is deep in slumber, for several hours of continuous, undisturbed sleep are of greater value than the same number of hours taken separately; and such a relaxation of the watch-keeping routine tends to strengthen the give-and-take attitude which is so important; but it must not be indulged in to excess on one side, or the sleeper will lose his self-respect and the other feel a martyr.

A light sleeper's rest can easily be spoilt by unnecessary noise, particularly if it is intermittent; the loose bottle or can of food rolling in the locker, the clinking of plates in the pantry, the slapping of the idle headsail sheet on deck. During the afternoon sleep, which is an important part of the routine when night watches are kept, bright light should be subdued as much as possible, and the tinkerbell, caused by a shaft of sunlight from a deck- or port-light, flashing to and fro across the cabin as the yacht rolls, should be extinguished. This can be done by having at each decklight a hinged plywood flap, a 'tinker-trap', which normally is held back out of the way by a turn-button; to black out a decklight its flap is released and held over the glass by a second turnbutton (Plate 42D and E). For each uncurtained port-light in the coachroof coamings or topsides a piece of cardboard can be provided fastened at its centre to a small plastic suction cup; if this is moistened and pressed firmly on the glass it will adhere (Plate 41D and E).

Only aboard a large vessel is anyone likely to make a habit of sleeping on deck at sea, but in port in the tropics when the cabin may be un-comfortably hot, the deck or cockpit may be the most popular place; the essentials then will be a flat space of sufficient area without obstruc-tions in a position where the awning gives complete shelter from dew or rain; an inflatable mattress with a small, foot-operated pump incorporated, such as is sold for beach use, may be found convenient.

Health

If the yacht is to visit the U.S.A. or the Panama Canal Zone, her people must be vaccinated and must carry vaccination certificates, or they may not be permitted to land. It is as well to be inoculated against typhoid, and one may choose to be inoculated against other diseases, depending on the countries it is intended to visit.

In addition to the first-aid equipment listed in *Cruising Under Sail*, the ocean-going yacht, which will be out of reach of medical aid for long periods, should have in her medicine chest a supply of penicillin and a hypodermic syringe for administering it; this will be of great value in the event of blood-poisoning, and it might keep an appendicitis case alive until medical help can be reached; it would also be of assistance with severe cases of tropical sores and other infections. If malarial countries are to be visited, there should be a supply of tablets, such as Camoquin, for the suppression or treatment of malaria, and Chloromycetin for the treatment of dysentery, as well as the remedy for any particular complaint to which one may be prone. A plentiful supply of laxatives is important.

But new discoveries and improvements in the treatment of disease are continually being made, and the advice of one's doctor, and his instruction in the administration of drugs, should be sought; indeed, prescriptions from him will be needed in order that the drugs required may be obtained. A visit to one's dentist should also be made some time before sailing so that he may effect any necessary fillings or extractions, and it would be wise to have on board small quantities of oil of cloves and zinc oxide so that a cavity could be filled temporarily. It should be realized that some items in the medicine chest will deteriorate in hot weather and that others have a time limit to their efficiency.

There is such a variety of ills to which the human body is liable that a voyage out of reach of a doctor might at first sight appear to be a hazardous undertaking; but to judge by the few troubles which have been experienced by small-boat voyagers over the years, the risk is very small. Robinson in *Svaap* had appendicitis when at the Galapagos Islands, and was fortunate in meeting a vessel able to radio for air transport to hospital—perhaps the prudent mariner would have his appendix removed before starting—*Driac II* had a typhoid case on board during an Atlantic crossing, but got him to Barbados in time; and there have been a few cases of blood- and food-poisoning, and of malaria, but the majority of the voyages have been made with healthy crews. However, the greatest care must be taken to avoid serious injury, such as burns or fractures, which could be fatal, and sensible precautions should be taken against sunburn and heat. Perhaps it is not so well known as it might be that when one is sweating profusely in great heat it is necessary to increase the intake of salt, either by taking salt pills or drinking salt water, otherwise one may collapse. It is also important to take care in avoiding coral cuts; these frequently fail to heal and often become septic.

The usual toilet requisites, toothpaste, shaving cream, shampoo, etc., keep well enough for long periods in hot weather, but the black plastic caps fitted to some containers tend to grow sticky and discolour the contents. It is possible to wash oneself and even one's hair effectively in salt water if soapless shampoo is used instead of soap, and a small quantity of Cetavlon may be added to the water to soften it and ward off the skin infections leading to boils and sores with which the seaman is sometimes troubled; in my experience so-called 'salt-water soap' is worthless. Clothes can be washed in salt water with the usual household detergents which are used for dishes—the liquid type being better for this purpose than the flake or powder—but they will not dry properly unless they are given a rinse in fresh water; for that purpose it is worth catching rainwater, but see page 34. In case one has at some time to take water on board from a source which is believed to be contaminated, it is as well to carry a supply of water-purifying tablets, such as Halazone; a small bottle contains 1,000 tablets, which is enough to purify 250 gallons. Some people prefer to boil all drinking water taken aboard in foreign countries. My wife and I never did this, except at the Azores, where friends assured us that the water was dangerous, and we rarely used Halazone; we suffered no ill effects, but in that perhaps we were fortunate.

Pest control

If a yacht spends much time alongside, pests of one kind or another are bound to get aboard, and if the berth is close to one where copra is loaded, the infestation of cockroaches and copra fly can become appalling. I remember one evening visiting a yacht which had spent some years in the Pacific, and when I switched on a light in the galley, the whole place—bulkheads, deckhead, and deck—appeared to move; all were alive with cockroaches. But this kind of thing need not happen if all lockers and other places where insects might live and breed are sprayed occasionally with a 5 per cent solution of D.D.T. or Chlordane in paraffin; for several months after application this should kill all insects that touch it. But as improvements in insecticides are continually being made, it might be as well before starting on a voyage to contact one of the firms that specialize in these, such as The Shepherd's Aerosols Company of 70 Jermyn Street, London, S.W.1.

The farther a vessel lies from the shore the less will she be troubled by insects, but sooner or later the foreign-going yacht will need insect screens to protect her people from flies and/or mosquitoes, and in this connection it should not be thought that insects are a bother only in hot places; indeed, my wife and I rarely found screens essential until we visited Maine, the northernmost East Coast state of the U.S.A.; but we did at times, notably in the northern part of Australia, and in New Caledonia, use mosquito netting over our bunks at night. For this we had wide pieces of netting rather longer than the bunks, and fastened them with common spring clothes-pegs to the handrails on the coachroof coamings above the bunks, from which they hung down to the cabin sole. They were easy to rig and were quite effective provided insects were not trapped inside them. I now feel sure that it is better to exclude insects from the entire accommodation by fitting portable screens, preferably of plastic netting, to all hatches, vents, and opening ports; and it will often be found that if these are shipped in position at sundown they can safely be removed when one turns in after the lights, which attract insects, have been extinguished, thus allowing a freer flow of air throughout the night.

But there may be occasions when the screens have to be used in daytime, as for example at Nukuhiva, Marquesas, where the *nono*— a small, black, sand-fly with a bite which sometimes becomes septic —is particularly troublesome. However, because screens of any sort reduce the flow of air, they need to be easy to ship and unship; the screen in the main hatchway might be hinged (Plate 42A and B) for ease of passage, and if there is a cat aboard one screen needs to be provided with a two-way cat-flap (Plate 41A). Ventilators and opening ports can easily be provided with removable screens in the following manner: Cut a length of stiff $\frac{1}{8}$-in. diameter wire (welding rod serves well) one inch greater than the circumference of the inside of the open port, and bend it in such a way that to fit the port it must be compressed a little. Turn out at right angles half an inch of each of its ends to provide finger grips, and seize these ends together with marline so that when the ring of wire is not in position the ends cannot spring apart more than, say, half an inch. Cut a piece of netting rather larger than the area enclosed by the wire, turn it over the wire, and sew it to itself, but leave the two half-inch wire ends protruding. To ship the screen hold it by the protruding ends, press them together to reduce the circumference, insert the ring in the port recess, and then release (Plate 41B and C). Some small insects, such as the 'noseeums' of the Bahamas, can penetrate mosquito screens unless these are treated with paraffin.

Sometimes one may not suspect the presence of mosquitoes until it is too late to ship the screens. Then one will be glad to have a supply of mosquito coils. These come in boxes usually holding half a dozen, together with a little metal stand, and are available in many Pacific islands and some other places. A coil is rigged up on the stand on a plate to catch the ash, and its outer end is lighted. The coil will then smoulder for many hours, giving off a smell, harmless to humans and not unpleasant, which mosquitoes detest. On hot, still nights one may prefer to use one of these coils rather than ship the screens. A supply of the ointments and sprays used to keep insects off one's skin should certainly be carried, otherwise shore excursions in some places, and life in the open at sundown, will be a misery.

Even a yacht lying at anchor well away from the shore may embark cockroaches when loading stores. This can be prevented by never permitting cardboard cartons aboard, by picking over all vegetables, and searching all egg boxes before allowing them below, and by submerging stems of bananas in the sea for a few minutes. The occasional cockroach which gets through these defences can probably be killed by spraying, or by the use of an aerosol 'bomb', i.e. a fumigant in a canister which when opened emits an insect-killing vapour. I have known ants come off from the shore along the mooring lines quite a distance, so a foot or two of these might with advantage be treated with insecticide.

Only twice have we had rats aboard (a cat is good insurance against this), and never any mice, centipedes, or snakes; but there are places, such as the rivers in the Guianas and in northern parts of Australia, where according to the Admiralty *Pilot*, there could be a risk of water-snakes coming aboard by way of the anchor cable unless this is fitted with a rat-guard, a cone of metal with its apex pointing inboard.

PLATE 43

Top: *Walkabout*, a replica of Robinson's famous *Svaap*, was built in Australia by the Driscoll brothers, and was sailed by them across the Indian Ocean to Durban. *Bottom*: A replica of another famous little vessel, Frank Wightman's 34-foot, hard-chine yawl *Wylo*, which he built himself and sailed from Cape Town to Trinidad; she was built on the lines of *Islander*, which Harry Pidgeon twice sailed round the world.

Pets

I have no personal experience of dogs as shipmates on long voyages, but some I have met gave the impression that for them life afloat was not much fun. However, there are many places where a good watch-dog would be of value in discouraging unwelcome visitors. *Arthur Rogers*, *Omoo*, *Tzu Hang*, and *Widgee* all carried dogs of the smaller breeds, and in each yacht the dog learnt to use a part of the deck as the heads, and this was sluiced with seawater.

Most cats appear to be as content afloat as they are ashore if they start the life young, and a large number of voyaging yachts carry cats. Some people consider they are of no practical use except for catching vermin, but they can be good company, and help to keep up morale by their indifference to bad weather and anxious situations. When visiting the shore in some places they readily pick up fleas; my wife and I used to dust ours when cruising coastwise with anti-flea powder every few weeks, but later made him wear an anti-flea collar; the powder drugs the fleas so that they can be popped with a pair of small tweezers. For the cat's heads we started with the traditional sand-box, but soon abandoned it as damp paws carried the sand all over the ship. The Carrs in *Havfruen III*, who had two Siamese, advised us to try wood planings; this, which can be got at any boat-yard, was an improvement, but we failed to train our cat to use a small basin like the Carrs' cats did, for he was a powerful digger and flung the planings all over the forepeak. So we rigged him up with a shallow plastic container to hold the planings in the bottom of a high-sided bread-bin. No doubt each cat has its particular likes and dislikes regarding food, though most enjoy fresh fish. Ours could not keep down for more than a few minutes most of the cat foods sold in cans, but he approved of all raw pieces of fresh chicken and meat, and of the canned foods for humans preferred tuna (in water, not oil), and if there was nothing better around would share with us an occasional can of corned beef. There are available packets of pea-size crisp biscuits, which at times are popular, and a raw yolk of egg (uncooked white of egg is indigestible) is of value when other fresh food is not available, and is relished by some felines.

PLATE 44

A famous ocean voyager, the brigantine *Yankee*, while owned by Irving Johnson, made four 18-month voyages west-about round the world with crews of up to 20 young amateurs. Originally a North Sea pilot vessel, she was steel built and could carry, including stun'sails, an area of 7,775 square-feet of canvas. Here her main gaff topsail is being set as she leaves Cape Town during her third voyage. Under new ownership she was lost on the fringing reef of Raratonga, Cook Islands.

Quarantine regulations make the carrying of dogs or cats in voyaging yachts difficult for owners who have to return to the shore to live. In the U.K., New Zealand, and some other countries, but not in the U.S.A., there is a compulsory quarantine period of up to 12 months, irrespective of the time taken on the last passage, or an assurance that the animal has had no contact with the shore for the same period. This quarantine period is much too long for an animal to be separated from the people with whom it has shared the confines of a small vessel. However, provided the animal does not set foot ashore, it may be permitted to spend the quarantine period aboard the yacht. No doubt it would be wise to have one's pet inoculated against rabies and distemper, and to have a vet's certificate to that effect.

Shooting and fishing

During our voyages my wife and I did very little fishing, and although we carried a 12-bore shotgun we regarded it purely as a weapon of defence, and used it only in attempts to drive off sharks, and once against a spearfish which attacked the yacht in the South Atlantic; for such a purpose a pistol would be more effective. The chance of damage being done by a large fish must surely be very small, but collision with a whale (perhaps asleep) does occasionally happen; both *Moonraker* and *Widgee* struck whales by night in the North Atlantic, but no serious damage was sustained.

Worth was keen on firearms, and in *Beyond* carried the following: 12-bore shotgun, ·22 rifle, 8-mm. rifle, ·22 pistol, ·32 pistol, and 9-mm. pistol—quite an armoury. He wrote to me about these as follows:

The object of the firearms is for the pot, sport, and possibly defence. The most useful for the pot is the 12-bore, though occasionally one gets a chance of goat, pig, or something larger. The point is that you seldom know what you may come up against. If only one gun is to be taken, perhaps the most useful would be a double-barrel affair, with one barrel 12-bore, the other a light rifle. Such weapons were popular on the Continent some years ago, but are now out of fashion. If a shotgun alone is carried, some buck-shot cartridges would take the place of the rifle for short range. If a shotgun and a rifle are to be carried, a high-velocity rifle is the most useful, though I doubt whether one would meet anything worthy of the 8-mm. Much native hunting is done with a ·22 rifle, but it may not stop even a goat, and I hate the idea of a wounded animal getting away. There are some parts of the world where, according to reputation, I would not like to sail unarmed, and so be perhaps at the mercy of any boatload of Arabs. Here again a shotgun is a fine weapon, particularly if loaded with buck-shot;

mine is a pump-action 12-bore Winchester carrying five rounds. The ·22 pistol and the ·22 rifle are old friends carried for cheap fun and practice.

Guns are difficult to keep in good order in a yacht, and need frequent cleaning. Do not take a 'best' gun for it is sure to get pitted somewhere, however careful you are. Do see that the guns have a proper home; leather cases will go mouldy and rust the guns; canvas cases are better. Ammunition is often difficult to get and very expensive in many parts of the world. Wherever asked we declared what arms we had on board, and we had no trouble. We did not shoot animals unless we could eat at least a portion of the meat, or unless there were people to whom we could give it.

I entirely agree with Worth's remarks on fishing. He wrote:

For one as unskilled in sea fishing as myself, it would be unwise to depend on catching fish [for the pot] in the Atlantic; in the Pacific there is a good chance of catching good fish near the coasts or islands, but not offshore. Off European coasts perhaps the best tackle is the ordinary mackerel line with spinner; say 12 feet of gut, a lead of about 3 lb. and 100 feet of $\frac{1}{8}$-inch diameter line. In the Pacific we used various coloured feather lures, 2 to 6 inches long and weighing 1 to 4 ounces, 15 feet of wire leader with a breaking strain of 150 lb., and 100 feet of $\frac{1}{4}$-inch diameter hemp line. A piece of shock-absorbing rubber was looped in the standing end of the line to take the shock if a big fish struck. We caught a fair amount of fish, mostly Spanish mackerel, and members of the bonito family up to about 15 lb., and occasionally up to 40 lb. The gear was broken by bigger fish several times. A rod is of no practical value, and is not a great deal of fun because you never know what you will catch. The cheapest place for fishing gear is in the Panama Canal Zone. Dictionary of Fishes by Rube Allyn, price $1.00, published by the Great Outdoors Association, St. Petersburg, Florida, is of great interest in identifying strange fish. We never hesitated to eat any fish that really looked like a fish should look, even if we could not name it, and we suffered no ill effects. Occasionally we caught some odd-shaped, ugly-looking fish which we would not touch.

Two other wild-life books of particular interest to the voyager are: Birds of the Ocean by W. B. Alexander (Putnam), and Giant Fishes, Whales and Dolphins by J. R. Norman and F. C. Fraser (Putnam).

11

IN PORT

*Ship's business—Duty-free stores—Mail—Hospitality—Yacht
clubs—Finance—The Panama Canal—The Suez Canal*

EVEN though he may have been looking forward to the occasion
with mounting enthusiasm, the voyager will probably find that
the transition from the sea to the city at the conclusion of a long
passage is far from enjoyable. It is likely that he will be tired, and a
little dazed by the noise, the heat, the smell, and the multitude of
strange faces after the silence and emptiness of the clean sea, and
would like more than anything else just then to be left in peace for a
little while, to rest and clean up, and adjust himself to the change.
But only in a few countries where small craft and their people are well
understood, will he be permitted to do so; elsewhere his yacht will
be boarded by officials who will want a number of forms completed
and may insist that an immediate visit be paid to the custom-house
or other office. This can give a wrong first impression of a country,
against which he should be on his guard.

Ship's business

It is advisable for the foreign-going British yacht to be registered
as a British ship, the certificate of registry being evidence of ownership
and the flag she is entitled to wear, and it establishes her identity.
The length of time she may remain in a country other than that in
which she was built is limited usually to a period not exceeding one
year, and if she overstays that time she may be subject to import duty,
which can range from 10 per cent (U.K.) to as much as 60 per cent of
her original cost. Possibly the payment of this could be postponed or
avoided by sailing to another country for a time and then returning,
but the customs and excise people should be consulted about this.

PLATE 45

The Panama Canal. *A*: In the locks, which are 1,000 feet long and 110 feet wide, large ships are
handled by electric locomotives known as mules. *C*: One of the three up-locks at Gatun, each
of which lifts a ship 28 feet in 8 minutes, and it is here that a yacht is most likely to get into
trouble because of the inrush of water. *B*: Each pair of huge gates is protected by a fender chain,
D, which is not lowered until the gates are fully open.

Each member of her crew should have a valid passport, but apart from visits to Egypt and the U.S.A., I have not found visas necessary; however, if anyone wishes to leave the yacht and travel overland, a visa would be desirable, and perhaps essential. When a yacht is about to depart from any country bound for another, the correct procedure is for her owner to telephone or call at the port office or the customhouse a day or so before sailing to inquire what are the regulations with which he must comply. Usually these are few and simple, but they vary from one country to another and are liable to alteration from time to time; the most important document to be obtained is the outward clearance, not that this is likely to be asked for elsewhere, but without it it may not be legal for the yacht to sail. Although by international agreement a bill of health—this refers not to the health of the people in the yacht, but to the state of health in the country she is leaving—is no longer required, it is still asked for occasionally. On one visit to the Panama Canal Zone I was asked for a deratting exemption certificate; as I did not have one, I had to pay $13 for that worthless piece of paper which, had I but known at the time, I could have obtained for nothing at Barbados. Sometimes an official from the immigration department may wish to visit the yacht just before she sails to make sure she has not left any of her people behind. If the requirements of the country for which she is bound are not known, her owner may inquire of the consul of that country if there are any special formalities which should be dealt with in advance; but in general I have found that the less I bother officials the less do they interfere with me.

The first stop in a foreign country should be at an official port of entry (such ports are usually mentioned in the text of the relevant *Pilot*), or failing that, at a port of some size where one can be certain of finding a customs officer, and code flag 'Q' should be flown. If the quarantine anchorage shown on the chart is too remote or exposed, as it frequently is, one should contact the pilot cutter or anchor in some conspicuous position, and not land until permission has been granted, but see below. Some officials do not like flag 'Q', as they associate it with the plague; nevertheless it is the correct signal to

---

PLATE 46

The Suez canal. *A*: Much of the 87-mile-long canal passes through featureless desert; this 28-mile stretch leading to Port Said is dead straight, and beside it run the railway and the Sweet Water canal. *B*: A yacht which cannot keep to convoy speed of $7\frac{1}{2}$ knots has to get used to meeting large ships at close quarters. The projector capable of lighting the canal 1,300 yards ahead can be seen at the stemhead of this tanker. *C*: A light beacon, and one of the small buoys which at frequent intervals mark the edge of the deep-water channel. *D*: The suction dredger *September Fifteenth* at work deepening the canal. *E*: One of the many signal stations.

S

make, its international meaning being: 'My vessel is healthy and I request free pratique', i.e. permission to land. Most likely an official will come off to grant pratique, or explain what must be done to comply with the regulations, but in some places no notice will be taken, and one may have to go ashore and visit the port office or custom-house, but several hours should be allowed to elapse before that step is taken. Incidentally, there is some confusion over the words 'clear' and 'enter'. A vessel 'clears' from one port and 'enters' at another, irrespective of the fact that in the process of entering she will be cleared by customs. In general avoid entering port with flag 'Q' flying after 1700 or before 0900, or at any time on Saturdays, Sundays, or other holidays, otherwise a charge may be made for overtime. A charge may be made at any time for transportation of the officials, though this is rare, and at a few places, such as Nassau in the Bahamas, no receipt for this is given, so one may guess what happens to it, especially as a single official may on one visit enter several yachts and ask of each the same sum for travelling. Possibly in the interest of others one should dispute this charge, but I question whether this is wise as port officials can, if they feel so inclined, be awkward and make the process of entry long and tiresome. But in my experience it is rare to encounter a dis-honest or discourteous port official, and usually if one welcomes these men aboard in a friendly manner they will not raise any difficulties; their chief job is to prevent smuggling, and this concerns not only wines, spirits, tobacco, and aliens, but many other items including cameras and radio sets, and in the interest of those who follow after, every facility should be given them.

The following notes on official procedure in a few places, based on my own experience over the past 18 years, may be of interest, but it should be remembered that the procedure is liable to be altered from time to time, and even to differ between one port and another in the same country.

In Spain, even in the smallest place, a yacht will usually be boarded by the captain of the port or members of the *guardia civil*, but these men rarely require more than the name of the yacht and her people, where from and whither bound; at Vigo, however, the secretary of the Real Club Nautico will ask for passports and get them stamped. In Portugal the regulations were altered some years ago so that passports no longer have to be surrendered to the international police; at Leixoes (the port for Porto) and at Cascais (the popular summer anchorage at the mouth of the Tagus) entry is done through the yacht clubs, and a green card, a sort of yacht's passport, is issued; thereafter the only formality is to get this stamped at each place visited, and to surrender it on leaving the country. This applies also in the Azores, but at Madeira in 1968, although a green card was issued, passports still had to be surrendered to the police and collected on the day of departure. In the Canary Islands yachts are not boarded, but a simple form must be completed at the port office on arrival.

The Windward and Leeward Islands are tiresome in that the yacht must enter at each, even though she may have come from another of the same nationality only a few miles away; but all that is usually required between these islands is a crew list in duplicate, and I have found it convenient to type a number of these in advance, leaving blank the date and ports of departure and arrival to be filled in later. At some of these islands one must land and visit the port office or police station, while at others one must wait to be visited; but there are usually other yachts around to inform one of the correct procedure. English Harbour has for some years now been a port of entry for Antigua, thus saving one the long beat back from St. Johns. St. Thomas is not a good place in which to enter or clear the American Virgin Islands, as the officials are unco-operative and insist that one must berth alongside on a lee shore. So the majority of yachts sailing between the British and American Virgins enter and clear at Cruz Bay, St. John, where the anchorage is safe and the officials are close and helpful; but a great deal of paper work is involved, and it is a help if the required forms can be obtained and partly completed in advance. At San Juan, Puerto Rico, the yacht is not boarded, but her arrival must be reported by telephone and a visit paid to the custom-house in the old (and distant) part of the city, where a fee is charged for entry and clearance. The Bahamas seem glad to have visiting yachts and make life easy for them; there are more than 25 ports of entry listed in that essential publication, *Yachtsman's Guide to the Bahamas* (obtainable from the Ministry of Tourism, Nassau), and on arrival at any one of them a *transire* will be issued, and the yacht will then be free to cruise where she wishes without further bother. But, if coming from the south and entering at Turks or Caicos, she should enter again farther north, as these two groups are under the jurisdiction of Jamaica.

For a visit to the U.S.A. passport visas and vaccination certificates (these are valid only for a period of three years) are required. During a 10-month period of cruising on the east coast of that country I had dealings with officials only twice, and found them helpful and friendly. The first stop should be at a port of entry, where a cruising licence for the yacht and an immigration card for each person on board will be issued free of charge. The former exempts the yacht from further formalities for a period of six months, after which it can be renewed for a further six-month period at any custom-house; but at the same time the immigration cards must also be renewed, and for this a charge of $10 each is made. On leaving the U.S.A. there are no formalities, but the cards and cruising licence must be sent by post to their office of issue. For the Panama Canal Zone see page 273.

On entering Mexico in 1969 a payment of about £6 had to be made, and at most places several crew lists (in Spanish) had to be furnished by the yacht and stamped by various port officials. In the same year no charge was being made in Costa Rica.

For a visit to the Galapagos Islands (Ecuadorean) 5 manifests, 3 crew lists, and 3 provision lists must be stamped by an Ecuadorean consul. In 1960 the consul at Colon did this for a charge of $8, while the consul in Panama City was charging $30 for the same service—but it is this kind of information that the voyager will pick up from other yachts as he goes along. I consider it would be unwise to go to the Galapagos Islands without having the ship's papers in order, and even then there may on arrival be a further charge of $10 or more. The treatment of yachts in those fascinating islands depends largely on the attitude of the Commandant in office at the time at Wreck Bay, port of entry, and in 1968 he was courteous and not demanding.

The Marqueses Islands can be entered at Taiohae Bay, but the only port of entry for the Society Islands and Tuamotu is Papeete on the island of Tahiti, and a permit from the Governor there must be obtained before visiting other islands in the group. Nevertheless there have been many instances of *gendarmes* granting permission for yachts to remain for a short time at the islands under their administration, but with the area being used by France for atomic experiments, it is likely that the regulations are now more strictly enforced. At Papeete pilotage is compulsory, and although the pilot is unlikely to do more than berth a yacht, and perhaps not that, his fee must be paid. On both our visits we found the business of entering and clearing time-consuming.

The government of Tonga is very conscious of the damage done by the coconut-destroying rhinoceros beetle, and a yacht wishing to go to Tongatapu, the capital at the south end of the group, should call there first, for if she has been to Vava'u in the north she will not be welcome, and may be ordered to lie a mile offshore during the hours of darkness when the beetle flies. Suva is the port of entry for

Fiji, and here again the authorities are strict because of the risk that a yacht calling elsewhere first might bring the beetle to an island which until then had been free of it.

New Zealand has little red tape for the visiting yacht. If coming from the north she may enter conveniently at the small town of Russell, Bay of Islands, where the local doctor will grant pratique, and is then free to cruise down the coast stopping where she likes, and on arrival at Auckland her owner should pay a brief visit to the custom-house. On arrival at Sydney, N.S.W., the yacht must anchor first in Watson Bay and contact the pilot cutter which makes that her headquarters; she will notify the customs. On no account may the yacht proceed farther up the harbour until she has been granted pratique, and after that her owner must visit the custom-house, where in 1954 the paper work was so formidable that I was glad to accept the services of a shipping agent who kindly did it for me without charge. In those days one was expected to enter and clear at every port visited; but that rule was such a burden that I ignored it and entered and cleared at my first and last port only. I understand that since then the paper work has been simplified to some extent, and certainly when I called at Darwin, in the Northern Territory, 8 years later, I encountered no difficulty. In South Africa the formalities were similar to those in Australia, but the rule about entering and clearing at every port was strictly enforced; I did not always find the office clerks civil.

During *Wanderer III's* circumnavigations harbour dues were charged only in a very few places and were small indeed. She had the services of official pilots only at the following places: twice through the Panama canal and once through the Suez canal; at Papeete twice, and at Mauritius and Durban, where friendly pilots offered their services free. It is as well to remember that a pilot does not necessarily understand the handling of small craft.

On arrival in any foreign harbour it is courteous and correct to call on the captain of the port. If he cannot be bothered with a yacht owner nothing will have been lost; but he may be pleased to see one, and he can sometimes do much to make one's stay more comfortable and convenient than it might otherwise be.

Duty-free stores

A foreign-going British yacht with a register tonnage of 40 or more has the right in British and Commonwealth ports to ship tobacco and spirits out of bond, that is duty-free and at the rate of 1 ounce of tobacco and $\frac{1}{4}$ pint of spirits per person per day. In practice, however, U.K.

customs will usually permit such stores, probably limited to a 13-week supply, to be shipped in a smaller yacht, provided they are satisfied that she is genuinely going foreign for a reasonable period of time. Application should be made in writing, giving the proposed route and major ports of call, to H.M. Customs and Excise, Kingsbeam House, Mark Lane, London, E.C.3.

Such stores can be obtained only through shipping merchants, who will do the required paper work and deliver on board. The cases should be opened only in the presence of a customs officer, when their contents must be stowed in a locker which he can seal. If, after leaving a British port with duty-free stores on board, the yacht puts into another British port, the ensign should have a knot tied in it as a signal for customs officers to come off and see that the seal has not been broken, or, if stores have been consumed, to collect the duty payable on them. On arrival at a foreign port with duty-free stores on board, the officials should be told; often that will be the end of the matter, the locker may be kept unsealed and its contents used *on board* without payment. I have only twice had stores resealed, in Fiji and California; but on arrival at Sydney a tally was taken of the stores, and duty had to be paid at the final Australian port on the stores that had been consumed while in Australian waters.

On earlier voyages I found it was possible to replenish stores out of bond at a number of places in the Commonwealth, but in recent years I have noticed that the officials are becoming reluctant to allow this, partly no doubt because of the increasing number of voyaging yachts, but chiefly because some of them have taken advantage of the privilege to smuggle. Therefore a yacht requiring cheap tobacco or spirits would do well to go to one of the so-called 'free ports', where these things are available duty-free, or nearly so. Examples of such places are: La Luz (Canary Islands), St. Barts (the French smuggling port of the West Indies), and Gibraltar; any quantity may be bought in the shops and taken aboard without formality. At Christmas Island and Keeling Cocos (both in the Indian Ocean), and at Ascension (South Atlantic), the inhabitants kindly invited me to buy small quantities from their private stocks.

Mail

Practically everyone has to make some arrangements for collecting mail, and it will be best if all mail at home goes to a bank or the office of some firm, where there will always be someone to attend to it, whereas a friend may fall ill or go away on holiday or business. Brief forwarding instructions can then be sent from time to time. A number of strategic ports can be selected for the collection of mail, but these should not be too close together or they might limit one's movements too much. I have had mail sent to banks, consulates, my publisher's offices, and yacht clubs, and have never had anything but a few magazines go adrift; but I have always written some time in advance saying approximately when I expect to arrive, asking if they will be kind enough to hold mail for me, and offering to pay any fees or expenses incurred. I have never been refused, though I have heard of British consulates declining to accept mail, nor have I ever been asked to pay anything. Perhaps yacht clubs are the least responsible, for often all mail there is left in a public place for every new arrival to sort through; but this at least means that one is expected and that one can gain some idea of what other yachts are on the way. On leaving the port a forwarding address can be given in case some mail arrives late. It is not wise to use poste restante; some post-offices will not keep mail for more than a short time, others make a charge, and nobody is personally responsible for seeing that the mail is handed over or forwarded on. Mail should be addressed: Mr. ..., British Yacht ..., Care of .... The yacht's name is important, especially on parcels, for apart from helping to identify one, it may make the difference between having to pay duty or not. If any gear for the yacht has to be sent out from England, it is best to enlist the services of a shipping agent, otherwise, and in spite of the fact that it should be duty-free for a yacht in transit, duty or a bribe may have to be paid.

Letter-writing can become a burden. One naturally wishes to keep in touch with friends and family at home, and as the cruise progresses one makes more and more friends along the route. During one three-year voyage my wife and I wrote something more than 600 letters between us. A typewriter is a help in dealing with this correspondence, and perhaps there are circumstances in which it is permissible to send a stereotyped letter, though this naturally lacks the personal touch and is not so keenly appreciated.

Hospitality

I have never made use of letters of introduction, as I consider they place their recipients in the awkward position of feeling that they must offer some hospitality no matter how inconvenient that may be. Indeed, such letters proved to be unnecessary, for the spontaneous friendship and hospitality extended by English-speaking and some other strangers abroad to my wife and me during our voyages was very remarkable, and I believe this to be the common and delightful experience of most little-ship voyagers. There was no occasion when we had real need of evening dress, but some respectable clothes suitable for wear in hot weather are essential. These are best of the permanently pressed kind, and in a small yacht a good way of preserving them is in their own little fibre case; but unless they are washed or cleaned before being stowed away after use they will soon become fusty, and in warm weather may grow a crop of mildew.

One of the chief difficulties arising from accepting hospitality ashore in some places is guarding the yacht against theft. Of course it should be possible to lock her up, but she may be broken into, or gear may be stolen from on deck. One will usually be warned if the risk is great (usually the smaller the port the smaller the risk), and the only solution then will be to employ a watchman or a guard dog.

There is little one can do to return shore hospitality, except to invite one's hosts on board for drinks, and this does seem to be appreciated. For such an occasion it is essential that the berth be a quiet one, for even the slightest motion, which is scarcely noticed by those who live afloat, may be enough to upset the guests and spoil the party. It is best for the yacht to lie alongside so that the guests may come and go without dinghy work. For this and the many other occasions when the yacht lies alongside, if she is not fitted with a metal-shod rubbing strake, fender boards may often be required, one for each pair of fenders (Plate 42c), so that she can lie against piles or other projections without damage to her topsides. In some places, notably in Mediterranean ports, the yacht will lie to her anchor with her stern warped close in to the quay. A stern gangway rigged to the shore is then a convenience, and one could scarcely improve on *Rena*'s. This consists of a light alloy ladder (this is often a useful thing to have on board) with its inboard end provided with a stout, hinged pin to ship in a socket on the taffrail, thus making a gooseneck fitting; its shore end is provided with two small wheels. Any fore-and-aft or vertical movement of the yacht is therefore taken care of by the wheels, and any athwartship movement by the pin on the taffrail. A plank laid on the rungs for the full length of the ladder provides a foothold.

With duty-free drink aboard, a party such as was suggested above is not a very expensive affair in spite of the large quantity sometimes consumed. Perhaps it might not be out of place to give here the recipe for the famous rum-punch of the Caribbean: 'One of sour, two of sweet, three of strong, and four of weak.' In practice this means the juice of one lime, two heaped teaspoonfuls of sugar, three ounces of rum, and plenty of water or ice; a little nutmeg should be grated to float on top. Ice can often be had easily enough in hot places, and if there is no refrigerator or ice-box aboard, there should be a large thermos flask with a wide neck in which to collect a small supply from the shore; but a warm hand should not be thrust into the cold flask or the glass may shatter. Incidentally, a unique way of using an ice-box was that employed by Bob Kittredge in *Svea*. When loading the box for a passage he did not buy any ice; instead he bought a quantity of canned beer (to which he was partial) and had this frozen at the local ice plant. He then lined the sides and bottom of the ice-box with several layers of the cans, and put his supply of frozen meat in the remaining central space. Thus no space was wasted, and there was no water to slop about in the bottom of the box when the temperature rose.

Yacht clubs

Yacht clubs abroad, except some of the commercial ones, generally invite the visiting yachtsman to be an honorary member during his stay; some of these clubs have excellent mooring, docking, and slipping facilities which they may put at the visitor's disposal free of charge, but this should not be taken for granted. Sometimes the visitor may be asked to present the club with his own yacht club burgee, so it is as well to carry a supply for this purpose. In addition to her own burgee at the main truck and ensign on its staff aft (the plain red ensign is more widely understood than is the blue or a defaced ensign), the yacht when abroad should wear at the main starboard crosstree a courtesy ensign, i.e. the merchant ensign of the country in whose waters she is. It is important to remember that the merchant ensign sometimes differs from the national flag; when it does so it is incorrect to wear the latter. One should have on board such courtesy ensigns as will be needed, so that the yacht may wear the appropriate one as she enters port; in a few countries failure to do this could entail a fine. Incidentally, it has become a custom, and a pleasant one I think, for a yacht to wear the ensigns of all the countries she has visited on the day she returns to her home port, and perhaps on some special occasion in the same year, such as the annual meet or rally of the owner's yacht club.

Most yacht clubs are concerned mainly with racing and social events, and the voyaging man would gain little or no benefit from membership of these. But there are some notable exceptions, such as the Royal Cruising Club, the Cruising Association, and the Little Ship Club, all with headquarters in London, and the Cruising Club of America in the U.S.A.

The Royal Cruising Club has a limited membership and gathers within its rather exclusive ranks a fair proportion of the British voyagers of the day; it awards trophies to members for their cruises and voyages. The club publishes up-to-date port information sheets compiled by members, for most European cruising grounds, and these are available to members only; it also publishes a yearly *Journal*, a well-produced book containing accounts of members' cruises, but this is available to the public under the title *Roving Commissions*. The Cruising Association probably has the finest nautical library in the country, and a large selection of charts for study in its Baker Street premises. The Little Ship Club, with a fine club house on the Thames, provides instruction in navigation and seamanship, and appoints honorary port officers, to whom members may apply when seeking local knowledge, in many places abroad.

The Cruising Club of America is perhaps best known abroad for its rules under which some of the most important ocean races are sailed, yet, like the R.C.C., it has as members many famous cruising men, and although its membership is not limited, it is exclusive. *The Cruising Club News*, in newspaper form, containing accounts of members' activities and other matters of interest, is published quarterly. The Blue Water Medal of the C.C.A. is awarded annually 'for the year's most meritorious example of seamanship, the recipient to be selected from among the amateurs of all the nations'. This 5-inch diameter bronze medal (Plate 42F) was struck in 1923, and is generally regarded among the voyaging fraternity as the premier and most coveted award. As it has been awarded for some of the most remarkable feats of seamanship, many of them performed by people whose names are now household words, the complete list, given below by kind permission of the officers and members, may be of interest.

1923　*Firecrest*　Alain Gerbault　France
　　　Left Gibraltar June 7, 1923, and arrived Fort Totten, L.I. exactly 100 days later. Non-stop. Dixon Kemp-designed British cutter, 34 feet l.o.a. Single-handed.

1924　*Shanghai*　Axel Ingwersen　Denmark
　　　Departed Shanghai February 20, 1923, and arrived Denmark via Cape of Good Hope in May, 1924. Double-ended ketch, 47 feet l.o.a., built by native labourers. Crew of three.

1925   *Islander*   Harry Pidgeon   U.S.A.
        First circumnavigation—from Los Angeles via Cape and Panama
        Canal, November 18, 1921—October 31, 1925. Home built 34-foot
        l.o.a. yawl of *Sea Bird* type. Single-handed.

1926   *Jolie Brise*   E. G. Martin   England
        Double trans-Atlantic crossing, including Bermuda Race. Le Havre
        pilot cutter 56 feet l.o.a. April 3, 1926, from Falmouth, July 27 to
        Plymouth.

1927   *Primrose IV*   Frederick L. Ames   U.S.A.
        This 50-foot l.o.a. schooner had been sailed to England for the 1926
        Fastnet. Medal was awarded for her return passage, from Ports-
        mouth, northabout, Iceland, Labrador, Cape Breton Island, 58
        days to Newport, R.I.

1928   *Seven Bells*   Thomas F. Cooke   U.S.A.
        An eastbound trans-Atlantic passage, Branford, Conn., to Falmouth,
        July 5—July 31, 1928. Roué-designed 56-foot l.o.a. ketch.

1929   *Postscript*   F. Slade Dale   U.S.A.
        A 4,000-mile cruise in the West Indies with crew of two, from and
        to Barnegat Bay, N.J. The 23-foot l.o.a. cutter, designed by the
        owner, was subsequently lost with all hands under different owner-
        ship. No power.

1930   *Carlsark*   Carl Weagant   U.S.A.
        A 13,000-mile cruise of this 46-foot l.o.a. ketch from Ithaca, N.Y.,
        to Ithaca, Greece, and return to New York City. Started June 20,
        1929, completed May 30, 1930.

1931   *Svaap*   William A. Robinson   U.S.A.
        This 32-foot 9-inch l.o.a. Alden ketch departed New London June
        23, 1928, in the Bermuda Race of that year, and circumnavigated
        via Panama and Suez Canals with crew of two, except for period of
        race. Arrived N.Y. November 24, 1931.

With-   *Jolie Brise*   Robert Somerset   England
out   Award for remarkable feat of seamanship and courage in rescuing
date   all but one of 11-man crew of burning schooner *Adriana* in the 1932
        Bermuda Race.

1933   *Dorade*   Roderick Stephens, Jr.   U.S.A.
        A three-month, 8,000-mile trans-Atlantic crossing from New York
        to Norway and return, including victory in the Fastnet Race. The
        52-foot 3-inch Stephens-designed yawl returned home from England
        by the northern route in the remarkable time of 26 days.

1934   *May L.*   W. B. Reese   England
        A single-handed passage in a small double-ended ketch from England
        in the fall of 1933 to Nassau in January, 1934.

1935   Charles F. Tillinghast   U.S.A.
        'For his seamanship in the effort to save three members of the crew
        of the *Hamrah* who were overboard in the North Atlantic, and in

bringing the disabled and short-handed ketch safely into Sydney, N.S.'

1936 *Arielle* Marin-Marie France
A single-handed trans-Atlantic passage in a 42-foot 7-inch l.o.a. motor boat (July 23–August 10, 1936) with two self-steering devices. Marie had sailed the cutter *Winnibelle II* (without power) from Brest to New York in 1933.

1937 *Duckling* Charles W. Atwater U.S.A.
A voyage from New York to Reykjavik, Iceland, and return to Newport via Trepassey, Newfoundland, June 19–August 26, 1937. A 37½-foot l.o.a. Mower cutter.

With- *Igdrasil* Roger S. Strout U.S.A.
out   Circumnavigation in a *Spray*-type cutter (eventually re-rigged as
date   yawl) designed and built by owner. He and his wife circumnavigated via Panama and Cape between June, 1934, and May, 1937.

1938 *Caplin* Cdr. Robert D. Graham, R.N. England
Bantry Bay, Ireland, to Funchal and Bermuda, between April 20 and June 27, 1938, and then to West Indies. Graham's daughter completed crew of two in 35-foot l.o.a. yawl.

1939 *Iris* John Martucci U.S.A.
An 11,000-mile cruise from New York to Naples and return in a 36-foot l.o.a. MacGregor yawl. The return home, including a non-stop 35-day run from Tangier to Bermuda, was made after outbreak of World War II.

1940 British Yachtsmen at Dunkerque England
Awarded to British yachtsmen, living and dead, who had helped in the evacuation of the British Expeditionary Force in June, 1940.

1941 *Orion* Robert Neilson U.S.A.
Orion was a 30-foot auxiliary ketch of 10-foot beam and 4½-foot draft designed by John G. Hanna. On June 5, 1941, Neilson and one companion sailed from Honolulu and arrived San Pedro, Calif., on July 15. The medal was awarded for this passage, and *Orion* subsequently carried on through the Panama Canal to Tampa, Florida, a total distance of 7,978 miles.

1947 *Gaucho* Ernesto C. Uriburu Argentina
A cruise in a 50-foot ketch from Buenos Aires through the Mediterranean and to the Suez Canal and then to New York, following Columbus's route from Palos, Spain, to San Salvador.

1950 *Lang Syne* William P. & Phyllis Crowe U.S.A.
From Honolulu around the Cape to New England, from Easter Sunday, 1948, to the spring of 1950. After the award the 39-foot l.o.a., home-built Block Island type double-ended schooner completed her circumnavigation to Hawaii.

1952 *Stornoway* Alfred Petersen U.S.A.
A circumnavigation from and to New York via the two major

canals in a 33-foot double-ended cutter. Single-handed, June, 1948–August 18, 1952.

1953 *Omoo* L. G. Van De Wiele Belgium
A circumnavigation by owner and wife and one other, plus dog, from Nice, France, to Zeebrugge, Belgium, July 7, 1951–August 2, 1953, via Canal and Cape of Good Hope. Steel 45-foot l.o.a. gaff-rigged ketch. Said to be first steel yacht and first dog to circumnavigate.

1954 *Viking* Sten & Brita Holmdahl Sweden
A circumnavigation by Canal and Cape of Good Hope by owner and wife from Marstrand to Gothenburg, Sweden, between June 17, 1952, and June 22, 1954. A double-ended 33-foot ketch converted by owner and wife from a fishing boat.

1955 *Wanderer III* Eric & Susan Hiscock England
Circumnavigation by Canal and Cape of Good Hope by owner and wife, July 24, 1952–July 13, 1955, in 30-foot Giles-designed sloop.

1956 *Mischief* H. W. Tilman England
20,000-mile voyage of 50-year-old Bristol pilot cutter from England through Strait of Magellan, up west coast of South America, through Panama Canal and return to England, July 6, 1955–July 10, 1956.

Without date Carleton Mitchell U.S.A.
'For his meritorious ocean passages, his sterling seamanship and his advancement of the sport by counsel and example.'

1957 *Landfall II* Dr. William F. Holcomb U.S.A.
Circumnavigation westabout from San Francisco of Schock-designed 46-foot 6-inch l.o.a. schooner via the Suez and Panama Canals, with side trips to South America, England, North Africa, and New York: September 18, 1953–September 15, 1957.

1958 *Les 4 Vents* Marcel Bardiaux France
Single-handed circumnavigation westabout around Cape Horn and the Cape of Good Hope, in home-built sloop 30-feet 9-inches l.o.a. From Ouistreham, France, May, 1950, to Arcachon, France, July 25, 1958.

1959 *Trekka* John Guzzwell Canada
Single-handed circumnavigation in home-built yawl 20-foot 10-inch l.o.a. via the Cape of Good Hope and the Panama Canal. From Victoria, B.C., to Victoria, September 10, 1955, to September 10, 1959.

Without date *Lehg I, Lehg II, Sirio* Vito Dumas Argentine
Global circumnavigation in *Lehg II*, 1942–43. Other phenomenal single-handed voyages in *Lehg I*, 1931–32; *Lehg II*, 1945–47; *Sirio*, 1955.

1960 *Gipsy Moth III* Francis Chichester England
Winner of the first single-handed trans-Atlantic Race in 1960, from east to west across the Atlantic.

With- *Seacrest* Dr. Paul B. Sheldon U.S.A.
out Extended cruises along the coasts of Nova Scotia, Newfoundland,
date and Labrador.
1962 *Adios* Thomas S. Steele U.S.A.
Two circumnavigations in a 32-foot ketch; one in 1950-55, the
other in 1957-63.
1963 Not awarded.
1964 *Pen Duick II* Eric Tabarly France
Winner of the second single-handed trans-Atlantic race, time
27 days, 2 hours. Ketch 45 feet l.o.a.
1965 *Delight* Wright Britton U.S.A.
Cruise to Greenland and return with his wife in 40-foot yawl.
1966 *Joshua* Bernard Moitessier France
Non-stop voyage from Moorea to Alicante by way of Cape Horn
with wife. Ketch 40 feet l.o.a.
1967 *Gipsy Moth IV* Sir Francis Chichester England
Single-handed voyage round the world from Plymouth, England,
to Plymouth, by way of the Cape and the Horn, with only one stop,
at Sydney, N.S.W.
1968 *Lively Lady* Sir Alec Rose England
Single-handed voyage round the world from Portsmouth, England,
to Portsmouth, by way of the Cape and the Horn, with only two
stops, at Melbourne and at Bluff, N.Z.
1969 *Elsie* Frank Casper U.S.A.
1970 *Carina* Dick Nye U.S.A.

Among the smaller clubs mention should perhaps be made of the
Seven Seas Cruising Association. This has no club premises, and
membership is limited to those who live continuously afloat. Each
month it publishes its *Bulletin*, the contents of which are letters (often
verbose and many dealing with the west coast of North America)
sent in by cruising members. From this source certain useful and
up-to-date information which cannot easily be obtained elsewhere
is available. *Bulletins* can be had by those who are not members from
P.O. Box 6354, San Diego, California, U.S.A. at a cost of $5.00 a year.

Finance

The voyager should apply to his bank for the latest information on
currency restrictions. He will no doubt take with him travellers'
cheques to the value permitted, and these will be cashed by banks or
tourist agencies, or sometimes by hotels or shops, in the countries
endorsed on them, or in any country if marked 'world wide'. If the
cheques bought are issued by American Express Company, which

has offices in many countries, including the U.K., their value will be refunded in the event of theft or loss, provided the instructions given with them are followed. Alternatively a letter of credit may be used, but I consider this has no advantage, and may be inconvenient in that money can only be drawn on it at a bank. In the U.S.A. the credit card system is widely used, and along the Intracoastal Waterway, for example, it is rare indeed for anyone to pay cash for the fuel he buys.

I presume that few people would start on a long voyage without having sufficient cash or credit to complete a large portion of it, but a few people have succeeded in earning a part or the whole of their living as they went along. For example, Dwight Long sometimes took paying guests in *Idle Hour*; W. Howell practised as a dental surgeon during his voyage from England to British Columbia in *Wanderer II*; L. F. Champion of *Little Bear* worked as a shipwright while his wife did secretarial work; the Carrs in *Havfruen III* took parties of Americans cruising among the West Indies; and others, of whom I am one, earned their way by writing articles for the press and illustrating them with photographs processed on board.

One of the best examples of a voyaging life being successfully financed by the efforts of the owners is that of Tom Hepworth and his wife Diana in the ex-Brixham-trawler *Arthur Rogers*. First they did trawl fishing off Guiana, then photographic work in the Panama Canal Zone; this was followed by a charter party trip to the Galapagos Islands and some paying guests now and then, until finally they became traders in the New Hebrides. I quote from one of their letters:

We seem to be settled for the foreseeable future in the New Hebrides, which are governed by the monstrous Franco-British condominium, whose single virtue is, however, that it interferes with nobody. For two and a half years now we have made a reasonable living as genuine south sea island traders. We take the good ship *Arthur Rogers* round the islands of the group, buying copra from the natives, and selling fish-hooks, stick tobacco, peroxide (to bleach the hair); torches, soap, ammunition, tea; saucepans, aspirin, pencils, rice; fountain pens, bright cottons, tinned milk, scent—a thousand and one things. Often we take back in the store all the money we have paid out for the copra, a highly satisfactory state of affairs. Sometimes the game is hazardous, as when we load copra off the rocks into open boats, with the ocean swell running. Then one backs in on a small anchor on the top of a wave, a bag weighing some 150 lb. is chucked in, and one pulls out frantically to avoid the next wave. Sometimes there is no anchorage for the ship—50 fathoms 50 yards from the shore—

and then Diana has to do the boat work while Tom stands off and on all day under motor. One works all the daylight hours, but mostly there is a quiet anchorage at night.

The old ship was eventually lost on a reef in the Solomons, but the Hepworths went on trading.

Mention should also be made here of the seven voyages Irving Johnson and his wife have made round the world. The first three were in the schooner *Yankee*, and the last four in the beautiful brigantine of the same name (Plate 44). These trips, each of 18 months' duration, were financed by the payments made by the crews, which numbered up to as many as twenty young men and women. Unfortunately this fine ship, under new ownership, was lost on the fringing reef of Rarotonga, one of the Cook islands.

The Panama Canal (Plate 45 and Fig. 64)

Much has been written and said about the difficulties and dangers attendant on a small yacht's transit of the Panama Canal. Perhaps the following notes may help to reassure the owner who plans to pass that way, and give him some idea of the procedure to be followed. The information is based on the three transits I have made, but it is possible that some changes in the regulations may have been made since then.

The American-occupied Canal Zone is a strip of territory about 10 miles wide, with the canal running approximately through the centre of it; but the city of Colon at the north (Caribbean) end, and Panama at the south (Pacific) end belong to the Republic of Panama. The general trend of the canal from Caribbean to Pacific is south-east. The rainy season is from May to December; January is a suitable month for passing through into the Pacific, because there is then a good chance of getting a fair wind to take one out of the Gulf of Panama and away to the south and west.

The harbour of Cristobal/Colon is protected by two massive breakwaters, each about 2 miles in length. On arrival the yacht should anchor on the flats at the east side of the harbour south of No. 9 pier and west of the oiling plant (marked 'coaling plant' on the Admiralty

PLATE 47
Voyaging people. *A*: Van and Jo Vancil, *Rena* (U.S.A.). *B*: George and Sonny Cook, *Wind's Song* (Canada). *C*: John and Mary Caldwell, *Outward Bound* (Australia). *D*: Sadun and Oda Boro, *Kismet* (Turkey). *E*: George and Elsa Koch, *Kairos* (Germany). *F*: David Guthrie, *Widgee* (U.K.). *G*: Annie Van de Wiele, *Hierro* (Belgium). To judge by their expressions, these voyagers from seven different countries all have something in common—contentment, a sense of achievement, or just plain happiness?

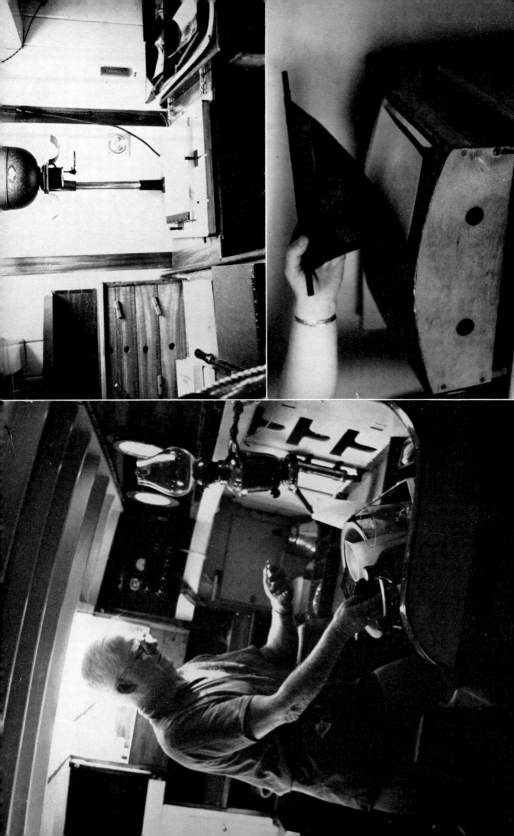

chart), and fly flag 'Q'. Officials representing customs, health, and the canal authorities will board her there, complete the paper work for entry, and compute the tonnage on which dues will be paid—the canal tonnage is about the same as the net tonnage. The owner must then row to Pier 9 (the use of outboard motors is forbidden) to obtain permission from customs to berth his yacht at the Panama Canal Yacht Club, which is situated at Cristobal on the east bank of the French canal in behind the oiling plant, where she may lie alongside for a small charge. The tide rises only about 1 foot, and the berth is comfortable, convenient, and well policed, but is sometimes foul with fuel oil. Showers, meals, and drinks can be had at the club, and all stores can be bought at Colon. The Panama Canal Yacht Club is a convenient place at which to scrub and paint the bottom before entering the Pacific, where slipping facilities are few and far between and are rather expensive. The club has two slips; the smaller one is able to haul out a yacht drawing 5 feet and with a displacement of 15 tons; the larger one can accommodate a draught of $6\frac{1}{2}$ feet and a displacement of 35 tons. The charge in 1969 was \$1.00 per overall foot for the first day and \$10.00 per day thereafter. In no circumstances may either slip be used for more than 5 days, and because of the demand it would be wise to book in advance. The address is: Box 5041, Cristobal, Panama Canal Zone.

To arrange for the transit, the owner should go first to the Admeasurer's office for the tonnage certificate, and pay the bill at the Collector's office in the adjoining building—the charge is only 72 cents a ton, and this includes the services of a pilot, pilotage being compulsory. The receipt is then taken to the Port Captain's office and arrangements will be made for the pilot to board the yacht at a pre-arranged date and time at or off the club. A yacht without an engine may not proceed under sail, but must be towed through, unless she can borrow an outboard motor of sufficient power. Arrangements for towage can usually be made very cheaply with the skipper of one of the banana boats, vessels of about 80 tons, and one may secure alongside her in the locks to avoid damage.

PLATE 48

Many of the photographs reproduced in this book were processed on board *Wanderer III*. *Left*: With tank and chemicals safe on the swinging cooker, I time the development of a film as the yacht rolls on her way across the Atlantic. *Top right*: But only when the yacht lay quietly in port was it possible to make enlargements, and for this the forepeak was used. The enlarger is in position on its temporary platform beneath the extended coachroof; benches each side hold paper and processing dishes. *Bottom right*: A home-made print drier and glazer.

T

The distance from the yacht club at Cristobal to the yacht club at Balboa is 46 sea miles. There are three up-locks grouped together at

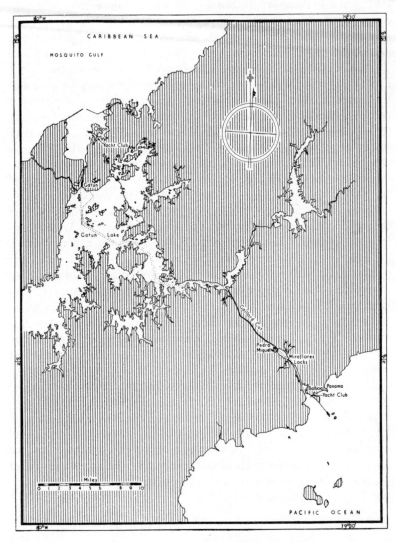

FIG. 64. PLAN OF THE PANAMA CANAL

Gatun, 5 miles from Cristobal; these lift ships a total height of 85 feet to the level of Gatun Lake. It is in these locks that a small vessel is most likely to get into difficulties; for when the water is let in through the huge holes in the floors of these chambers, which each measure 1,000 feet by 110 feet, and fill in 8 minutes, the disturbance is similar to that in a tide race, and a small vessel may get out of hand and be thrown against the wall, a bowsprit, bumkin, or crosstree being particularly vulnerable. The safest method of negotiating these locks is to arrange in advance with Traffic Control for the yacht to be held in the centre of each by four lines, one from each bow and one from each quarter. The lock-master will be notified in advance so that he may have lock-hands standing by with heaving lines, but if the wind is strong the men may not be able to get the lines out to the yacht, and she will then have to lie against the wall or another small vessel. She must provide her own lines, each not less than 20 fathoms in length; she must also provide a separate hand for each line, otherwise she will be compelled to employ canal seamen at high cost. If one is short-handed it is usually easy enough to get members of the yacht club to make the trip as a joy-ride, for many of them, although employed on the canal, have never been through it. At Gatun the lock-hands will carry the warps along from one lock to the next, but the yacht must move ahead under her own power; the mules—electric locomotives—are only used for large ships. If one has to share a lock with a big ship, and this is usual, she will go in first in the up-locks and last in the down-locks, and one must be prepared for a strong back-wash if she should use her propeller.

The channel through Gatun Lake—a flooded jungle—is about 14 miles long, and it is well marked with buoys and lighthouses. At the far end of the lake 12 miles of canal take one on through Gaillard Cut, the narrowest part of the Continental Divide, to the first of the down-locks at Pedro Miguel (Americans call this Peter MacGill); a mile farther on at Miraflores, and placed together, are the remaining two down-locks. The down-locks present no difficulty, as the water in them subsides quite gently, but in the last of them—the sea-lock—there is apt to be a disturbance when the gates are opened and fresh water and sea water mingle. As all locks are duplicated, north-bound and south-bound ships are independent.

The pilot will ask the master of the last lock to telephone the Balboa Yacht Club, giving the estimated time of arrival, and he will take the yacht to the club, which is situated on the north-east side of the canal approach, where a boatman will show her to a mooring which is free for the first 10 days, and a launch will call to take the pilot off. This berth is, by reason of the prevailing winds, far better than it looks on the chart, and one suffers little inconvenience from the wash of passing ships, but the tide rises 15 feet at springs, and the stream runs hard. The yacht club has a pier with water and fuel laid on, and a daylight launch service, but no shore facilities except a cold shower and a bar; however, visitors have the privilege of using the American Legation Club in the same building. Foreign yachts are no longer permitted to buy anything from the wonderfully stocked canal commissariat, not even through a canal employee, so any stores which have not already been bought at Colon (that in my opinion is the better end for shopping) must be bought (with American dollars) at Panama city 2 miles away.

The channel from Balboa out to the Gulf of Panama is wide, straight, and well marked, and in 1969 a yacht was not required to employ a pilot there.

The Suez Canal (Plate 46, and Fig. 65)

At the time of writing no one can say when the Suez Canal will be reopened to traffic or what the regulations for yachts using it are likely to be; but perhaps a few notes on it may be of interest, and I hope of some practical value in the future. It is, however, not so much the transit of the canal as the passage of the Red Sea which presents difficulties to a sailing vessel, and this is why most circum-navigating yachtsmen in the days before auxiliary engines became reliable and powerful preferred to take the Cape route.

In the northern part of the Red Sea, from Suez down to about Port Sudan, the wind blows from NNW. throughout the year. In the southern portion the prevailing wind is SSE., but with this important exception: from June to September (the very hot months) the winds there are light northerlies mixed with calms. Clearly, therefore, it is easiest for a sailing vessel to make the passage from north to south, for provided she does it at the right time of year she should have a fair wind all the way. The north-bound yacht, even at the best time of year, is sure to have a headwind for about 500 miles of the trip.

When I passed northward through the Suez Canal in January 1962, a yacht approaching either end of it from the sea had to make her number with the pilot cutter, and was usually directed to the yacht basin at Port Fouad, opposite the town of Port Said, or if approaching Suez was told to anchor off the monument at the west side of the canal entrance. This last can be a dangerous berth in a strong onshore wind, and if caught there in such conditions the yacht should at once go without waiting for permission into the mouth of Suez creek, which is

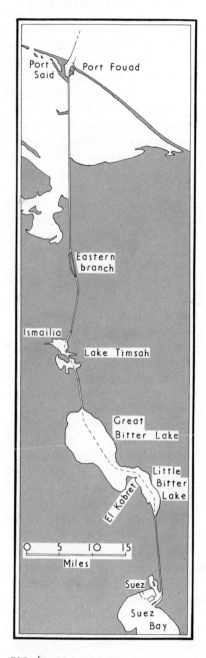

FIG. 65. PLAN OF THE SUEZ CANAL

safe but inconvenient. When I was there one was not permitted to make any use of the yacht club a little way up the creek, but soon after that restriction was withdrawn, see opposite page.

In advance of my arrival I had asked a British firm of shipping agents, Hull, Blyth & Company, to handle the paper work for me, and this they kindly did without charge. At that time there were no dues for the transit of the canal by yachts, but I paid about £5, which included stamp duty and a £2 permit to permit me to buy a few gallons of fuel for the engine. Some other yachts have not been so fortunate, and a few have been charged fantastic sums by Egyptians calling themselves agents.

The canal is 87 miles long, and as it passes through flat desert there are no locks. The difference in tidal level between the Gulf of Suez and the Mediterranean is taken care of by the fact that the canal runs through three lakes, the Little Bitter, the Great Bitter, and Timsah, in that order from the south. Between Suez and the Little Bitter lake tidal streams of up to 2½ knots are felt; elsewhere there is no tide, but a current may be experienced. Pilotage was compulsory, one pilot going as far as Ismailia on Lake Timsah, and another taking over there; no charge was made for their services. Merchant ships went through in convoys, one north-bound, and two south-bound each day, the former going through without a stop; one south-bound convoy tied up in the branch canal 30 miles south of Port Said, the other anchored in the Great Bitter lake, to let the north-bound convoy pass. A yacht had to start off at the tail end of a convoy, but could not keep up unless she was capable of 7½ knots, and she could not travel after dark because of the impossibility of carrying the required projector capable of lighting up the canal 1,300 yards ahead. So a yacht normally took two days over her transit, spending a night at Ismailia, a green oasis in the brown desert, where the yacht club was hospitable, but one was not permitted to leave its grounds. Another possible night anchorage was at El Kabret on the Little Bitter lake.

The edges of the deep channel were marked by buoys; there were many signal stations, but only two bridges, one a pontoon, the other a swing. Mostly the banks were too high to see over from the low deck of a small vessel, but even from aloft there was little to be seen but sand, and the only excitements were meeting the convoys going the other way. In the narrow canal each ship raised the water level ahead of herself, then sucked it from the banks as she went by; it was important not to be too close to the bank at such moments, or to make fast to the bank. I did not notice any tendency of passing ships to push or suck the yacht off her course, but it was essential to keep plenty of steerage way.

Shortly after my account of the transit of the canal had been published in *Yachting World*, that magazine received a friendly and encouraging letter from Mr. M. A. Moursy, President of the Suez Canal Yachting Club, part of which is quoted below:

Having read Eric Hiscock's article . . . we would like to assure *Wanderer III* and all similar cruising yachts that we are doing our best to facilitate their passage and mooring in the Suez Canal and its ports.

The Suez Canal Authority is getting in touch with the Authorities concerned to reduce the amount of paper and delay associated with entering the U.A.R. ports, for cruising yachts.

The club premises at Suez and anchorage will be facilitated, and steps will be taken to get the responsible authorities in Ismailia to allow captains and crews of yachts in transit to land, visit the town, and pay a visit to Cairo when requested.

We would like to see *Wanderer III* again and so many other cruising yachts as possible, they are all welcome, and we hope that they, too, will enjoy their trips through the canal.

12

PHOTOGRAPHY

Cameras—Colour photography—Processing black-and-white films—Enlarging—Movie work

MOST voyaging yachts have at least one camera aboard for pleasure or profit, and the general rules for photography, which will be found in a good text book such as Newcombe's *35mm. Photo Technique* (Focal Press), will apply just as well afloat as ashore. But on a long voyage, and particularly in the tropics, the matter is complicated by the fact that high temperatures and humidities are unfavourable to the life and performance of all sensitized materials and of some types of apparatus, and these conditions favour the growth of moulds. Also the processing of films and the making of enlargements aboard a small seagoing yacht present special difficulties, but these are not insuperable.

Cameras

As many of the exposures made during a voyage will be on or near the water, the light will generally be bright, so a cheap camera with a lens of small aperture may serve well enough for average subjects, and will have the merit of simplicity in operation. But whatever the type of camera, it is important that the range of shutter speeds should extend to about 1/250th of a second, because the motion of the yacht will render unsharp pictures made with a longer exposure. My preference is for the precision miniature camera, using 35-mm. film, and having interchangeable lenses. Because of its short focal length, the standard lens of such a camera gives considerable depth of field even at a wide aperture, and the ability to change it at a moment's notice for a wide-angle lens to include more in the picture, or for a lens of longer focal length to enable a distant subject to fill the frame (this is particularly important when using colour reversal film), is a great advantage. Although black-and-white reproductions can be made from colour transparencies, it is more convenient and satisfactory to make them direct from negatives in the usual way. So if both colour and black-and-white photographs are required, two cameras are desirable. Here the camera with interchangeable lenses has the merit that one need buy only a second body instead of a complete camera, because the lens or lenses one already possesses will serve in either. Colour film is more readily available in the 35-mm. than in larger sizes, and as most of it is slow film, the large aperture of the more expensive camera will permit short exposures to be made.

There are on the market completely watertight cameras, and for some normal cameras there are available watertight cases with arrangements for working the controls from the outside; these are intended for under-water photography, and would no doubt be of value if one had to make a landing through surf and there was a risk of the dinghy filling or capsizing. But for protection against rain or spray this expensive equipment is not required, and the following procedure will serve almost as well. Envelop the camera in a loose-fitting polythene bag with a hole cut for the lens, and secure the bag to the lens mount with an elastic band. If the bag is sufficiently thin and loose-fitting, all the controls of the camera can be worked from the outside. Protect the front element of the lens with an ultra-violet filter, which is almost colourless and calls for no increase of exposure; rain or spray on this filter can be wiped off, leaving the lens untouched, and if in time the filter becomes scratched it can be renewed for a few shillings. I keep one of these filters permanently on each lens for protection against rain, spray, dust, and fingermarks. If by mischance salt water does get on to the front element of a lens it should at once be wiped off with a very soft cloth, such as an old linen handkerchief, moistened with fresh water, and the lens then dried with an equally soft cloth.

A camera ought not to be left exposed to direct sunlight, and if its case or strap is stitched with cotton thread, this should be tested occasionally as it can become rotten without visible sign. In a warm and humid climate mould will grow rapidly on leather bellows, cases, and the like; the best deterrent appears to be wax polish. In some places the inhabitants are afraid of or object to photographs being taken, while at others they are almost embarrassing in their wish to be photographed (for a fee), so it is as well when ashore to carry the cameras in a shopping bag to prevent their appearance being too obvious. A Polaroid camera, which produces a finished print in a few moments, might be of value when making friends with natives in some places.

To protect the cameras and the partly exposed films in them from damp when not in use, I keep them in their hold-all case, and the case in a plastic bag together with a supply of indicator silica gel. This desiccating agent is in the form of small crystals; when it is dry it has a bright blue colour, but as it absorbs water the colour changes to a pale pink; it can then be given a new lease of life by drying it on the hotplate of the cooking stove or in the oven. Clearly the amount of drying required will depend on the humidity, the size of the camera container, and the frequency with which the latter is opened; perhaps once or twice a week. Silica gel can be bought in perforated cans about 3 inches in diameter and half an inch thick, each can having a small window so that the colour of the crystals can be seen.

Colour photography

For colour work during our voyages my wife and I used Kodachrome exclusively, as this reversal film makes excellent transparencies and lends itself to block-making, as is evidenced by the magnificent reproductions in *National Geographic*; we were not interested in colour prints, for which other types of film are preferable. Kodachrome film, which must be processed by a laboratory, is slow —A.S.A. 25—and as it has relatively little latitude, the exposure needs to be judged fairly closely; so we always used an exposure meter with it, and indeed regarded that instrument as essential until it went out of business in mid-Pacific, and could not be replaced for many months. During that period we made use of the exposure guide supplied with the film, and the results were almost the equal of those we had got when using the meter, but the latter is much quicker and more convenient. With Kodachrome over-exposure results in a pale, colourless transparency; under-exposure gives a dark, heavy image lacking shadow detail, with a tendency to a mauve or bluish cast, but is the preferable error. When exposed at or near the sea it has a tendency to be too blue because of the abundance of ultra-violet rays, but this can be obviated by using an ultra-violet filter.

Colour film is most likely to suffer climatic damage after exposure, so it is inadvisable to leave a partly exposed film in the camera for any length of time. Aboard *Wanderer III* after a film had been exposed and wound back into its cassette, this was placed in the maker's container, but without the cap, and the container placed in a storage can together with silica gel. When a place with air-mail service was reached, we put the cap on and posted the film at once either to Kodak in England or to the next processing address along our route. As it is sometimes difficult to get films sent out of the country in which they have been processed, necessitating as it may an exchange transaction requiring government sanction, we arranged for the film to be posted by the laboratory to some address in that country to await our arrival. Some countries, Italy, for example, do not permit films to travel by post out of the country, and although a post-office may accept them they may be destroyed; so inquiries should first be made. The price paid for Kodachrome film bought in the U.S.A. and some other countries does not cover the cost of processing as it does in the U.K. But prepaid processing mailers can be bought at the same time as the films, and these will be honoured by Kodak processing laboratories throughout the world.

Some of the film bought in England for our first long voyage, and not used until we were in the western Pacific, returned from processing with an overall green tinge, believed to be due to the high temperature in which, inevitably, it had been stored while on board. We therefore jettisoned the remaining unused stock and bought fresh in New Zealand, and we did not experience that trouble again although temperatures were high on the Australian coast and in the Indian Ocean. On subsequent voyages we started with a smaller quantity and restocked at more frequent intervals abroad.

On receipt of transparencies we stowed them away with silica gel until there was a suitable opportunity for mounting them between glass in metal slides; for this it is essential that the atmosphere be dry, otherwise damp patches may appear beneath the glass to the detriment of the transparencies. The type of slide in which adhesive must be made damp is unsuitable.

During the later stages of the first voyage we carried on board a 500-watt projector, and were then able to give illustrated talks ashore in some places. After returning home we gave a large number of $1\frac{1}{2}$-hour shows, using a selection of about 300 slides, to audiences exceeding 60,000.

Processing black-and-white films

As the commercial processing of miniature black-and-white films is often so badly done, I always do my own developing even when at sea; this is not so difficult as is often believed, for everything, except perhaps the loading of the developing tank, can be done in full daylight. Originally I used the common type of tank, the spiral of which must be loaded in total darkness, and this I did in a zip-fastened changing bag of double black satin. But in hot weather my hands sweated in the confines of the bag, and although I endeavoured to hold the film by its edges only, there were occasions when it got marked by finger prints. More recently I have been using the German made

Rondinax tank, which can be loaded in daylight and obviates the need to handle the film. With this tank, constant instead of intermittent agitation has to be employed, and this calls for a reduction of development time by about one-quarter.

Having studied the Kodak booklet, *Photography in the Tropics*, I was aware of the risk of reticulation—a crinkled appearance of the sensitized layer—and frilling or stripping—lifting of the sensitized layer from its base—which may attend development at high temperatures. For the first voyage I therefore laid in a supply of tropical developer DK.15, which is loaded with sodium sulphate to reduce these risks, together with pre-development hardener. However, the resulting negatives were not to my liking, being rather grainy and sometimes too contrasty. In all probability this was not the fault of the developer, but as I happened to have on board a supply of Meritol-metol, the developer I had been accustomed to using in moderate temperatures, I experimented with it. Although it is not recommended for use above 70 °F., I used it with complete success from then on, and sometimes at a temperature of 88 °F., and I soon gave up pre-hardening as this appeared to be unnecessary. With increased temperature the time of development must, of course, be reduced, so I made a series of experiments to ascertain the correct developing times for each 5° step from 70° to 90°, but there is no point in giving the figures here as the film I was then using is no longer made. During a recent voyage in less extreme climatic conditions, I developed Kodak Plus-X film in Microdol-X with continuous agitation, and obtained good results with the following times:

70, 72, 75, 77, 80 degrees F.
$8\frac{1}{4}$, $7\frac{3}{4}$, $6\frac{1}{2}$, 6, $5\frac{1}{4}$ minutes

Having developed several hundred films at temperatures considerably above normal without once having experienced reticulation or stripping, I conclude that these troubles may be due not so much to processing at high temperatures as to differences in temperature of the various solutions. If, for example, one uses a refrigerator to cool the developer to, say, 68° but uses a fixing solution at 85°, I think there will be a risk. But as I had no refrigerator or ice-box aboard, developer, rinse, fixing solution, and washing water were all at approximately the same temperature. But if processing has to be done at anything above 90° (this is most unlikely afloat) the advice in the Kodak booklet should certainly be followed. Incidentally, the thinner the sensitized layer the smaller is the risk of reticulation, so slow films with thin emulsions should be more suited to use in tropical conditions than rapid films with thicker layers. I have not used films faster than A.S.A. 125.

At sea I made use of the swinging galley stove to hold the tank and the chemicals safe and level (Plate 48, *left*), and timed with a stop-watch. Development was followed by a quick rinse in fresh water, followed by 10 minutes in an acid hardening fixing bath, made up on board from powder. Because of the limited supply, one may not wish to wash films on board in fresh water, though if they are given 5-minute soakings in 8 tank-fulls of water, which is adequate, the quantity used is less than one gallon. However, sea water can be used with complete confidence so long as it contains no sediment or other foreign matter— indeed, for this purpose it is more efficient than fresh. A final rinse must be given in fresh water, preferably with a few drops of wetting lotion added, and the film may then be stretched between spiked clips to dry. In quiet weather I used to put the film in the rigging, but dur-ing strong winds, or in anchorages where insects abounded, drying was done in the screened forepeak; in wet or very humid weather I have dried films in the hot air above the cooker, keeping them con-stantly on the move so that they could not become too hot. Kodak recommend that before drying the film should be rinsed in fungicide to prevent moulds growing on it. No doubt this is sound advice, but I found it was not necessary, provided the film was wrapped in dry photographic paper and stowed in its unsealed container in the storage can along with silica gel, for moulds are not able to grow in a dry atmosphere. My aim throughout was to use the minimum number of solutions and make processing as simple and speedy as possible, and any step that experience showed to be unnecessary was eliminated. Each container was numbered, and an index to negatives compiled so that any one needed could be found in a few moments.

Enlarging

As I wished to make enlargements with which to illustrate articles, the forepeak was arranged in such a way that it could be used as a darkroom when in port; during one voyage I made some 300 whole plate ($8\frac{1}{2} \times 6\frac{1}{2}$ in.) enlargements there, and many of the photographs reproduced in this book were also made there. The door and hatch were sufficiently light-tight for bromide paper, and a mushroom ventilator in the deck provided a little air, but in the tropics the dark-room temperature sometimes exceeded 95 °F, even with my wife pouring water over the foredeck; a small electric fan would have been a help. There was a bench each side, one for dishes of chemicals, the other for paper, etc. A hinged flap could be bolted in place horizontally between them (Plate 48, *top right*) to make a platform for the enlarger, which being beneath the extended coachroof had ample headroom.

The enlarger (a Leitz Valoy) was intended to run off a mains supply, using a full-size opal lamp; I fitted it with a 12-volt 30-watt lamp to run off the yacht's supply, but as such a lamp could only be obtained in a small size, it was necessary to insert a piece of opal glass between it and the condenser to give even illumination, and this absorbed a lot of the already limited available light. Since then, in a larger yacht, I use a dynamotor to step up the 24-volt supply to 240 volts so that the correct mains lamp can be used.

Bromide paper was packed specially for the trip. Each batch of 50 sheets was wrapped in tinfoil before being put in its box, and each box was sealed up in its own can. This kept the paper in good condition, but once a box had been opened its contents deteriorated after a week or so with slightly degraded enlargements as a result. I understand that Kodak paper can no longer be had specially packed for the tropics. I used D. 163 paper developer, one can of powder making 20 ounces of stock solution, which for use was diluted one part with three parts of water. After a quick rinse in fresh water the developed paper was fixed in the same solution as was used for films, and eight changes of sea water were used for washing with a final rinse in fresh. As the enlargements were required purely for reproduction in magazines and papers, all were made on glossy, single-weight paper. After washing they were squeegeed face down on a chromed glazing plate, which was then placed on a drier of my own construction. This consisted of a wood box with an open bottom and a curved metal top (Plate 48, *bottom right*) on which the glazing plate was placed, and an apron of bunting was drawn over the enlargements and fastened down. With the drier placed in sunlight the enlargements cracked off the plate dry and highly glazed in about 10 minutes, or with the drier placed across the fiddles of the cooker with one burner alight, in about 2 minutes. Enlargements kept on board for several months in the tropics lost some of their glaze and became a little degraded.

Movie work

On our second world voyage my wife and I for the first time tried our hands at movie-making, our intention being to produce a colour film fit to be shown to large audiences; we also hoped we might be able to sell the television rights. These requirements called for 16-mm. film instead of 8-mm., and because a sound track would have to be put on it, the film had to be shot (and projected) at 24 frames per second instead of the more usual 16 or 18 f.p.s.; this meant that the exposure of each frame would be shorter and that a larger stop would have to be used. The camera we chose was a second-hand Bell & Howell with a turret holding 3 lenses, a 1-inch (normal), a ¾-inch (wide angle), and a 2-inch (telephoto). We had to learn to express ourselves in the new medium, something like learning a new language; we had to find out how to tell a story with a sequence of linked shots, and how to portray a passage of time—a fascinating, but a time-consuming occupation.

On the 3-year voyage we shot 8,300 feet of Kodachrome, and cut this by a half during editing to leave a film with a screening time of 1½ hours. Through Drayton Film Productions of Abingdon Road, London, W.8, we then had a colour copy made on single-perforated film, a magnetic stripe was put on, and using our own sound projector we put our own sound on it, using my commentary and some of the tape recordings we had made. We showed this film to some large audiences, and on one occasion in Canada the picture had to fill a 21-foot screen, which it did satisfactorily. Certainly the high first cost of camera, projector, and editor, and the purchase of so much film (about £350 worth), and the making of the copy, would not have been justified by this alone, and I therefore question the wisdom of attempting to make a movie for profit during a voyage. But if one can in advance persuade the B.B.C., or some other television concern, to sponsor the film to the extent of providing the film stock, it may be worth doing, but in my experience British television people are not very interested in this kind of thing, and if they want a film of far-away places and unusual people, they can send a professional unit to do the job so very much better than we can; but they cannot film the day-by-day life aboard an ocean-crossing yacht.

However, my wife and I were fortunate in that the B.B.C. did buy the television rights in our film. They made a black-and-white copy, cut it to run for one hour, got me to put a commentary on it, and televised it in two half-hour shows in the Adventure series. These appear to have been popular, and at the time of writing have been televised in the U.K. three times, and in many countries overseas, a fee being paid on each occasion.

On board we took the precautions against humidity that are mentioned in connection with still cameras and films on page 281, but all the film went to England for processing, and then to a friend who examined it and advised us on how to improve our technique. We saw none of it until the voyage was over, but the first foot or so of each

100-foot spool was exposed on a code number for reference, and we kept a list of all shots and a note of the lighting and the clothes we were wearing, so that we could if necessary repeat a faulty shot. We learnt that the best training for the handling of a movie camera at sea is the use of the sextant, which teaches one always to hold the instrument vertical, and as with still photography, we learnt to fill the frame, and not to shoot until we 'could see the whites of their eyes'. We realized the need for cutaways to bridge the gaps between shots—a hand on the tiller, a flag in the wind, the bow-wave, the wake—but during the editing we could have done with many more. We also discovered that it is not worth while to make fades or dissolves in the field, as one cannot tell where they will be required until the editing is being done, and then they are best kept to a minimum and made by a laboratory.

APPENDIX I
PASSAGE TIMES

WHEN an ocean passage is being planned, it is of some value to know how many days other yachts have taken over it, and at the time the passage is being made such information is of particular interest. The time taken will depend on the strength and direction of the wind—which in turn may depend on the time of year—and to some extent on the size of the yacht, especially if the wind is strong; but when the wind is moderate and fair, it sometimes happens that the passage time of a small yacht will compare very favourably with that of a larger one.

The times taken on a variety of passages in the Atlantic, Pacific, and Indian Ocean, by some of the yachts that have been mentioned in this book, will be found in the following pages, together with the dates of departure. Most of the distances given are those arrived at by adding together the days' runs; they are not necessarily the shortest distances. Where such figures are not available, the distances, measured along the usual sailing-ship routes, have been taken from the small-scale ocean charts; such measurements are only approximate, and are therefore followed by the letter 'a'.

Particulars of the yachts mentioned below, with their owners' names, number of crew, and the years in which the passages were made, will be found in Appendix II.

Destination	Port of departure	Yacht	Date of departure	Days	Sea miles
ATLANTIC OCEAN, south-bound					
Gibraltar	Falmouth	*Wanderer II*	17 Sept.	13	1,100 a
Madeira	Dublin	*Saoirse*	20 June	13	1,350 a
,,	Plymouth	*Solace*	Sept.	13	1,200 a
,,	Douarnanez, Fr.	*Winnibelle*	10 May	16	1,080
Las Palmas	Falmouth	*Mischief*	6 July	17	1,500
Dakar, Senegal	Canary Is.	*Driac II*	8 Nov.	8	824
Montevideo, Uruguay	Las Palmas	*Mischief*	29 July	64	4,600
(Crossed equator in 28° W.)					
Bermuda	St. John's, Newfoundland	*Emanuel*	3 Nov.	23	1,072
(13 days of gales and 4 days of calm).					
Haiti, W.I.	Bermuda	*Svaap*	19 July	10	1,012
Recife, Brazil	Barbados, B.W.I.	*Speedwell*	20 Dec.	32	2,448
(Mostly close-hauled.)					
Recife	Madeira	*Saoirse*	6 July	37	3,000 a
(Crossed equator in 22° W.)					
Tristan da Cunha	Recife	*Speedwell*	28 Jan.	26	2,050 a

U

Destination	Port of departure	Yacht	Date of departure	Days	Sea miles

ATLANTIC OCEAN, west-bound

Destination	Port of departure	Yacht	Date of departure	Days	Sea miles
Ponta Delgada, Azores	La Coruña	*Driac II*	28 Aug.	11	860
,,	Vigo	*Wanderer II*	22 June	13	847
Canary Is.	Gibraltar	*Wanderer II*	Feb.	10	760
,,	,,	*Svaap*	19 Sept.	8	,,
,,	,,	*Speedwell*	5 Dec.	14	,,

Crossing by northern route

Destination	Port of departure	Yacht	Date of departure	Days	Sea miles
St. John's, Newfound-land	Bantry Bay, Eire	*Emanuel*	26 May	24	1,765

(Made latitude 53° N. in longitude 32° W.)

Destination	Port of departure	Yacht	Date of departure	Days	Sea miles
New York	Falmouth	*Vertue XXXV*	15 April	47	3,669

(Passed 100′ NW. of Flores and to 35° N. in 48° W., thence to 36° N. in 66° W., and direct to destination.)

Crossing by southern, trade wind, route

Destination	Port of departure	Yacht	Date of departure	Days	Sea miles
Antigua	Madeira	*Widgee*	27 Sept.	26	2,988
Grenada	,,	*Rena*	11 Nov.	26	3,000 a
Barbados, B.W.I.	,,	*Moonraker*	9 Oct.	30	2,960
,, ,,	,,	*Kochab*	2 Nov.	33	,,
,, ,,	,,	*Havfruen III*	11 Nov.	23	,,
,, ,,	Canary Is.	*Sopranino*	11 Jan.	28	2,700 a
,, ,,	,,	*Wanderer II*	Feb.	29	,,
,, ,,	,,	*Buttercup*	5 May	29	,,
,, ,,	,,	*Wanderer III*	11 Oct.	26	,,
,, ,,	,,	*Viking*	15 Oct.	32	,,
,, ,,	,,	*Omoo*	17 Oct.	23	,,
,, ,,	,,	*Diotima*	29 Oct.	44	,,

(7 days of calm, and 4 of headwinds.)

Destination	Port of departure	Yacht	Date of departure	Days	Sea miles
Barbados, B.W.I.	Canary Is.	*Beyond*	6 Nov.	21	,,
,, ,,	,,	*Speedwell*	22 Dec.	25	,,

The recommended route for this passage is to make 25° N. in 25°–30° W., then to 18° N. in 40° W., and thence to destination; but the season of the year must be taken into consideration when judging how far south it will be necessary to go to ensure holding the trade wind.

Destination	Port of departure	Yacht	Date of departure	Days	Sea miles
Barbados, B.W.I.	Cape Verde Is.	*Driac II*	1 Dec.	19	2,040
Demerara R.	,,	*Wanderer III*	19 Oct.	19	2,052
New York	Madeira	*Karin III*	5 April	43	3,980
,,	Canary Is.	*Svaap*	4 Oct.	38	4,000
,,	Bermuda	*Lang Syne*	18 May	6	700 a
Cristobal, Panama	St. Thomas	*Treasure*	April	8	1,000 a
,, ,,	Antigua	*Wanderer III*	28 Dec.	10	1,160

ATLANTIC OCEAN, north-bound

Destination	Port of departure	Yacht	Date of departure	Days	Sea miles
St. Helena	Cape Town	*Diddikai*	3 Mar.	18	1,700
,,	,,	*Omoo*	4 Mar.	16	,,
,,	,,	*Wanderer III*	17 Mar.	16	,,
,,	,,	*Hurricane*	7 May	17	,,

Destination	Port of departure	Yacht	Date of departure	Days	Sea miles
St. Helena	Cape Town	*Driac II*	30 Jan.	16	1,700
Falmouth	,,	*Viking*	23 Feb.	78	7,500 a
(Crossed equator in 20° W., and left Azores to starboard.)					
Ascension	St. Helena	*Omoo*	28 Mar.	6	704
,,	,,	*Wanderer III*	6 April	6	,,
,,	,,	*Diddikai*	7 April	8	,,
Barbados	,,	*Hurricane*	29 May	30	3,800
,,	Ascension	*Trekka*	15 Mar.	37	3,000 a
(Crossed equator in 35° W.)					
Horta, Azores	Ascension	*Wanderer III*	17 April	52	3,444
(Crossed equator in 23° W., then close-hauled to 31° N., 37° W.)					
Horta	Ascension	*Diddikai*	28 April	41	,,
(Crossed equator in 20° W.)					
Ponta Delgada, Azores	Ascension	*Omoo*	9 April	41	3,550
(Crossed equator in 20° W.)					
Trinidad, B.W.I.	Ascension	*Driac II*	3 Mar.	29	3,072
(Steered the direct course.)					
R. Plate	Magellan Strait	*Waltzing Matilda*	24 Mar.	12	1,100
Rio de Janeiro	R. Plate	,,	12 May	14	1,052
Recife	Falkland Is.	*Saoirse*	28 Feb.	27	3,340
,,	Rio de Janeiro	*Lang Syne*	28 Dec.	12	1,170 a
Trinidad	Recife	,,	13 Jan.	14	2,000 a
Horta, Azores	,,	*Saoirse*	7 April	28	3,180
(Crossed equator in 30° W. then to 22° N. 38° W.)					
Cowes, Isle of Wight	Recife	*Waltzing Matilda*	17 June	46	4,400
Bermuda	Grenada, B.W.I.	*Driac II*	22 April	17	1,200 a
,,	Cristobal, Panama Canal Zone	*Mischief*	1 May	30	2,100 a
Liverpool, Nova Scotia	Abaco, Bahamas	*Wind's Song*	June	12	1,270
Mt. Desert I, Maine	,, ,,	*Wanderer III*	18 May	12	1,177
Studland Bay	Gibraltar	,,	17 July	21	1,130

ATLANTIC OCEAN, east-bound

Bermuda	Charleston, Florida	*Driac II*	11 May	12	816
,,	,, ,,	*Moonraker*	4 June	8	,,
Dartmouth, Eng.	Dartmouth, N.S.	*Nova Espero*	5 July	43	2,450
Hamble R.	Newport, R.I.	*Jester*	6 July	34	3,030 a
Flores, Azores	Bermuda	*Moonraker*	20 June	20	1,730
Horta, Azores	,,	*Emanuel*	24 April	17	1,879
,, ,,	,,	*Diotima*	8 July	24	,,
Falmouth	,,	*Widgee*	Aug.	28	3,295
Fowey	,,	*Moonraker*	19 July	40	3,300 a
Portsmouth	,,	*Driac II*	16 May	35	3,400 a
Scilly Isles	,,	*Mischief*	7 June	32	3,200 a
,, ,,	Horta	*Emanuel*	18 May	17	1,254

Destination	Port of departure	Yacht	Date of departure	Days	Sea miles
Falmouth	Horta	*Wanderer III*	22 June	14	1,260
,,	,,	*Diddikai*	29 June	19	,,
Fowey	,,	*Diotima*	12 Aug.	19	1,280
Portsmouth	,,	*Driac II*	24 Sept.	24	1,390
Dublin	,,	*Saoirse*	3 June	17	1,375
Falmouth	San Miguel, Azores	*Wanderer II*	15 July	13	1,185
Tangier	,, ,,	*Driac II*	2 July	16	950 a
Gibraltar	Teceira ,,	*Rena*	Aug.	10	1,080
Cape Town	Recife	*Saoirse*	1 Sept.	36	3,550 a
,,	Tristan da Cunha	*Speedwell*	26 Feb.	13	1,500 a

PACIFIC OCEAN, passages with a westerly component

Destination	Port of departure	Yacht	Date of departure	Days	Sea miles
Galapagos Is.	Balboa, Panama Canal Zone	*Moonraker*	12 Feb.	12	1,000 a
,,	,,	*Kochab*	Feb.	12	,,
,,	,,	*Arthur Rogers*	5 Mar.	10	,,
,,	,,	*Amaryllis*	29 Mar.	12	,,
,,	,,	*Hurricane*	Sept.	23	,,
,,	,,	*Svaap*	27 Sept.	16	,,
Marquesas Is.	,,	*Wanderer III*	26 Jan.	37	3,970
,,	,,	*Omoo*	10 Feb.	45	,,
,,	,,	*Viking*	Feb.	51	,,
,,	San Diego, Calif.	*Tropic Seas*	May	42	3,000 a
Mangareva, Gambier Islands	Galapagos Is.	*Arthur Rogers*	3 May	31	,,
,,	,,	*Varua*	15 Sept.	18	,,
Marquesas Is.	,,	*Moonraker*	1 Mar.	37	3,185
,,	,,	*Amaryllis*	14 Apr.	26	3,057
,,	,,	*Beyond*	15 April	22	3,010
,,	,,	*Hurricane*	Oct.	30	,,
Tahiti	,,	*Svaap*	6 Dec.	32	3,600
,,	,,	*Wanderer II*		34	,,
Los Angeles, Calif.	,,	*Kochab*	19 Mar.	47	2,900
Tahiti	Marquesas Is.	*Wanderer III*	19 Mar.	7	770
,,	,,	*Viking*	Mar.	17	,,
,,	,,	*Omoo*	15 April	9	,,
Hilo, Hawaii	San Francisco	*Wanderer II*	19 Aug.	32	2,160
,,	,,	*Tzu Hang*	15 Oct.	24	,,
,,	Tahiti	*Wanderer II*	3 Mar.	32	2,670

(Crossed equator in 150° W., instead of 147° W. as recommended.)

Hilo	Takaroa, Tuamotu	*Stortebecker III*	8 Dec.	25	2,145

(Crossed equator in 145° W.)

Honolulu	Los Angeles	*Kochab*	4 July	15	2,300
,,	Bora-Bora, Society Islands	*Moonraker*	14 June	32	2,678

(Crossed equator in 146° W.)

Honolulu	Balboa, Panama Canal Zone	*Little Bear*		49	5,300

(Motored for 800 miles.)

Destination	Port of departure	Yacht	Date of departure	Days	Sea miles
Honolulu	Balboa, Panama Canal Zone	*Trekka*	21 May	62	5,400
Pago Pago, American Samoa	Bora-Bora, Society Islands	*Wanderer III*	29 May	13	1,070
Apia, Samoa	Honolulu	*Kochab*	8 Aug.	15	2,300
,,	Fanning I., Line Islands	*Trekka*		14	1,300 a
Russell, N.Z.	Suva, Fiji	*Wanderer III*	31 Aug.	17	1,097
,,	Kandavu, Fiji	*Little Bear*		19	950
Auckland, N.Z.	Ndravuni, Fiji	*Beyond*	27 Aug.	11	1,160
,,	Vavau, Tonga Is.	*Kochab*	14 Sept.	24	1,300
Sydney, N.S.W.	Whangaroa, N.Z.	*Wanderer III*	12 Jan.	10	1,156
,,	Suva, Fiji	*Treasure*	13 Oct.	11	1,700
Port Moresby, Papua, New Guinea	Noumea, New Caledonia	*Wanderer III*	5 May	15	1,368

PACIFIC OCEAN, passages with an easterly component

Picton, N.Z.	Melbourne, Victoria	*Saoirse*	30 Mar.	17	1,500 a
Auckland, N.Z.	Sydney, N.S.W.	*Waltzing Matilda*	29 Oct.	16	1,300
,,	,,	*Amaryllis*	17 Dec.	15	,,
Rapa	Auckland	*Stortebecker III*	30 Aug.	24	2,260
Stanley, Falkland Is. (By way of Cape Horn.)	,,	*Saoirse*	22 Oct.	46	5,800
Port Slight, Chile	Kawau I., N.Z.	*Waltzing Matilda*	31 Dec.	45	5,000 a
Port Corral, Chile	Rapa	*Varua*	11 Jan.	37	4,740
Port San Juan, B.C.	Honolulu	*Stortebecker III*	1 April	31	2,350
Victoria, B.C.	,,	*Moonraker*	30 July	29	2,630
,,	Hawaii	*Wanderer II*	Aug.	41	2,430 a

INDIAN OCEAN, passages with a westerly component

S. Goulburn I., Northern Territory	Thursday I., Torres Strait	*Wanderer III*	9 June	5	532
Christmas I.	,,	*Beyond*	18 Feb.	29	2,190
(Called at Crocker I., where 1 day was spent; engine used considerably.)					
Christmas I.	S. Goulburn I.	*Wanderer III*	19 June	15	1,658
,,	Bali, Indonesia	*Driac II*	9 July	6	585
,,	Darwin, Northern Territory	*Viking*	Aug.	17	1,500 a
Keeling Cocos	Port Moresby, Papua, New Guinea	*Omoo*	9 Sept.	31	3,000 a
,,	Christmas I.	*Wanderer III*	13 July	4	527
,,	,,	*Driac II*	21 July	6	527
Rodriguez I.	Keeling Cocos	*Driac II*	6 Aug.	18	1,986
,,	,,	*Wanderer III*	19 Aug.	19	,,
,,	Fremantle, W. Australia	*Walkabout*	Aug.	44	3,000 a
Mauritius	Keeling Cocos	*Omoo*	16 Oct.	29	2,400

Destination	Port of departure	Yacht	Date of departure	Days	Sea miles
Mauritius	Keeling Cocos	*Viking*		17	2,400
Durban, S. Africa	Mauritius	*Wanderer III*	2 Oct.	19	1,655
„ „	„	*Omoo*	20 Nov.	15	„
Cape Town	„	*Viking*	4 Nov.	26	2,400 a
Fort Dauphin, Madagascar	La Réunion	*Driac II*	21 Sept.	6	512
Durban	„	*Walkabout*	Nov.	20	1,550
Lourenço Marques, Mozambique	Fort Dauphin	*Driac II*	6 Oct.	11	803
Chagos	Keeling Cocos	*Beyond*	10 April	12	1,420
Seychelles	Chagos	„	23 April	9	1,020
„	Keeling Cocos	*Wanderer III*	1 Sept.	26	2,561
Aden	Seychelles	„	6 Oct.	25	1,419
„	„	*Beyond*	11 May	14	1,500
(Made the African coast in 4° N. Engine much used.)					
Colombo, Ceylon	Penang, Malaya	*Svaap*	Jan.	9	1,282
Zanzibar	Colombo	*Hurricane*	31 Dec.	27	3,000 a
Mombasa	Seychelles	*Tzu Hang*	Aug.	10	1,100 a

In the Indian Ocean very few eastward passages have been made by yachts.

APPENDIX II

PARTICULARS OF YACHTS REFERRED TO IN THIS BOOK

IN the following alphabetical list will be found brief particulars of most of the yachts mentioned or illustrated in this book, and of the more notable voyages made by them. No attempt has been made to list all the ports called at, but when known sufficient have been included to indicate the route taken where there are alternatives. Each yacht's name is followed by the name of her owner/skipper at the time the voyage referred to was made; his nationality is British unless stated otherwise. The number of crew includes the owner/skipper. Construction is of wood unless stated to be otherwise.

Altair, Charles F. Grey (American). Centreboard cutter, designed by R. P. Derring and owner, and built of steel by Wm. Schwarz in Wisconsin, U.S.A., in 1961. L.o.a. 36 ft., b. 11·5 ft., d. with plates up 3 ft. Diesel engine. Sailed from Chicago to Bahamas in 1964, thence to Nova Scotia and by great circle course to U.K.; via canals to Mediterranean, and then by southern route to West Indies and Florida. Time 2½ years, crew the owner and his wife.

Amaryllis, G. H. P. Muhlhauser. Gaff yawl, built by A. E. Payne in 1882. L.o.a. 62 ft., l.w.l. 52 ft., b. 13 ft., d. 10 ft. Paraffin engine 12/15 h.p. Left Plymouth, Devon, in September 1920, sailed round the world by way of Panama, Australia, New Zealand, Torres Strait, Singapore, and Suez, returning to Dartmouth in July 1923. Crew, three and four.

Arthur Rogers, Mr. and Mrs. T. Hepworth. Gaff ketch, built on trawler lines by R. Jackman & Sons at Brixham, Devon, in 1922. L.o.a. 68 ft., b. 15·8 ft., d. 8·8 ft. Gray oil engine. Left England in September 1947, crossed Atlantic by southern route to British Guiana, and through Panama Canal to south Pacific islands and New Zealand, doing fishing and charter work and trading. Crew, three to eight.

Awanhee, Dr. R. Griffith (American). Cutter designed by Uffa Fox and built of ferro-cement by owner in New Zealand. L.o.a. 52 ft. Left New Zealand June 1965 and made a circumnavigation west-about by way of the Cape and the Horn, 35,000 miles in 291 sailing days. Crew, the owner and his wife and 11-year old son, but sometimes more.

Beyond, Mr. and Mrs. T. C. Worth. (For description and plans see pages 100–4.) Cutter, designed by Laurent Giles & Partners and built of aluminium alloy by the Sussex Shipbuilding Co. at Shoreham, Sussex, in 1951. L.o.a. 43 ft., l.w.l. 32 ft., b. 10·7 ft., d. 7 ft., s.a. 690 sq. ft. Coventry Godiva diesel engine, 22 h.p. Left Helford, Cornwall, in July 1952 and sailed round the world by way of Panama, New Zealand, Australia, Torres Strait, Seychelles, and Suez, and returned to Helford in September 1954. Crew, four to New Zealand, thence the two owners only.

Buttercup, Major Ian Major. Sliding gunter sloop, turtle deck and twin keels; designed by Robert Clark, and built by Rowhedge Ironworks, Colchester, Essex, in 1936, but subsequently altered by her owner. L.o.a. 25 ft., b. 7 ft., s.a. 236 sq. ft. Air-cooled Lister diesel engine, 3½ h.p. Sailed from south France to West Indies by southern route, 1956, using an early type of vane steering gear. Crew, two.

Cardinal Vertue. Sloop designed by Laurent Giles & Partners and built by Elkins at Christchurch, Hants, in 1948. L.o.a. 25·3 ft., l.w.l. 21·5 ft., b. 7·5 ft., d. 4·5 ft., s.a. 294 sq. ft. Engine removed in 1962. Owner Dr. David Lewis competed in first single-handed transatlantic race. Owner W. Nance (Australian) circumnavigated east-about single-handed by way of the Cape and the Horn, leaving England in September 1963 and reaching Florida in August 1965.

Coimbra, N. B. Redfern (South African). Cutter with reverse sheer, designed by Laurent Giles & Partners and built of aluminium alloy by McLean & Sons, Gourock, Scotland, in 1952. L.o.a. 40 ft., l.w.l. 32 ft., b. 10·5 ft., d. 6·25 ft., s.a. 685 sq. ft. Parsons petrol engine. While bound from England to South Africa she was wrecked on Tristan da Cunha in July 1953.

Diddikai, W. Wild (South African). Ketch designed by Francis L. Herreshoff and built by her owner at Cape Town in 1953. L.o.a. 36 ft., l.w.l. 31·5 ft., b. 9 ft., d. 5·5 ft., s.a. 500 sq. ft. Engine. Sailed from Cape Town to Hamble River, Hants, in 1957, by way of St. Helena, Ascension, and Horta, Azores. Crew, three. Later sailed to West Indies.

Diotima, Vice-Admiral Sir Lennon Goldsmith. Gaff cutter, with raised deck amidships; designed by J. G. Hanna and built by Philip & Son at Dartmouth, Devon, in 1949. L.o.a. 30·5 ft., l.w.l. 28·5 ft., b. 10 ft., d. 5 ft., s.a. 470 sq. ft. Stuart Turner petrol engine, 8 h.p. Left Helford, Cornwall, in October 1949, and sailed to West Indies by the southern route, thence to Bermuda and Azores, and reached Fowey, Cornwall, in August 1951. Crew, owner and Miss Adams.

Director, B. and S. Fahnestock (American). Ex Portland, Maine, pilot schooner, gaff rigged, built in 1912. L.o.a. 60 ft., b. 16 ft., d. 10·5 ft. Engine. Sailed from New York to Panama, crossed South Pacific to New Guinea, thence to China, 1936–7. Crew, five and six.

Driac II. First owner A. G. H. Macpherson, second owner W. Leng. Cutter with steel frame and wood planking; designed by S. N. Graham and built by Feltham at Portsmouth, Hants, in 1932. L.o.a. 32·2 ft., l.w.l. 25·5 ft., b. 8·6 ft., d. 5·5 ft., s.a. 546 sq. ft. Thornycroft petrol engine.

Sailed from England to Azores and back in 1932; England to Iceland and back in 1933; England to West Indies and Mexico by southern route, and thence via Charleston, Bermuda, and Azores to Gibraltar in 1934–5; Gibraltar to India and Indonesia by way of Suez, thence to South Africa, 1935–8; mostly with crew of three. In 1939 W. Leng sailed the yacht to England by way of the West Indies and Bermuda; crew, three.

Elsie, ex *Liberia*. Cutter designed as North Sea fishing research vessel and built near Hamburg in 1958. L.o.a. 30 ft., b. 10½ ft. First owner Dr. H. Lindermann (German) sailed from Germany to the Congo and Bahamas. Second owner, Frank Casper (American), circumnavigated single-handed via Panama, New Zealand, Torres Strait, and the Cape, leaving Miami, Florida, December 1963 and taking three years.

Emanuel, Commander R. D. Graham. Gaff cutter with raised deck amidships; designed and built by A. Anderson & Son at Penarth, Wales, in 1928, L.o.a. 30 ft., l.w.l. 25 ft., b. 8·5 ft., d. 5 ft., s.a. 475 sq. ft. No engine. Sailed single-handed from Poole, Dorset, to Bantry Bay, Eire, thence to St. John's, Newfoundland, by northern route, and to Labrador and Bermuda, in 1934. Returned to England by way of the Azores in 1935. Crew, two.

Felicity Ann, Mrs. Ann Davison. Sloop, designed and built by Mashford Bros. at Plymouth in 1949. L.o.a. 23 ft., l.w.l. 19 ft., b. 7 ft., d. 4·7 ft., s.a. 300 sq. ft. Coventry Victor diesel engine. Sailed single-handed from Plymouth, Devon, to West Indies by southern route in 1952, thence to New York. Mrs. Davison is the first woman to have crossed the Atlantic single-handed.

Finisterre, Carleton Mitchell (American). Centreboard yawl designed by Sparkman & Stephens, and built by Seth Persson at Saybrook, U.S.A., in 1953. L.o.a. 38·6 ft., l.w.l. 27·5 ft., b. 11·25 ft., d. 4·25 ft., with centreboard 7·6 ft., s.a. 713 sq. ft. Gray four-cylinder petrol engine. Crossed Atlantic from Bermuda to Gibraltar by way of the Azores in 1956. Crew, five.

Galway Blazer, Commander W. D. Æ. King. Ketch, designed by Laurent Giles & Partners and built by Camper & Nicholsons at Gosport, Hants, in 1949. L.o.a. 31 ft., l.w.l. 24 ft., b. 7·4 ft., d. 6 ft. No engine. Sailed from England to West Indies by southern route in 1949. Crew, two.

Gesture, R. McIlvride (New Zealand). Cutter designed by Woollacott and built in New Zealand by her owner. Sailed in 1953 from New Zealand to Lord Howe Island, where she was dismasted by a hurricane, thence to Sydney, N.S.W., under jury rig. Crew, five. On the return trip to New Zealand she was again dismasted.

Gipsy Moth III, Francis Chichester. Cutter designed by Robert Clark and built by Tyrrell at Arklow, Eire, in 1959. L.o.a. 39·6 ft., l.w.l. 29 ft., b. 10 ft., d. 6·4 ft., s.a. 679 sq. ft. Coventry petrol engine. Won the first single-handed transatlantic race, and was second in the second race.

Gipsy Moth IV, Sir Francis Chichester. Ketch designed by Illingworth & Primrose and built of laminated wood by Camper & Nicholsons at Gosport, Hants, in 1966. L.o.a. 53 ft., l.w.l. 38·5 ft., b. 10·5 ft., d. 7·7 ft., displacement 11·5 tons, s.a. 750 sq. ft. Perkins diesel engine. Left Plymouth, Devon, in 1966 and sailed single-handed round the world south of the three stormy capes, a distance of 29,617 miles in 226 days, making only one stop, at Sydney, N.S.W. Now on permanent exhibition near the National Maritime Museum, Greenwich.

Havfruen III, Group-Captain T. H. Carr. Gaff ketch, designed by Colin Archer and built at Porsgrund, Norway, in 1897. L.o.a. 60 ft., l.w.l. 53 ft., b. 17·8 ft., d. 8·5 ft., s.a. 1530 sq. ft. Thornycroft diesel engine. Sailed from England to West Indies by southern route in 1954, returning to England by way of Bermuda. Left England in 1956 and sailed round the world by way of Panama, Torres Strait, and South Africa. Crew, two to six.

Hurricane, R. Kauffman (American). Gaff ketch, designed by owner and S. Krebbs and built by S. Krebbs near New Orleans, U.S.A. in 1935. L.o.a. 45 ft., b. 14 ft. Engine. Sailed round the world from New Orleans by way of Panama, Pacific islands, Sydney, Torres Strait, Singapore, Ceylon, Zanzibar, and Cape of Good Hope, 1936-8. Crew, three.

Ice Bird, Dr. J. I. Cunningham. Vertue class cutter designed by Laurent Giles & Partners and built by Aero Marine at Emsworth, Hants, in 1952. L.o.a. 26 ft., l.w.l. 21·5 ft., b. 7·2 ft., d. 4·5 ft., s.a. 360 sq. ft. No engine. Sailed single-handed from Northern Ireland to West Indies by southern route in 1952, and subsequently returned to Ireland.

Imogen, Captain Otway Waller. Gaff yawl, designed by Albert Strange. L.o.a. 26 ft., l.w.l. 22·5 ft., b. 8 ft., d. 3·5 ft. Ailsa Craig petrol engine. Sailed single-handed from the Shannon, Eire, to Vigo, Spain, thence to Madeira and Canary Islands. Probably the first yacht to be rigged with self-steering twin running sails.

Islander, Harry Pidgeon (American). Gaff yawl, hard chine, designed by Fleming Day, and built by owner at Los Angeles, California, in 1918. L.o.a. 34 ft., b. 10·75 ft., d. 5 ft., s.a. 630 sq. ft. No engine. Left Los Angeles in November 1921, sailed single-handed round the world by way of Pacific Islands, Torres Strait, Indian Ocean islands, Cape of Good Hope, and Panama, returning to Los Angeles in October 1925. A second single-handed circumnavigation was made during 1932-7, but during a third attempt, with his wife on board, was wrecked in the New Hebrides.

Island Girl, J. S. Letcher, Jr. (American). Hard-chine cutter. L.o.a. 20·3 ft., l.w.l. 18·5 ft., b. 7 ft., d. 3·5 ft., s.a. 250 sq. ft. In 1963 sailed single-handed from Los Angeles, California, to Hawaii, and in 1964 from Hawaii to Alaska. Was run down while on passage from Sitka to Santa Barbara, but reached port.

Jester, Colonel H. G. Hasler. Designed by owner and built by H. Feltham at Portsmouth, Hants, in 1953. L.o.a. 25 ft., l.w.l. 20 ft., b. 7 ft., d. 4 ft., Chinese rig 250 sq. ft. Four single-handed Atlantic crossings including the first two single-handed races, during which a much more northerly route was taken than by any other competitor. She got second place in the first race, although much the smallest yacht.

Joshua, Bernard Moitessier (French). Steel ketch designed by Jean Knocker and built by J. Fricaud in 1961. L.o.a. 39·6 ft., b. 12 ft., d. 5·2 ft., s.a. 960 sq. ft. Sailed from France to Tahiti via Panama. Left Moorea, Society Islands, in November 1965, and sailed non-stop to Alicante, Spain, by way of the Horn, 14,216 miles in 126 days. Crew, owner and his wife.

Kairos, Mr. and Mrs. G. Koch (German). Centreboard sloop built of steel at Hamburg in 1955. L.o.a. 32 ft., l.w.l. 25 ft., b. 9·5 ft., d. 3 ft. and 6 ft. with board down. Left Hamburg in May 1964 and circumnavigated by way of Panama, Torres Strait, and the Cape, and is believed to be the first German yacht to have done so. Crew, the two owners.

Kochab, Dr. I. J. Franklen-Evans. (For description and plans see pages 104-10.) Yawl, designed by Arthur C. Robb and built by Herbert Woods at Potter Heigham, Norfolk, in 1956. L.o.a. 39·5 ft., l.w.l. 29 ft., b. 10·7 ft., d. 6 ft., s.a. 824 sq. ft. Coventry Victor diesel engine, 9/11 h.p. Left England in 1956 and sailed to Australia by way of Panama, Galapagos, Los Angeles, Hawaii, Samoa, Tonga, and New Zealand. In 1960-2 returned to the U.K. by way of the west coast of the U.S.A. and Panama, and in 1964 sailed out to New Zealand again. Crew, two and three.

Kon Tiki, Thor Heyerdahl and others (Norwegian). A balsa wood raft, 45 ft. long and 18 ft. beam, built by owners and rigged with a single square-sail. Sailed from Callao, Peru, on 28 April 1947, and was beached on Raroia Reef, Tuamotu, on 7 August. Crew, six.

Kurun, J. Y. le Toumelin (French). Gaff cutter, designed by Dervin and built by Moullec at Le Croisic, France, in 1947. L.o.a. 33 ft., l.w.l. 27·8 ft., b. 11·8 ft., d. 5·4 ft. No engine. Left Le Croisic in September 1949 and sailed single-handed round the world by way of Panama, South Pacific islands, Torres Strait, Cape of Good Hope, and returned to Le Croisic in July 1952.

Lang Syne, Mr. and Mrs. Crowe (American). Two-masted schooner, gaff-rigged on the fore, jib-headed on the main, built by her owners at Hawaii in 1936. L.o.a. 39 ft., l.w.l. 34 ft., b. 14 ft., d. 6 ft., s.a. 1,068 sq. ft. Scripps diesel engine. In 1946 sailed from Hawaii to Los Angeles and South Pacific islands, returning to Hawaii in 1947. Left again in April 1948 and sailed round the world by way of Pacific islands, east coast of Australia, Torres Strait, Singapore, Ceylon, Zanzibar, Cape of Good Hope, River Congo, Rio de Janeiro, West Indies, New York, Panama, and Los Angeles, returning to Hawaiian islands in March 1952. Crew, Mr. and Mrs. Crowe.

Lehg II, Vito Dumas (Argentinian). Ketch designed and built by Manuel M. Campos in 1934. L.o.a. 31·2 ft., b. 10·8 ft., s.a. 454 sq. ft. In 1942 sailed single-handed from Buenos Aires east-about round the world south of the three stormy capes, 20,420 miles in 272 days with 7 ports of call.

Little Bear, L. F. Champion (American). (For description and plans see pages 110–14.) Gaff ketch, designed by J. G. Hanna and built by the owner and his wife at San Francisco. L.o.a. 36·7 ft., l.w.l. 32·9 ft., b. 12 ft., d. 4·5 ft., s.a. 700 sq. ft. Universal petrol engine, 50 h.p. Left Balboa (Panama) in 1955 and sailed to east coast of Australia by way of Hawaiian and South Pacific islands and New Zealand; returned by way of South Pacific islands. Crew, Mr. and Mrs. Champion.

Lively Lady. Yawl (mizzen used for mizzen staysail only) designed by first owner, S. J. P. Cambridge, O.B.E., and F. Shepherd, and built by owner in Calcutta in 1948. L.o.a. 36 ft., l.w.l. 27 ft., b. 8·9 ft., d. 7·4 ft., s.a. 550 sq. ft. Morris paraffin engine. Sailed single-handed by second owner, Sir Alec Rose, from Portsmouth, Hants, to Melbourne, Australia, east-about in 150 days, then by way of the Horn to Portsmouth. Also sailed in the second single-handed transatlantic race, and return.

Madalèna, Vice-Admiral Sir Lennon Goldsmith. Gaff cutter, designed by owner and built by Philip & Son at Dartmouth, Devon, in 1934. L.o.a. 54·75 ft., l.w.l. 50·75 ft., b. 14·7 ft., d. 8 ft. Diesel engine 30 h.p. Sailed from England in October 1935 to West Indies by the southern route, thence to Bermuda, and returned to England in July 1936. Crew, six.

Mischief, H. W. Tilman. Gaff cutter, ex-Bristol Channel pilot cutter, built by Baker at Cardiff in 1906. L.o.a. 46 ft., l.w.l. 40·6 ft., b. 13 ft., d. 7·5 ft. Petrol motor. Left Falmouth, Cornwall, in 1955 and sailed round South America by way of Magellan Strait and Panama and back to England. In 1957-8 sailed from England to Cape Town and towards Kergulen I., returning via Suez. Subsequently made several voyages to Greenland, and in 1966 to South Georgia. Crew, five and six. Lost in the Arctic in 1968.

Moonraker, Dr. and Mrs. E. A. Pye. Gaff cutter, converted Looe fishing boat, built by Ferris at Looe, Cornwall, in the 1890s. L.o.a. 29 ft., l.w.l. 29 ft., b. 9·7 ft., d. 6 ft., s.a. 700 sq. ft. Bergius paraffin engine. Sailed from Fowey, Cornwall, by the southern route to the West Indies and Florida, and returned to Fowey by way of Bermuda and Azores, 1949–50. Sailed from Fowey by southern route to West Indies, Panama, Tahiti, Hawaii, British Columbia, and Alaska, and returned to England by way of Panama and Bermuda, 1952-4. Crew, three. Now on permanent exhibition at Exeter Maritime Museum.

Nova Espero, Stanley and Colin Smith. Gunter sloop, designed and built by owners at Halifax, Nova Scotia, in 1949. L.o.a. 20 ft., l.w.l. 16 ft., b. 6·5 ft., d. 3 ft., s.a. 153 sq. ft. No engine. Sailed from Dartmouth, Nova

Scotia, to Dartmouth, Devon, in 1949, and from London to New York by northern route, for which she was converted to yawl rig, in 1951. Crew, two.

Omoo, L. G. Van de Wiele (Belgian). (For description and plans see pages 114-18.) Gaff ketch, designed jointly by owner and F. Mulder, and built of steel by Meyntjens at Antwerp in 1948. L.o.a. 45·3 ft., l.w.l. 37 ft., b. 12·1 ft., d. 6·2 ft., s.a. 976 sq. ft. Kermath diesel engine, 27 h.p. Leaving Nice, France, in July 1951, sailed round the world by way of Panama, South Pacific islands, Torres Strait, and Cape of Good Hope, and reached Zeebrugge, Belgium, in August 1953. Crew, three.

Outward Bound, John and Mary Caldwell (Australian). Ketch designed by L. Francis Herreshoff and built by owners at Sydney, N.S.W., in 1958. L.o.a. 45·3 ft., l.w.l. 38 ft., b. 12·5 ft., d. 5·5 ft., s.a. 1,100 sq. ft. Perkins diesel engine 52 h.p. Sailed from Sydney to West Indies by way of Torres Strait and Suez. Crew, owners and their two sons.

Rehu Moana, Dr. David Lewis (New Zealander). Plywood catamaran sloop, designed by Colin Mudie and built by Prout Bros. at Canvey Island, Essex, in 1963. L.o.a. 40 ft., l.w.l. 35 ft., b. 17 ft., d. 3 ft. and 5 ft. with centreboards down; displacement 8 tons, s.a. 700 sq. ft. Engine 4 h.p. Seagull outboard. Took part in second single-handed transatlantic race, then with wife and two baby daughters circumnavigated by way of Magellan Strait, New Zealand, and the Cape.

Rena, Commander and Mrs. Vancil (American). (For description and plans see pages 119-22.) Ketch, designed by John Alden and built by owners at Great Bridge, Virginia, U.S.A., in 1962. L.o.a. 45·6 ft., l.w.l. 34·8 ft., b. 12·7 ft., d. 5·9 ft., s.a. 1,062 sq. ft. Sheppard 50 h.p. diesel engine. In 1962 sailed from Norfolk, Virginia, to Azores and Mediterranean, returning to Florida by southern route. Crew, the owners and one occasional friend.

Salmo, Commander A. G. Hamilton. Vertue class sloop, designed by Laurent Giles & Partners; built by Elkins at Christchurch, Hants, in 1948. L.o.a. 25·4 ft., l.w.l. 21·5 ft., b. 7·2 ft., d. 4·5 ft., s.a. 294 sq. ft. Morris petrol engine. Sailed single-handed from Scotland to Quebec by northern route in 1956. Then with wife sailed to Los Angeles via Panama and Society Islands.

Sandefjord, Erling Tambs (Norwegian). Gaff ketch, ex-Norwegian lifeboat; designed and built by Colin Archer at Risör, Norway, in 1913. L.o.a. 47·1 ft., l.w.l. 40·8 ft., b. 26·2 ft., d. 8 ft., s.a. 889 sq. ft. Bergius paraffin engine. Crossed Atlantic from east to west in 1935, capsizing *en route*, and subsequently sailed to Cape Town. In 1965, when owned by Pat and Barry Cullen, sailed from Durban, South Africa, and circumnavigated by way of the Cape, Panama, and Torres Strait, returning to Durban 21 months later.

Saoirse, Conor O'Brien (Irish). (For description and plans see pages 122–6.) Gaff ketch with gunter mizzen, designed by owner and built by the Fishery School at Baltimore, Eire, in 1922. L.o.a. 42 ft., l.w.l. 37 ft., b. 12·2 ft., d. 6·8 ft., s.a. 1,000 sq. ft. No engine. Left Dublin in June 1923 and sailed round the world east-about by way of Recife, Cape of Good Hope, Melbourne, New Zealand, and Cape Horn, and returned to Dublin in June 1925. Crew, two, three, and four.

Sea Queen, Stone and Vincent, and skippered by J. C. Voss (probably Canadian). Gaff yawl, hard chine; designed by Fleming Day and built by Stone at Yokohama, Japan, in 1912. L.o.a. 25·7 ft., l.w.l. 19 ft., b. 8·25 ft., d. 3·5 ft., s.a. 400 sq. ft. No engine. In 1912, while attempting to sail east across the Pacific from Japan, she was capsized and dismasted in a typhoon, but righted herself and reached port under jury rig. Crew, three.

Solace, Commander V. C. F. Clark. Ketch designed and built by David Hillyard at Littlehampton, Sussex, in 1929. L.o.a. 33·7 ft., l.w.l. 27 ft., b. 9 ft., d. 5·2 ft., s.a. 620 sq. ft. Coventry diesel engine. Left England in September 1953 and circumnavigated by way of Panama, New Zealand, Torres Strait, and the Cape. Was driven ashore on Palmerston Atoll and severely damaged. Crew, two.

Sopranino, P. J. Ellam. Clinker-built cutter designed by Laurent Giles & Partners, and built by Wootten at Cookham Dean, Berks., in 1950. L.o.a. 20 ft., l.w.l. 17·75 ft., b. 5·3 ft., d. 3·7 ft., s.a. 200 sq. ft. No engine. Sailed by the southern route from England to the West Indies and U.S.A. in 1951–2. Crew, two.

Speedwell of Hong Kong, first owner Commander A. G. Hamilton, second owner J. Goodwin. Vertue class sloop designed by Laurent Giles & Partners; built by Wing On Shing at Hong Kong in 1952. L.o.a. 25·3 ft., l.w.l. 21·5 ft., b. 7·4 ft., d. 4·75 ft., s.a. 330 sq. ft. No engine. Under first owner sailed from Singapore to England by way of the Cape of Good Hope in 1952; crew of two. Under second owner, sailed from England to West Indies single-handed in 1956; thence to Cape Town by way of Recife and Tristan da Cunha in 1958; crew of two.

Spray, Joshua Slocum (American). Gaff cutter, later converted to yawl, rebuilt by owner near Boston, U.S.A., in 1892. L.o.a. 36·75 ft., b. 14·2 ft., d. 4 ft. No engine. Left Boston in April 1895 and sailed single-handed round the world by way of Gibraltar, Magellan Strait, South Pacific islands, east coast of Australia, Torres Strait, Indian Ocean islands, and Cape of Good Hope, returning to Boston in June 1898. The first single-handed circumnavigation.

Stortebecker III, Dr. I. J. Franklen-Evans. Yawl, designed and built by Rasmussen at Bremerwerder, Germany, in 1937. L.o.a. 33·5 ft., l.w.l. 24·5 ft., b. 8·4 ft., d. 5·25 ft. Penta petrol engine, 10 h.p. Sailed from England to New Zealand by way of Panama and South Pacific islands in 1949–50; crew of two. Sailed from New Zealand to Victoria, B.C., by way of Pacific islands in 1952–3; crew of three.

Svaap, W. A. Robinson (American). Ketch, designed by Alden and built in the U.S.A. L.o.a. 32·5 ft., l.w.l. 27·5 ft., b. 9·5 ft., d. 5·5 ft., s.a. 660 sq. ft. Kermath petrol engine, 10 h.p. Left New York in June 1928, and sailed round the world by way of Bermuda, Panama, Pacific islands, Malaya, and Suez, and returned to New York in November 1931. Crew, two.

Svea, R. Kittredge (American). Ketch designed by Aage Utzon, and built by Randers in Denmark 1946. L.o.a. 38 ft., l.w.l. 35 ft., b. 10 ft., d. 5·75 ft., s.a. 900 sq. ft. Diesel engine, 45 h.p. Left San Diego, California, in March 1961, and sailed by way of Pacific islands, New Zealand, Torres Strait, and Suez to Fort Lauderdale, Florida, taking four years. Crew, two to four.

Tilikum, J. C. Voss (probably Canadian). Dugout canoe, decked and rigged as a three-masted schooner, gaff-rigged on fore and main, jib-headed on mizzen. L.o.a. 38 ft., b. 5·5 ft., d. 2·25 ft. Left Victoria, B.C. in 1901, sailed by way of Pacific islands, Australia, New Zealand, Torres Strait, Cape of Good Hope, Pernambuco, and Azores to London, the voyage taking three years and three months. Crew, two.

Treasure, John Guzzwell. Cutter designed by Laurent Giles & Partners with modifications by owner, and built (three-skin construction sheathed with g.r.p.) by owner near Lymington, Hants, in 1965. L.o.a. 45·3 ft., l.w.l. 34 ft., b. 12 ft., d. 7 ft., displacement 15·2 tons, s.a. 1,020 sq. ft. Parsons 56 h.p. diesel engine. Left Falmouth, England, December 1965, and sailed to Sydney, N.S.W., by way of Panama and Pacific islands. Crew, the owner, wife and two small sons.

Trekka, John Guzzwell. (For description and plans see pages 126–9.) Ketch, designed by Laurent Giles & Partners and built by owner at Vancouver, B.C., in 1955. L.o.a. 20·8 ft., l.w.l. 18·5 ft., b. 6·5 ft., d. 4·5 ft., s.a. 184 sq. ft. No engine. Sailed single-handed round the world by way of New Zealand, Torres Strait, the Cape, and Panama. Later, under new ownership, made a second circumnavigation.

Tzu Hang, Brigadier and Mrs. Miles Smeeton. Ketch, designed by H. S. Rouse and built by Hop Kee at Hong Kong in 1939. L.o.a. 46·2 ft., l.w.l. 36 ft., b. 11·7 ft., d. 7 ft. Gray petrol motor. Sailed from England to Vancouver, B.C., by way of Panama in 1951–2. Sailed from Vancouver to New Zealand and Melbourne, then when on passage towards England, and before reaching Cape Horn, was dismasted and made Coronel, Chile, under jury rig; on a second attempt to round the Horn she was again dismasted, and again made a Chilian port (Valparaiso) under jury rig; 1955–7. Was shipped to U.K. and then made a circumnavigation east-about by way of Suez,

E. Africa, Japan, Aleutian Islands, Panama, and Greenland. Crew, the owners and an occasional third hand. Left England in 1968 and sailed to Vancouver, B.C., via the Horn and Hawaii. Crew, three.

Varua, W. A. Robinson (American). Brigantine of composite construction, designed by Starling Burgess and owner, and built by owner's yard at Ipswich, Massachusetts, U.S.A., in 1942. L.o.a. 70 ft., l.w.l. 60 ft., b. 16 ft., d. (laden) 8 ft., s.a. 2,700 sq. ft. Deutz diesel engine, 47 h.p. In 1951 sailed south from Tahiti into the roaring forties and so to Valdivia, Chile; cruised north to Panama, and returned to Tahiti by way of Galapagos and Gambier Islands. Crew, five. Subsequently made many inter-island voyages in the west Pacific.

Vertue XXXV, Humphrey Barton. Vertue class sloop designed by Laurent Giles & Partners; built by Elkins at Christchurch, Hants, in 1950. L.o.a. 25·25 ft., l.w.l. 21·5 ft., b. 7·2 ft., d. 4·5 ft., s.a. 314 sq. ft. No engine. Sailed from Falmouth, Cornwall, to New York by the intermediate route in 1950. Crew, two.

Viking, Mr. and Mrs. S. Holmdahl (Swedish). Ketch, converted from fishing vessel by her owners. L.o.a. 33·5 ft., l.w.l. 30 ft., b. 12 ft., d. 6·5 ft. No engine. Left Sweden in 1952 and sailed round the world by way of southern route to West Indies and Panama, thence by South Pacific islands, Torres Strait, Indian Ocean islands, and Cape of Good Hope back to Sweden, taking two years for the voyage. Crew, Mr. and Mrs. Holmdahl. Lost in Galapagos Islands under new ownership.

Walkabout, R. and M. Driscoll (Australian). Ketch, designed by Alden and built by owners at Fremantle, Western Australia, in 1952. L.o.a. 33·2 ft., b. 9·5 ft., d. 5·5 ft. Stuart Turner petrol engine, 8 h.p. Sailed from Fremantle, W.A., across the Indian Ocean to Durban, South Africa, in 1954. Crew, the two owners.

Waltzing Matilda, Philip Davenport (Australian). (For description and plans see pages 129–34.) Cutter, designed and built by J. Muir at Hobart, Tasmania, in 1949. L.o.a. 46·4 ft., l.w.l. 36 ft., b. 12 ft., d. 6·5 ft., s.a. 1,050 sq. ft. Universal petrol engine, 15/20 h.p. Sailed from Sydney, N.S.W., to Cowes, Isle of Wight, by way of New Zealand, Magellan Strait, and South American ports in 1950. Crew, four. Lost off St. Lucia under new ownership in 1969.

Wanderer II, first owner Eric Hiscock, second owner W. Howell (Australian), third owner R. Jones (American). Gaff cutter, designed by Laurent Giles & Partners and built by Napier at Poole, Dorset, in 1936. L.o.a. 24 ft., l.w.l. 20·75 ft., b. 7·1 ft., d. 5 ft., s.a. 495 sq. ft. No engine. With first owner sailed from England to Azores and back in 1950; crew of two. With second owner sailed from England to Tahiti, taking the southern route to West Indies and Panama, in 1952–3, crew of two; from Tahiti to Victoria, B.C., by way of Hawaii in 1953 single-handed. With third owner sailed from Seattle, U.S.A., to Hawaii in 1956; crew of two.

Wanderer III, Eric Hiscock. (For description and plans see pages 134–9.) Sloop, designed by Laurent Giles & Partners and built by Wm. King at Burnham-on-Crouch, Essex, in 1952. L.o.a. 30·3 ft., l.w.l. 26·3 ft., b. 8·4 ft., d. 5 ft., s.a. 600 sq. ft. Stuart Turner petrol engine. Left Yarmouth, Isle of Wight, in July 1952 and circumnavigated by way of Panama, New Zealand, E. Australia, Torres Strait, and the Cape. Left on second circumnavigation July 1959 and returned via Suez. In 1965–7 sailed to U.S.A. and back. Crew, owner and his wife.

Wanderer IV, Eric and Susan Hiscock. Ketch designed by S. M. Van der Meer and built of steel in Holland in 1968. L.o.a. 49·5 ft., l.w.l. 39·8 ft., b. 12·8 ft., d. 6·2 ft., s.a. 1,050 sq. ft. Engine 61 h.p. Ford diesel. Sailed from England in 1968 to west coast of North America via West Indies and Panama. Crew, the two owners.

Widgee, Captain David Guthrie. Sloop designed by Laurent Giles & Partners and built by Colne Marine at Wivenhoe, Essex, in 1965. L.o.a. 29·5 ft., l.w.l. 24·5 ft., b. 9·3 ft., d. 5 ft., s.a. 533 sq. ft. Volvo Penta 15 h.p. diesel engine. In 1965 sailed single-handed from U.K. to West Indies and returned following year.

Wind's Song, Commander George Cook (Canadian). Wishbone, centre-board ketch, designed by owner and built by him at Hackett's Cove, Nova Scotia, 1963. L.o.a. 50 ft., l.w.l. 42 ft., b. 13·5 ft., draught with centre-boards down 7·75 ft., s.a. 1,000 sq. ft. Has made several voyages between Nova Scotia and Bahamas. Crew, two to four.

Winnibelle, Marin-Marie (French). Gaff cutter, built at Boulogne, France, in 1931; l.o.a. 36 ft. Diesel engine, 9 h.p. Sailed single-handed from Douarnenez, France, by the southern route to West Indies and New York in 1933.

Wylo, F. A. Wightman (South African). Gaff yawl, built by owner at Cape Town in 1937. L.o.a. 34 ft., l.w.l. 27 ft., b. 10·8 ft., d. 5·25 ft. No engine. Sailed from Cape Town to Trinidad, B.W.I., in 1947. Crew, two.

Yankee, Commander Irving Johnson (American). Brigantine, ex-North Sea pilot vessel of steel construction converted in England in 1947. L.o.a. 96 ft., l.w.l. 81 ft., b. 21 ft., d. 11 ft. Two G.M. diesel engines of 56 h.p. each. Since 1948 has made four voyages, each of 18 months' duration, west-about round the world, all of them by way of Panama and the Cape of Good Hope, and all of them starting and finishing at Gloucester, U.S.A. Crew, up to twenty-two. Under new ownership was lost on Rarotonga, Cook Islands.

BOOKS

So many accounts of voyages made in small craft have been published that it is possible to include here only some of those which have been written by people mentioned in this book, and from which some of my information has been obtained.

Author	Vessel	Title and publisher
Anson, Lord George	*H.M.S. Centurion*	*A Voyage Round the World* (Knapton, 1748)
Barton, Humphrey	*Vertue XXXV*	*Vertue XXXV* (Hart-Davis)
„ „	Various	*Atlantic Adventurers* (Adlard Coles)
Bruce, Erroll	„	*Deep-Sea Sailing* (Stanley Paul)
Chichester, Francis	*Gipsy Moth IV*	*Gipsy Moth Circles the World* (Hodder)
Clark, Victor	*Solace*	*On the Wind of a Dream* (Hutchinson)
Coles, K. Adlard	Various	*Heavy Weather Sailing* (Adlard Coles)
Crealock, W. I. B.	*Arthur Rogers*	*Towards Tahiti* (Hart-Davis)
Crowe, Bill and Phyllis	*Lang Syne*	*Heaven, Hell and Salt Water* (Adlard Coles)
Davenport, Philip	*Waltzing Matilda*	*The Voyage of Waltzing Matilda* (Hutchinson)
Davison, Ann	*Felicity Ann*	*My Ship is so Small* (Peter Davies)
Dumas, Vito	*Lehg II*	*Alone Through the Roaring Forties* (Adlard Coles)
Ellam, Patrick, and Mudie, Colin	*Sopranino*	*Sopranino* (Hart-Davis)
Fahnestock, Bruce and Sheridan	*Director*	*Stars to Windward* (Robert Hale)
Franklen-Evans, Dr. I. J.	*Stortebecker III*	*R.C.C. Journal*, 1952, 1953
„ „	*Kochab*	*R.C.C. Journal*, 1957
„ „	„	*Roving Commissions*, 1964 (R.C.C. Press)
Goldsmith, M. L.	*Diotima*	*R.C.C. Journal*, 1951
Graham, R. D.	*Emanuel*	*Rough Passage* (Hart-Davis)
Guzzwell, John	*Trekka*	*Trekka Round the World* (Adlard Coles)
Hasler, H. G.	*Jester*	*Roving Commissions*, 1964 (R.C.C. Press)
Heyerdahl, Thor	*Kon-Tiki*	*The Kon-Tiki Expedition* (Allen & Unwin)

Author	Vessel	Title and publisher
Hiscock, Eric	*Wanderer III*	*Around the World in Wanderer III* (Oxford)
„	„	*Beyond the West Horizon* (Oxford)
Howell, William	*Wanderer II*	*White Cliffs to Coral Reefs* (Odhams)
Hughes, John Scott (Edited by)	*Driac II*	*Macpherson's Voyages* (Methuen)
Johnson, Irving and Electa	*Yankee*	*Yankee's Wander World* (Norton)
Johnson, Irving and Electa, and Edes, Lydia	„	*Yankee's People and Places* (Norton)
Kauffman, Ray	*Hurricane*	*Hurricane's Wake* (Herbert Jenkins)
le Toumelin, Jacques-Yves	*Kurun*	*Kurun Around the World* (Hart-Davis)
Lewis, David	*Rehu Moana*	*Daughters of the Wind* (Gollancz)
Long, Dwight	*Idle Hour*	*Sailing All Seas in the Idle Hour* (Hart-Davis)
Marin-Marie	*Winnibelle*	*Wind Aloft, Wind Alow* (Peter Davies)
Martyr, Weston	*Southseaman*	*The Southseaman* (Hart-Davis)
Muhlhauser, G. H. P.	*Amaryllis*	*The Cruise of the Amaryllis* (Hart-Davis)
O'Brien, Conor	*Saoirse*	*Across Three Oceans* (Hart-Davis)
„	Various	*The Small Ocean-Going Yacht* (Oxford)
Pidgeon, Harry	*Islander*	*Around the World Single-handed* (Hart-Davis)
Pye, Dr. E. A.	*Moonraker*	*Red Mains'l* (Herbert Jenkins)
„	„	*The Sea is for Sailing* (Hart-Davis)
Robinson, W. A.	*Svaap*	*Deep Water and Shoal* (Hart-Davis)
„	*Varua*	*To the Great Southern Sea* (Peter Davies)
Slocum, Joshua	*Spray* and *Liberdade*	*Sailing Alone Around the World* (Hart-Davis)
Smeeton, Miles	*Tzu Hang*	*Once is Enough* (Hart-Davis)
„	„	*Sunrise to Windward* (Hart-Davis)
Smith, Stanley, and Violet, Charles	*Nova Espero*	*The Wind Calls the Tune* (Adlard Coles)
Tilman, H. W.	*Mischief*	*Mischief in Patagonia* (Cambridge)
„	„	*Mischief Among the Penguins* (Hart-Davis)

Author	Vessel	Title and publisher
Van de Wiele, Annie	*Omoo*	*The West in My Eyes* (Hart-Davis)
Voss, J. C.	*Tilikum* and *Sea Queen*	*The Venturesome Voyages of Captain Voss* (Hart-Davis)
Wightman, Frank A.	*Wylo*	*The Wind is Free* (Hart-Davis)
Worth, T. C.	*Beyond*	*R.C.C. Journal*, 1953, 1954

Also numerous issues of *Yachting*, *Yachting Monthly*, *Yachting World*, and *The Yachtsman*.

INDEX

Figures in italic indicate the facing pages of plates

PRINTED IN GREAT BRITAIN
AT THE UNIVERSITY PRESS, OXFORD
BY VIVIAN RIDLER
PRINTER TO THE UNIVERSITY

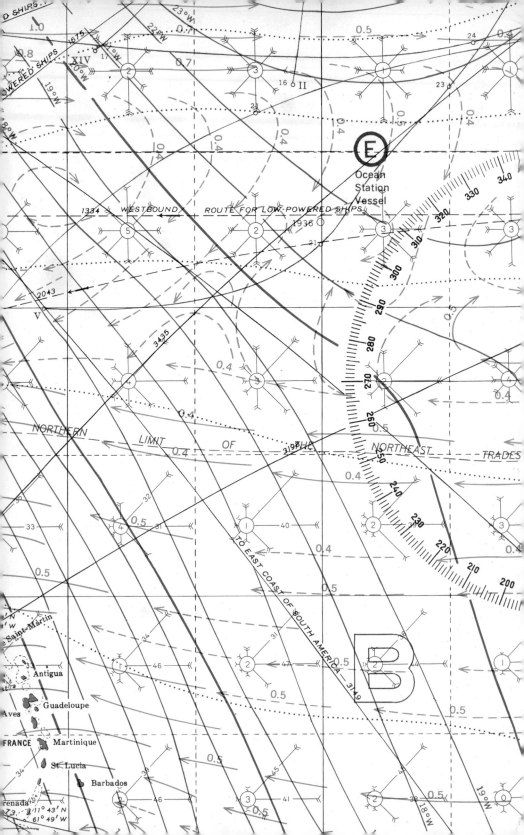